$$\left(\frac{f}{g}\right)' = \frac{gf' - g'f}{g^2} \quad p\ 3?$$

Fundamentals of Complex Analysis

for
Mathematics, Science, and Engineering

Fundamentals of Complex Analysis

for Mathematics, Science, and Engineering

E. B. SAFF

A. D. SNIDER

University of South Florida

Prentice-Hall, Inc. Englewood Cliffs, New Jersey

Library of Congress Cataloging in Publication Data

SAFF, E B (date)
    Fundamentals of complex analysis for mathematics,
science, and engineering.

    Includes bibliographies and index.
    1.–Mathematical analysis. 2.–Functions of complex
variables. I.–Snider, Arthur David (date),
    II.–Title.
QA300.S18   515'.9   75–25823
ISBN 0–13–332148–7

© 1976 by PRENTICE-HALL, INC., Englewood Cliffs, New Jersey.

Printed in the United States of America.

10 9 8 7

PRENTICE-HALL INTERNATIONAL, INC., *London*
PRENTICE-HALL OF AUSTRALIA, PTY. LTD., *Sydney*
PRENTICE-HALL OF CANADA, LTD., *Toronto*
PRENTICE-HALL OF INDIA PRIVATE LIMITED, *New Delhi*
PRENTICE-HALL OF JAPAN, INC., *Tokyo*
PRENTICE-HALL OF SOUTHEAST ASIA (PTE.) LTD., *Singapore*

To Loretta
whose encouragement, typing, and coffee made it possible

Contents

* A star denotes sections which can be omitted without interrupting the logical flow.

Preface

The *raison d'existence* for this book is that we feel that mathematics, science, and engineering undergraduates who have completed the calculus sequence are capable of understanding the basics of complex analysis and applying its methods to solve physical problems. Accordingly, we have addressed ourselves to this audience in our attempt to make the fundamentals of the subject more easily accessible to readers who have little inclination to wade through the rigors of the axiomatic approach. To accomplish this goal we have modeled the text after standard calculus books, both in level of exposition and layout, and have incorporated physical applications throughout the text so that the mathematical methodology will appear less sterile to the reader.

To be more specific about our mode of exposition, we begin by addressing the question most instructors ask first: To what extent is the book self-contained, i.e., which results are proved and which are merely stated? Let us say that we have elected to include all the proofs that reflect the spirit of analytic function theory and to omit most of those which involve deeper results from real analysis. For example, we do not prove the convergence of Riemann sums for complex integrals, nor do we prove the Cauchy criterion for convergence of a series, Goursat's generalization of Cauchy's Theorem, or the Riemann Mapping Theorem. Moreover, in keeping with our philosophy of avoiding pedantics, we have shunned the ordered-pairs interpretations of complex numbers and retained the more intuitive approach (which is based on algebraic field extensions).

Cauchy's Theorem is given two alternative presentations in Chapter 4. The first is based on the deformation of contours, or what is known to topologists as homotopy. We have taken some pains to make this approach understandable and transparent to the unsophisticated reader because it is easy to visualize and to apply in specific situations. The second treatment interprets contour integrals in terms of line integrals and invokes Green's Theorem to complete the argument. These parallel developments constitute the two parts of Section 4 in Chapter 4; either one may be read, and the other omitted, without disrupting the exposition (although it should not be difficult to discern our preference, from this paragraph).

The well-known applications of analytic function theory to two-dimensional problems in potential theory appear throughout the text but receive special emphasis in Chapter 7 on conformal mapping. We draw the distinction between direct methods, wherein a mapping must be constructed to solve a specific problem, and indirect methods which appear to pull a mapping "out of thin air" and then investigate what problems it solves. In doing so we hope to dispel the impression given in many other books that all applications of the technique fall in the latter category. For convenience, Appendix II contains a list of some useful mappings, and Appendix I presents the physical basis for the appearance of harmonic functions in electrostatics and equilibrium flow of heat and fluids. Section 6 of Chapter 2 discusses a visualization of harmonic functions which lends insight into their properties.

The analysis of linear systems is another concept which recurs in the text. The basic idea of frequency analysis is introduced in Chapter 3 following the study of the exponential function, and the development culminates in Chapter 8 with the exposition of Fourier series and integrals and the Laplace transform. (One might say that harmonic functions and linear system analysis constitute two "sub-plots" in the overall treatment.) In keeping with the philosophy stated earlier, we offer proofs of the Fourier and Laplace representations for those cases where analytic function theory applies. Other features of the book are optional sections (marked *) which enrich either the mathematical development or the physical interpretation, a thorough coverage of the uses of residue theory in evaluating integrals, Summary and Suggested Reading sections at the end of each chapter, and a wealth of worked-out examples illustrating the theorems.

We have adopted the conventional use of decimal notation to refer to chapters and sections: thus "Sec. 5.2" indicates Section 2 of Chapter 5.

The material in the book can be covered fairly leisurely in a two quarter course. A one semester course would probably dictate the omission of the optional sections. For a one quarter course designed to teach methods for applications, we recommend that the instructor deemphasize Sections 1.5, 2.1, 2.2, 3.1, 3.2, 3.3, 5.4, 5.7, 5.8, 6.2, 6.6, 6.7, and 7.4 in favor of the other topics.

A number of people have contributed to the successful production of this book. We would like to express our gratitude to Professors Kenneth Hoffman of M.I.T., Ken Gross of the University of North Carolina, and Jack Britton, James Gard, and David Horowitz of the University of South Florida for their helpful criticisms, to USF Dean Donald Saff for the superb jacket design, and to the staff of Prentice-Hall for their co-operation. Finally, we wish to acknowledge our thesis directors, Joseph L. Walsh and Paul Garabedian, for their contribution to our mathematical education.

Tampa, Florida

EDWARD BARRY SAFF
ARTHUR DAVID SNIDER

Fundamentals of Complex Analysis

for Mathematics, Science, and Engineering

1

Complex Numbers

1.1 THE ALGEBRA OF COMPLEX NUMBERS

To achieve a proper perspective for studying the system of complex numbers, let us begin by reviewing briefly the construction of the various numbers used in computation.

We start with the rational numbers. These are ratios of integers and are written in the form m/n, $n \neq 0$, with the stipulation that all rationals of the form n/n are equal to 1. The arithmetic operations of addition, subtraction, multiplication, and division with these numbers can always be performed in a finite number of steps, and the results are, again, rational numbers. Furthermore, there are certain simple rules which involve the order in which the computations can proceed. These are the familiar commutative, associative, and distributive laws:

Commutative law of addition

$$a + b = b + a$$

Commutative law of multiplication

$$ab = ba$$

Associative law of addition

$$a + (b + c) = (a + b) + c$$

Associative law of multiplication

$$a(bc) = (ab)c$$

Distributive law

$$(a + b)c = ac + bc,$$

for any rationals a, b, and c.

Notice that the rationals are the only numbers we would ever need to solve equations of the form

$$ax + b = 0.$$

The solution, for nonzero a, is $x = -b/a$, and since this is the ratio of two rationals, it is itself rational.

However, if we try to solve quadratic equations we find that some of them have no solution; for example, the simple equation

(1) $$x^2 = 2$$

cannot be satisfied by any rational number (see Prob. 20 at the end of this section). Therefore, to get a more satisfactory number system, we extend the concept of "number" by appending to the rationals a new symbol, mnemonically written as $\sqrt{2}$, which is defined to be a solution of Eq. (1). Our revised concept of a number is now an expression in the standard form

(2) $$a + b\sqrt{2},$$

where a and b are rationals. Addition or subtraction is performed according to

(3) $$(a + b\sqrt{2}) \pm (c + d\sqrt{2}) = (a \pm c) + (b \pm d)\sqrt{2}.$$

Multiplication is defined via the distributive law with the proviso that the square of the symbol $\sqrt{2}$ can always be replaced by the rational number 2. Thus we have

(4) $$(a + b\sqrt{2})(c + d\sqrt{2}) = (ac + 2bd) + (bc + ad)\sqrt{2}.$$

Finally, using the well-known process of "rationalizing the denominator" we can put the quotient of any two of these new numbers into the standard form

(5) $$\frac{a + b\sqrt{2}}{c + d\sqrt{2}} = \frac{a + b\sqrt{2}}{c + d\sqrt{2}} \frac{c - d\sqrt{2}}{c - d\sqrt{2}} = \frac{ac - 2bd}{c^2 - 2d^2} + \frac{bc - ad}{c^2 - 2d^2}\sqrt{2}.$$

This procedure of "calculating with radicals" should be very familiar to the reader, and the resulting arithmetic system can easily be shown to contain no inconsistencies and to satisfy the commutative, associative, and distributive laws. However, observe that the symbol $\sqrt{2}$ has not been absorbed by the rational numbers painlessly. Indeed, in the standard form (2) and in the algorithms (3), (4), and (5) its presence stands out like a sore thumb. Actually, we are only using the symbol $\sqrt{2}$ to "hold a place," while we compute around it using the rational components, except for those occasional opportunities when it occurs squared and we are temporarily relieved of having to carry it. So the inclusion of $\sqrt{2}$ as a number is a somewhat artificial process, devised solely so that we might have a richer system in which we can solve the equation $x^2 = 2$.

With this in mind, let us jump to the stage where we have appended all the real numbers to our system. Some of them, such as $\sqrt[4]{17}$, arise as solutions of more complicated equations, while others, such as π and e, come from certain limit processes. Each irrational is absorbed in a somewhat artificial manner, but once again the resulting conglomerate of numbers and arithmetic operations contains no contradictions and satisfies the commutative, associative, and distributive laws.†

At this point we observe that we still cannot solve the equation

$$(6) \qquad\qquad x^2 = -1.$$

But now our experience suggests that we can expand our number system once again by appending a symbol for a solution to Eq. (6); customarily the symbol used is i.‡ Next we imitate the model of expressions (2) through (5) (pertaining to $\sqrt{2}$) and thereby generalize our concept of number as follows:

DEFINITION 1   *A **complex number** is an expression of the form $a + bi$, where a and b are real numbers. Two complex numbers $a + bi$ and $c + di$ are said to be equal $(a + bi = c + di)$ if and only if $a = c$ and $b = d$.*

The operations of addition and subtraction of complex numbers are given by

$$(a + bi) \pm (c + di) = (a \pm c) + (b \pm d)i.$$

In accordance with the distributive law and the proviso that $i^2 = -1$, we postulate the following:
The multiplication of two complex numbers is given by

$$(a + bi)(c + di) = (ac - bd) + (bc + ad)i.$$

---

†The algebraic aspects of extending a number field are discussed in Ref. [7] at the end of this chapter.

‡Engineers often use the letter j.

To compute the quotient of two complex numbers, we again rationalize the denominator:

$$\frac{a + bi}{c + di} = \frac{a + bi}{c + di} \cdot \frac{c - di}{c - di} = \frac{ac + bd}{c^2 + d^2} + \frac{bc - ad}{c^2 + d^2} i.$$

Thus we formally postulate the following:

The division of complex numbers is given by

$$\frac{a + bi}{c + di} = \frac{ac + bd}{c^2 + d^2} + \frac{bc - ad}{c^2 + d^2} i \qquad (\text{if } c^2 + d^2 \neq 0).$$

These are rules for computing in the complex number system. The usual algebraic properties are easy to verify and they appear in the exercises.

EXAMPLE 1   Find the quotient

$$\frac{(6 + 2i) - (1 + 3i)}{(-1 + i) - 2}.$$

SOLUTION

$$\frac{(6 + 2i) - (1 + 3i)}{(-1 + i) - 2} = \frac{5 - i}{-3 + i} = \frac{(5 - i)}{(-3 + i)} \frac{(-3 - i)}{(-3 - i)}$$

$$= \frac{-15 - 1 - 5i + 3i}{9 + 1}$$

$$= -\frac{8}{5} - \frac{1}{5} i. \qquad \blacksquare^{\dagger}$$

Historically, i was considered as an "imaginary" number because of the blatant impossibility of solving Eq. (6) with any of the numbers at hand. With the perspective we have developed, we can see that this label could also be applied to the numbers $\sqrt{2}$ or $\sqrt[4]{17}$; like them, i is simply one more symbol appended to a given number system to create a richer system. Nonetheless, tradition dictates the following designations:

DEFINITION 2   The **real part** *of the complex number $a + bi$ is the (real) number a; its* **imaginary part** *is the (real) number b. If a is zero, the number is said to be a* **pure imaginary number**.

For convenience we frequently use a single letter, usually z, to denote a complex number. Its real and imaginary parts are then written Re z and Im z, respectively. With this notation we have $z = $ Re $z + i$ Im z.

Observe that the equation $z_1 = z_2$ holds if and only if Re $z_1 = $ Re z_2 and Im $z_1 = $ Im z_2. Thus any equation involving complex numbers can be interpreted as a pair of real equations.

†A slug marks the end of solutions or proofs throughout the text.

We wish to point out that unlike the real number system, there is no natural ordering of the complex numbers; it is meaningless, for example, to ask whether $2 + 3i$ is greater than or less than $3 + 2i$. (See Prob. 21.)

Exercises 1.1

1. Verify that $-i$ is also a root of Eq. (6).

2. Verify the commutative, associative, and distributive laws for complex numbers.

3. Notice that 0 and 1 still retain their "identity" properties as complex numbers; i.e., $0 + z = z$ and $1 \cdot z = z$ when z is complex.
   (a) Verify that complex subtraction is the inverse of complex addition (i.e., $z_3 = z_2 - z_1$ if and only if $z_3 + z_1 = z_2$).
   (b) Verify that complex division, as given in the text, is the inverse of complex multiplication (i.e., if $z_2 \neq 0$, $z_3 = z_1/z_2$ if and only if $z_3 z_2 = z_1$).

4. Prove that if $z_1 z_2 = 0$, then $z_1 = 0$ or $z_2 = 0$.

In Problems 5–10, write the number in the form $a + bi$.

5. (a) $-3\left(\dfrac{i}{2}\right)$ (b) $(8 + i) - (5 + i)$

   (c) $(-1 + i)^2$ (d) $\dfrac{2 - i}{\frac{1}{3}}$

   (e) $\dfrac{8i - 1}{i}$ (f) $\dfrac{-1 + 5i}{2 + 3i}$

6. $\dfrac{(8 + 2i) - (1 - i)}{(2 + i)^2}$

7. $\dfrac{2 + 3i}{1 + 2i} - \dfrac{8 + i}{6 - i}$

8. $\left[\dfrac{2 + i}{6i - (1 - 2i)}\right]^2$

9. $i^3(i + 1)^2$

10. $(2 + i)(-1 - i)(3 - 2i)$

11. Prove that $\text{Re}(iz) = -\text{Im } z$.

12. Let k be an integer. Find i^{4k}, i^{4k+1}, i^{4k+2}, and i^{4k+3}.

13. Use the answers to Problem 12 to evaluate

$$3i^{11} + 6i^3 + \frac{8}{i^{20}} + i^{-1}.$$

14. Show that the complex number $z = -1 + i$ satisfies the equation $z^2 + 2z + 2 = 0$.

15. Write the complex equation $z^3 + 5z^2 = z + 3i$ as two real equations.

16. Let z be a complex number such that $\operatorname{Im} z > 0$. Prove that $\operatorname{Im}(1/z) < 0$.

17. Let z_1, z_2 be two complex numbers such that $z_1 + z_2$ and $z_1 z_2$ are each negative real numbers. Prove that z_1 and z_2 must be real numbers.

18. Verify that $\operatorname{Re}(\sum_{j=1}^{n} z_j) = \sum_{j=1}^{n} \operatorname{Re} z_j$ and that $\operatorname{Im}(\sum_{j=1}^{n} z_j) = \sum_{j=1}^{n} \operatorname{Im} z_j$. [The real (imaginary) part of the sum is the sum of the real (imaginary) parts.] Formulate, and then *disprove*, the corresponding conjectures for multiplication.

19. Prove the Binomial Theorem for complex numbers:

$$(z_1 + z_2)^n = z_1^n + \binom{n}{1} z_1^{n-1} z_2 + \cdots + \binom{n}{k} z_1^{n-k} z_2^k + \cdots + z_2^n,$$

where n is a positive integer, and $\binom{n}{k} = n!/k!\,(n-k)!$.

20. Prove that there is no rational number x which satisfies $x^2 = 2$. [HINT: Show that if p/q were a solution, where p and q are integers, then 2 would have to divide both p and q. This contradicts the fact that such a ratio can always be written without common divisors.]

21. The definition of the relation denoted by $>$ in the real number system is based upon the existence of a subset $\mathscr{P}$ (the positive reals) having the following properties:

(i) For any number $\alpha \neq 0$, either α or $-\alpha$ (but not both) belongs to $\mathscr{P}$.

(ii) If α and β belong to $\mathscr{P}$, so does $\alpha + \beta$.

(iii) If α and β belong to $\mathscr{P}$, so does $\alpha \cdot \beta$.

When such a set $\mathscr{P}$ exists we write $\alpha > \beta$ if and only if $\alpha - \beta$ belongs to $\mathscr{P}$.† Prove that the *complex* number system does not possess a nonempty subset $\mathscr{P}$ having properties (i), (ii), and (iii). [HINT: Argue that neither i nor $-i$ could belong to such a set $\mathscr{P}$.]

†Readers familiar with computers may note that this is, in fact, the method by which the statement $\alpha > \beta$ is tested.

7

*Section 1.2
Point
Representation of
Complex Numbers;
Absolute Value and
Complex
Conjugates*

1.2 POINT REPRESENTATION OF COMPLEX NUMBERS; ABSOLUTE VALUE AND COMPLEX CONJUGATES

It is presumed that the reader is familiar with the Cartesian coordinate system (Fig. 1.1) which establishes a one-to-one correspondence between points in the xy-plane and ordered pairs of real numbers. The ordered pair $(-2, 3)$, for example, corresponds to that point P which lies two units to the left of the y-axis and three units above the x-axis.

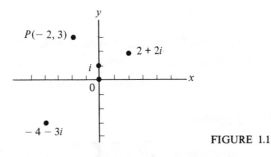

FIGURE 1.1

The Cartesian coordinate system suggests a convenient way to represent complex numbers as points in the xy-plane. Namely, to each complex number $a + bi$ we associate that point in the xy-plane which has the coordinates (a, b). The complex number $-2 + 3i$ is therefore represented by the point P in Fig. 1.1. Also shown in Fig. 1.1 are the points which represent the complex numbers 0, i, $2 + 2i$, and $-4 - 3i$.

When the xy-plane is used for the purpose of describing complex numbers it is referred to as the *complex plane* or *z-plane*.† Since each point on the x-axis represents a real number, this axis is called the *real axis*. Analogously, the y-axis is called the *imaginary axis* for it represents the pure imaginary numbers.

Hereafter, we shall refer to the point which represents the complex number z as simply *the point z*; i.e., the point $z = a + bi$ is the point with coordinates (a, b).

EXAMPLE 2   Suppose that n particles with masses m_1, m_2, ..., m_n are located at the respective points $z_1, z_2, ..., z_n$ in the complex plane. Show that the center of mass of the system is the point

$$\hat{z} = \frac{m_1 z_1 + m_2 z_2 + \cdots + m_n z_n}{m_1 + m_2 + \cdots + m_n}.$$

SOLUTION   Write $z_1 = x_1 + y_1 i$, $z_2 = x_2 + y_2 i$, ..., $z_n = x_n + y_n i$, and let M be the total mass $\sum_{k=1}^{n} m_k$. Presumably the reader will recall

†The term Argand plane is sometimes used.

that the center of mass of the given system is the point with coordinates $(\hat{x}, \hat{y})$, where

$$\hat{x} = \frac{\sum_{k=1}^{n} m_k x_k}{M}, \qquad \hat{y} = \frac{\sum_{k=1}^{n} m_k y_k}{M}.$$

But clearly $\hat{x}$ and $\hat{y}$ are, respectively, the real and imaginary parts of the complex number $(\sum_{k=1}^{n} m_k z_k)/M = \hat{z}$. ∎

Absolute Value　By the Pythagorean theorem the distance from the point $z = a + bi$ to the origin is given by $\sqrt{a^2 + b^2}$. Special notation for this distance is given in

DEFINITION 3　*The* **absolute value** *or* **modulus** *of the number* $z = a + bi$ *is denoted by* $|z|$ *and is given by*

$$|z| = \sqrt{a^2 + b^2}.$$

In particular,

$$|0| = 0, \qquad \left|\frac{i}{2}\right| = \frac{1}{2}, \qquad |3 - 4i| = \sqrt{9 + 16} = 5.$$

The reader should note that $|z|$ is always a nonnegative real number and that the *only* complex number whose modulus is zero is the number 0.

Let $z_1 = a_1 + b_1 i$ and $z_2 = a_2 + b_2 i$. Then

$$|z_1 - z_2| = |(a_1 - a_2) + (b_1 - b_2)i| = \sqrt{(a_1 - a_2)^2 + (b_1 - b_2)^2},$$

which is the distance between the points with coordinates (a_1, b_1) and (a_2, b_2). Hence *the distance between the points* z_1 *and* z_2 *is given by* $|z_1 - z_2|$. This fact is useful in describing certain curves in the plane. Consider, for example, the set of all numbers z which satisfy the equation

(7) $$|z - z_0| = r,$$

where z_0 is a fixed complex number and r is a positive real number. This set consists of all points z whose distance from z_0 is r. Consequently Eq. (7) is the equation of a circle.

EXAMPLE 3　Describe the locus of points z which satisfy the equations

(a)　$|z + 2| = |z - 1|$.　　　　　　　　(b)　$|z - 1| = \text{Re } z + 1$.

SOLUTION　A point z satisfies Eq. (a) if and only if it is equidistant from the points -2 and 1. Hence Eq. (a) is the equation of the perpendicular bisector of the line segment joining -2 and 1; i.e., Eq. (a) describes the line $x = -1/2$.

9

Section 1.2
Point
Representation of
Complex Numbers;
Absolute Value and
Complex
Conjugates

The geometric interpretation of Eq. (b) is less obvious. However, if we replace z by $x + iy$ in Eq. (b), we obtain the Cartesian equation $\sqrt{(x-1)^2 + y^2} = x + 1$, or $y^2 = 4x$, which describes a parabola. ∎

Complex Conjugates The reflection of the point $z = a + bi$ in the real axis is the point $a - bi$ (see Fig. 1.2). As we shall see, the relationship between $a + bi$ and $a - bi$ will play a significant role in the theory of complex variables. We introduce special notation for this concept in

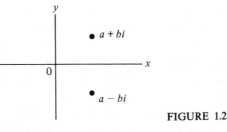

FIGURE 1.2

DEFINITION 4 *The* **complex conjugate** *of the number $z = a + bi$ is denoted by $\bar{z}$ and is given by*

$$\bar{z} = a - bi.$$

Thus

$$\overline{-1 + 5i} = -1 - 5i, \qquad \overline{\pi - i} = \pi + i, \qquad \bar{8} = 8.$$

It follows from Definition 4 that $z = \bar{z}$ if and only if z is a real number. Also it is clear that the conjugate of the sum (difference) of two complex numbers is equal to the sum (difference) of their conjugates; i.e.,

$$\overline{z_1 \pm z_2} = \bar{z}_1 \pm \bar{z}_2.$$

Perhaps not so obvious is the analogous property for multiplication:

EXAMPLE 4 Prove that the conjugate of the product of two complex numbers is equal to the product of the conjugates of these numbers.

SOLUTION It is required to verify that

(8)
$$\overline{(z_1 z_2)} = \bar{z}_1 \bar{z}_2.$$

Write $z_1 = a_1 + b_1 i$, $z_2 = a_2 + b_2 i$. Then

$$\overline{(z_1 z_2)} = \overline{a_1 a_2 - b_1 b_2 + (a_1 b_2 + a_2 b_1)i}$$

$$= a_1 a_2 - b_1 b_2 - (a_1 b_2 + a_2 b_1)i.$$

On the other hand,

$$\bar{z}_1 \bar{z}_2 = (a_1 - b_1 i)(a_2 - b_2 i) = a_1 a_2 - b_1 b_2 - a_1 b_2 i - a_2 b_1 i$$
$$= a_1 a_2 - b_1 b_2 - (a_1 b_2 + a_2 b_1)i.$$

Thus Eq. (8) holds. ▌

There is another, possibly more enlightening, way to see Eq. (8). First, notice that taking conjugates simply amounts to changing the sign of i. Second, recall the role that i plays in computations; it merely "holds a place while we compute around it," replacing its square by -1 whenever possible. Except for these occurrences i is never really absorbed into the computations; we could just as well call it j, or λ, or any other nonnumerical symbol whose square we agree to replace by -1. *In fact, we could replace it throughout by the symbol* $-i$, *since the square of the latter is also* -1. In other words, if we replace i in the expressions for the numbers z_1 and z_2 by $-i$ and then multiply, the only thing different about the product will be the appearance of $-i$ instead of i. But, expressed in terms of conjugation, this is precisely what Eq. (8) claims.

In addition to Eq. (8) the following properties can be seen:

(9)
$$\overline{\left(\frac{z_1}{z_2}\right)} = \frac{\bar{z}_1}{\bar{z}_2} \quad (z_2 \neq 0);$$

(10)
$$z + \bar{z} = 2 \operatorname{Re} z;$$

(11)
$$z - \bar{z} = 2i \operatorname{Im} z;$$

(12)
$$\bar{\bar{z}} = z.$$

Property (10) implies that the sum of a complex number and its conjugate is a real number, while (11) states that the difference of these two complex numbers is a pure imaginary number. Property (12) means that the conjugate of the conjugate of a complex number is the number itself.

It is clear from Definition 4 that

$$|z| = |\bar{z}|;$$

i.e., the points z and $\bar{z}$ are equidistant from the origin. Furthermore, since

$$z\bar{z} = (a + bi)(a - bi) = a^2 + b^2,$$

we have

(13)
$$z\bar{z} = |z|^2.$$

This is a useful fact to remember: The square of the modulus of a complex number equals the number times its conjugate.

11

Section 1.2
Point
Representation of
Complex Numbers;
Absolute Value and
Complex
Conjugates

Observe that we have already employed complex conjugates in Sec. 1.1 in the process of rationalizing the denominator for the division algorithm.

Exercises 1.2

1. Show that the point $(z_1 + z_2)/2$ is the midpoint of the line segment joining z_1 and z_2.

2. Given four particles of masses 2, 1, 3, and 5 located at the respective points $1 + i$, $-3i$, $1 - 2i$, and -6, find the center of mass of this system.

3. Which of the points i, $2 - i$, and -3 is farthest from the origin?

4. Show that if r is a nonnegative real number, then $|rz| = r|z|$.

5. Prove that $|\text{Re } z| \le |z|$ and $|\text{Im } z| \le |z|$.

6. Let $z = 3 - 2i$. Plot the points z, $-z$, $\bar{z}$, $-\bar{z}$, and $1/z$ in the complex plane.

7. Prove that if $|z| = \text{Re } z$, then z is a nonnegative real number.

8. Show that the points 1, $-1/2 + i\sqrt{3}/2$, and $-1/2 - i\sqrt{3}/2$ are the vertices of an equilateral triangle.

9. Show that the points $3 + i$, 6, and $4 + 4i$ are the vertices of a right triangle.

10. Describe the set of points z in the complex plane which satisfy each of the following.
    (a) $\text{Im } z = -2$ (b) $|z - 1 + i| = 3$
    (c) $|2z - i| = 4$ (d) $|z - 1| = |z + i|$
    (e) $|z| = \text{Re } z + 2$ (f) $|z - 1| + |z + 1| = 7$
    (g) $|z| = 3|z - 1|$ (h) $\text{Re } z \ge 4$
    (i) $|z - i| < 2$ (j) $|z| > 6$

11. Verify properties (9), (10), and (11).

12. Prove that if $(\bar{z})^2 = z^2$, then z is either real or pure imaginary.

13. Show that $|z - 1| = |\bar{z} - 1|$.

14. Suppose that four particles each of mass 1 are located at the points $z_1, \bar{z}_1, z_2$, and $\bar{z}_2$. Show that the center of mass of this system lies on the real axis.

15. Prove that $|z_1 z_2| = |z_1||z_2|$. [HINT: Use Eqs. (13) and (8) to show that $|z_1 z_2|^2 = |z_1|^2 |z_2|^2$.]

16. Prove that $(\bar{z})^k = \overline{(z^k)}$ for every integer k.

17. Let $a_0, a_1, \ldots, a_n$ be real constants. Show that if z_0 is a root of the equation $a_0 z^n + a_1 z^{n-1} + \cdots + a_n = 0$, then so is $\bar{z}_0$.

18. Prove that if $|z| = 1$ $(z \neq 1)$, then $\mathrm{Re}[1/(1 - z)] = \frac{1}{2}$.

1.3 VECTORS AND POLAR FORMS

With each point z in the complex plane can be associated a *vector*, namely, the directed line segment from the origin to the point z. Recall that vectors are characterized by length and direction, and that a given vector remains unchanged under translation. Thus the vector determined by $z = 1 + i$ is the same as the vector from the point $2 + i$ to the point $3 + 2i$ (see Fig. 1.3). Note that every vector parallel to the real axis corresponds to a real number, while those parallel to the imaginary axis represent pure imaginary numbers. Observe, also, that the length of the vector associated with z is $|z|$.

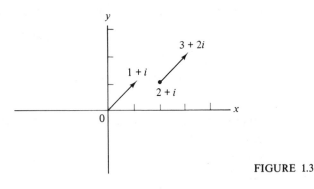

FIGURE 1.3

Let $\mathbf{v}_1$ and $\mathbf{v}_2$ denote the vectors determined by the points z_1 and z_2, respectively. The vector sum $\mathbf{v} = \mathbf{v}_1 + \mathbf{v}_2$ is given by the parallelogram law, which is illustrated in Fig. 1.4. If $z_1 = x_1 + iy_1$ and $z_2 = x_2 + iy_2$, then the terminal point of the vector $\mathbf{v}$ in Fig. 1.4 has the coordinates $(x_1 + x_2, y_1 + y_2)$; i.e., it corresponds to the point $z_1 + z_2$. Thus we see that the correspondence between complex numbers and planar vectors carries over to the operation of addition.

Hereafter, the vector determined by the point z will be simply called *the vector z*.

Recall the geometric fact that the length of any side of a triangle is less than or equal to the sum of the lengths of the other two sides. If we apply this theorem to the triangle in Fig. 1.4 with vertices 0, z_1, and $z_1 + z_2$, we

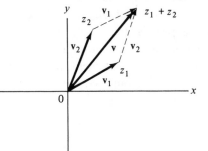

FIGURE 1.4

deduce a very important law relating sum and magnitudes of complex numbers:

The Triangle Inequality *For any two complex numbers z_1 and z_2 we have*

(14)
$$|z_1 + z_2| \le |z_1| + |z_2|.$$

The triangle inequality can easily be extended to more than two complex numbers, as requested in Prob. 20.

The vector $z_2 - z_1$, when added to the vector z_1, obviously yields the vector z_2. Thus $z_2 - z_1$ can be represented as the directed line segment

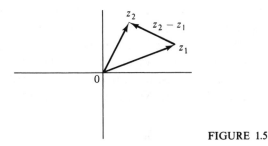

FIGURE 1.5

from z_1 to z_2 (see Fig. 1.5). Applying the geometric theorem to the triangle in Fig. 1.5, we deduce another form of the triangle inequality:

$$|z_2| \le |z_1| + |z_2 - z_1|$$

or

(15)
$$|z_2| - |z_1| \le |z_2 - z_1|.$$

Inequality (15) states that the difference in the lengths of any two sides of a triangle is no greater than the length of the third side.

EXAMPLE 5 Prove that the three distinct points z_1, z_2, and z_3 lie on the same straight line if and only if $z_3 - z_2 = c(z_2 - z_1)$ for some real number c.

SOLUTION Recall that two vectors are parallel if and only if one is a scalar multiple of the other. In the language of complex numbers, this says that z is parallel to w if and only if $z = cw$, where c is real. From Fig. 1.6 we see that the condition that the points z_1, z_2, and z_3 be collinear is equivalent to the statement that the vector $z_3 - z_2$ is parallel to the vector $z_2 - z_1$. Using our characterization of parallelism, the conclusion follows immediately. ∎

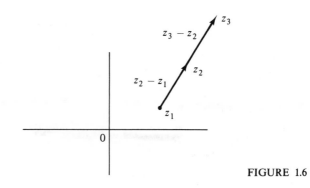

FIGURE 1.6

There is another set of parameters which characterize the vector from the origin to the point z (other, that is, than the real and imaginary parts of z) which more intimately reflects its interpretation as an object with magnitude and direction. These are the polar coordinates, r and θ, of the point z. The coordinate r is the distance from the origin to z, and θ is the angle of inclination of the vector z, measured positively in a counterclockwise sense from the positive real axis (and thus negative when measured clockwise) (see Fig. 1.7). Note that r is never negative, as we use it.

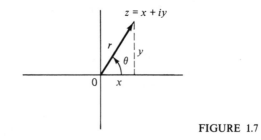

FIGURE 1.7

From Fig. 1.7 we deduce the relations

(16) $$x = r \cos \theta, \qquad y = r \sin \theta,$$

and

(17) $$r = \sqrt{x^2 + y^2} = |z|,$$

while θ, usually given in radians, is determined by the equations

$$(18) \qquad \cos \theta = \frac{x}{|z|}, \qquad \sin \theta = \frac{y}{|z|}.$$

Hence the complex number $z = x + iy$ can be written in the *polar form*

$$(19) \qquad z = r(\cos \theta + i \sin \theta).$$

Conversely, one should observe that whenever z is expressed as $\rho(\cos \phi + i \sin \phi)$ with ρ positive, then ρ must be equal to $|z|$. However, ϕ could differ from θ by an integer multiple of 2π.

In fact, using Eqs. (18) one can determine θ only up to a multiple of 2π. The value of any of these equivalent angles is called the *argument* or *phase* of z and is denoted by

$$\arg z.$$

Clearly, if θ_0 qualifies as a value of arg z, then so do

$$\theta_0 \pm 2\pi, \theta_0 \pm 4\pi, \theta_0 \pm 6\pi, \ldots,$$

and any value of arg z must be one of these. In particular, the values of arg i are

$$\frac{\pi}{2}, \frac{\pi}{2} \pm 2\pi, \frac{\pi}{2} \pm 4\pi, \ldots,$$

and we write

$$\arg i = \frac{\pi}{2} + 2k\pi \qquad (k = 0, \pm 1, \pm 2, \ldots).$$

The ambiguity arises, of course, because of the interpretation of θ as an angle. The argument of $z = 0$ cannot be defined.

Notice that any half-open interval of the real numbers of length 2π will contain one, and only one, value of arg z. For definiteness, the particular value of arg z which lies in the interval $(-\pi, \pi]$ is referred to as the *principal value of the argument* and is denoted by Arg z (note the capitalization). For example, $\text{Arg}(1 - i) = -\pi/4$. Observe that Arg z is a discontinuous function; it jumps by an amount 2π as we cross the negative real axis in the complex plane.

EXAMPLE 6   Find $\arg(1 + \sqrt{3}\,i)$ and write $1 + \sqrt{3}\,i$ in polar form.

SOLUTION   Note that $r = |1 + \sqrt{3}\,i| = 2$ and that the equations $\cos \theta = \frac{1}{2}$, $\sin \theta = \sqrt{3}/2$ are satisfied by $\theta = \pi/3$. Hence $\arg(1 + \sqrt{3}\,i) = \pi/3 + 2k\pi$, $k = 0, \pm 1, \pm 2, \ldots$ [in particular, $\text{Arg}(1 + \sqrt{3}\,i) = \pi/3$] and the desired polar form is $1 + \sqrt{3}\,i = 2(\cos \pi/3 + i \sin \pi/3)$.   ∎

Although the polar form is inconvenient when one is adding complex numbers, it lends a very interesting geometric interpretation to the process of multiplication. If we let

(20) $\quad z_1 = r_1(\cos\theta_1 + i\sin\theta_1), \qquad z_2 = r_2(\cos\theta_2 + i\sin\theta_2),$

then we compute

$$z_1 z_2 = r_1 r_2[(\cos\theta_1\ \cos\theta_2 - \sin\theta_1\ \sin\theta_2)$$
$$+ i(\sin\theta_1\ \cos\theta_2 + \cos\theta_1\ \sin\theta_2)],$$

and so

(21) $\quad z_1 z_2 = r_1 r_2[\cos(\theta_1 + \theta_2) + i\sin(\theta_1 + \theta_2)].$

From Eq. (21) we see that $\theta_1 + \theta_2$ is a particular value of arg $z_1 z_2$. Hence *the argument of the product is the sum of the arguments*, and we write symbolically

(22) $\qquad\qquad \arg z_1 z_2 = \arg z_1 + \arg z_2\,.$

To be precise, Eq. (22) means that if particular values are assigned to any two of the terms, then one can find a value of the third term such that Eq. (22) is satisfied. Also, from Eq. (21) we see that

(23) $\qquad\qquad |z_1 z_2| = |z_1|\,|z_2|\,.$

Geometrically, the vector $z_1 z_2$ has length equal to the product of the lengths of the vectors z_1 and z_2, and has angle of inclination equal to the sum of the angles of inclination of the vectors z_1 and z_2 (see Fig. 1.8). For instance, since the vector i has length 1 and angle of inclination $\pi/2$, it follows that the vector iz can be obtained by rotating the vector z through a right angle in the counterclockwise direction.

Observing that division is the inverse operation to multiplication, we are led to the following equations:

(24) $\qquad\qquad \dfrac{z_1}{z_2} = \dfrac{r_1}{r_2}[\cos(\theta_1 - \theta_2) + i\sin(\theta_1 - \theta_2)],$

(25) $\qquad\qquad \arg\left(\dfrac{z_1}{z_2}\right) = \arg z_1 - \arg z_2\,,$

and

(26) $\qquad\qquad \left|\dfrac{z_1}{z_2}\right| = \dfrac{|z_1|}{|z_2|}\,.$

Equation (24) can be proved in a manner similar to Eq. (21), and Eqs. (25) and (26) follow immediately. Geometrically, the vector z_1/z_2 has length equal to the quotient of the lengths of the vectors z_1 and z_2,

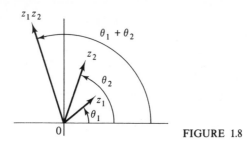

FIGURE 1.8

and has angle of inclination equal to the difference of the angles of inclination of the vectors z_1 and z_2.

As a special case of Eq. (25) we have

$$\arg\left(\frac{1}{z}\right) = -\arg z.$$

EXAMPLE 7 Write the quotient $(1 + i)/(\sqrt{3} - i)$ in polar form.

SOLUTION The polar forms for the numbers $1 + i$ and $\sqrt{3} - i$ are,
respectively, $\sqrt{2}(\cos \pi/4 + i \sin \pi/4)$ and $2[\cos(-\pi/6) + i \sin(-\pi/6)]$.
Hence from Eq. (24) we have

$$\frac{1 + i}{\sqrt{3} - i} = \frac{\sqrt{2}}{2}\left\{\cos\left[\frac{\pi}{4} - \left(-\frac{\pi}{6}\right)\right] + i \sin\left[\frac{\pi}{4} - \left(-\frac{\pi}{6}\right)\right]\right\}$$

$$= \frac{\sqrt{2}}{2}\left[\cos\left(\frac{5\pi}{12}\right) + i \sin\left(\frac{5\pi}{12}\right)\right]. \quad \blacksquare$$

EXAMPLE 8 Prove that the line l through the points z_1 and z_2 is perpendicular to the line L through the points z_3 and z_4 if and only if

$$(27) \qquad \arg\frac{z_1 - z_2}{z_3 - z_4} = \frac{\pi}{2} + 2k\pi \qquad (k = 0, \pm 1, \pm 2, \ldots)$$

or

$$(28) \qquad \arg\frac{z_1 - z_2}{z_3 - z_4} = \frac{3\pi}{2} + 2k\pi \qquad (k = 0, \pm 1, \pm 2, \ldots).$$

SOLUTION Note that the lines l and L are perpendicular if and only
if the vectors $z_1 - z_2$ and $z_3 - z_4$ are perpendicular. Let θ_0 be the particular value of $\arg[(z_1 - z_2)/(z_3 - z_4)]$ which satisfies $0 \le \theta_0 < 2\pi$. Since

$$\arg\frac{z_1 - z_2}{z_3 - z_4} = \arg(z_1 - z_2) - \arg(z_3 - z_4),$$

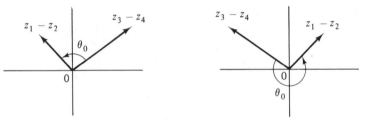

FIGURE 1.9

we deduce that θ_0 is the positive angle with initial side $z_3 - z_4$ and terminal side $z_1 - z_2$ (see Fig. 1.9). Thus the lines l and L are perpendicular if and only if $\theta_0 = \pi/2$ or $\theta_0 = 3\pi/2$, i.e., if and only if Eq. (27) or (28) holds. ∎

Recall that, geometrically, the vector $\bar{z}$ is the reflection in the real axis of the vector z (see Fig. 1.10). Hence we see that *the argument of the conjugate of a complex number is the negative of the argument of the number*; i.e.,

(29) $$\arg \bar{z} = -\arg z.$$

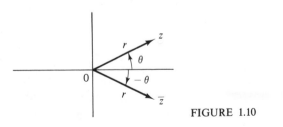

FIGURE 1.10

Exercises 1.3

1. Let $z_1 = 2 - i$ and $z_2 = 1 + i$. Use the parallelogram law to construct each of the following vectors.

    (a) $z_1 + z_2$ (b) $z_1 - z_2$ (c) $2z_1 - 3z_2$

2. Prove that the vector z_1 is parallel to the vector z_2 if and only if $\text{Im}(z_1 \bar{z}_2) = 0$.

3. Show that every point z on the line through the distinct points z_1 and z_2 is of the form $z = z_1 + c(z_2 - z_1)$, where c is a real number.

4. Show that $|z_1 z_2 z_3| = |z_1||z_2||z_3|$.

5. Prove that for any integer k, $|z^k| = |z|^k$.

6. Find the following.

(a) $\left| \dfrac{1 + 2i}{-2 - i} \right|$

(b) $|\overline{(1 + i)}(2 - 3i)(4i - 3)|$

(c) $\left| \dfrac{i(2 + i)^3}{(1 - i)^2} \right|$

(d) $\left| \dfrac{(\pi + i)^{100}}{(\pi - i)^{100}} \right|$

7. Prove that $|z_1 - z_2| \le |z_1| + |z_2|$.

8. Prove that $||z_1| - |z_2|| \le |z_1 - z_2|$.

9. Show geometrically that the nonzero complex numbers z_1 and z_2 satisfy $|z_1 + z_2| = |z_1| + |z_2|$ if and only if they have the same argument.

10. Find the argument of each of the following complex numbers and write each in polar form.

(a) $-1/2$

(b) $-3 + 3i$

(c) $-\pi i$

(d) $-2\sqrt{3} - 2i$

(e) $(1 - i)(-\sqrt{3} + i)$

(f) $(\sqrt{3} - i)^2$

(g) $\dfrac{-1 + \sqrt{3}\,i}{2 + 2i}$

(h) $\dfrac{-\sqrt{7}(1 + i)}{\sqrt{3} + i}$

11. Construct each of the following vectors.

(a) $7\left(\cos \dfrac{3\pi}{4} + i \sin \dfrac{3\pi}{4} \right)$

(b) $4\left[\cos\left(\dfrac{-\pi}{6} \right) + i \sin\left(\dfrac{-\pi}{6} \right) \right]$

(c) $\cos \dfrac{3\pi}{5} + i \sin \dfrac{3\pi}{5}$

(d) $3\left(\cos \dfrac{27\pi}{4} + i \sin \dfrac{27\pi}{4} \right)$

12. Prove that $\arg z_1 z_2 z_3 = \arg z_1 + \arg z_2 + \arg z_3$.

13. Prove that $\arg z_1 \overline{z_2} = \arg z_1 - \arg z_2$.

14. Prove that $\arg z_1 = \arg z_2$ if and only if $z_1 = c z_2$, where c is a positive real number.

15. Find the following.

(a) $\operatorname{Arg}(-6 - 6i)$

(b) $\operatorname{Arg}(-\pi)$

(c) $\operatorname{Arg}(10i)$

(d) $\operatorname{Arg}(\sqrt{3} - i)$

16. Decide which of the following statements are true.

    F (a) Arg $z_1 z_2 = $ Arg $z_1 + $ Arg z_2, $z_1 \neq 0$, $z_2 \neq 0$.
    T (b) Arg $\bar{z} = -$Arg z, z not a real number.
    F (c) Arg$(z_1/z_2) = $ Arg $z_1 - $ Arg z_2, $z_1 \neq 0$, $z_2 \neq 0$.
    T (d) arg $z = $ Arg $z + 2\pi k$, $k = 0, \pm 1, \pm 2, \ldots$, $z \neq 0$.

17. Given the vector z, interpret geometrically the vector $(\cos \phi + i \sin \phi)z$.

18. Let $z_1, z_2,$ and z_3 be distinct points and let ϕ be a particular value of arg$[(z_3 - z_1)/(z_2 - z_1)]$. Prove that

    $$|z_3 - z_2|^2 = |z_3 - z_1|^2 + |z_2 - z_1|^2$$
    $$- 2|z_3 - z_1||z_2 - z_1| \cos \phi.$$

    [HINT: Consider the triangle with vertices z_1, z_2, z_3.]

19. Translate the following geometric theorem into the language of complex numbers: The sum of the squares of the lengths of the diagonals of a parallelogram is equal to the sum of the squares of its sides. (See Fig. 1.4.)

20. Use mathematical induction to prove the *Generalized Triangle Inequality*:

    $$\left| \sum_{k=1}^{n} z_k \right| \leq \sum_{k=1}^{n} |z_k|.$$

21. Let $m_1, m_2,$ and m_3 be three positive real numbers and let $z_1, z_2,$ and z_3 be three complex numbers, each of modulus less than or equal to 1. Use the Generalized Triangle Inequality (Prob. 20) to prove that

    $$\left| \frac{m_1 z_1 + m_2 z_2 + m_3 z_3}{m_1 + m_2 + m_3} \right| \leq 1,$$

    and give a physical interpretation of the inequality.

22. Let $z_1, z_2,$ and z_3 be distinct points that lie on the circle of radius 1 about the origin. Prove that either

    $$\arg \frac{z_2 - z_3}{z_1 - z_3} = \frac{1}{2} \text{Arg} \frac{z_2}{z_1} + 2k\pi \qquad (k = 0, \pm 1, \pm 2, \ldots)$$

    or

    $$\arg \frac{z_2 - z_3}{z_1 - z_3} = -\frac{1}{2} \text{Arg} \frac{z_2}{z_1} + 2k\pi \qquad (k = 0, \pm 1, \pm 2, \ldots).$$

In this section we shall derive formulas for the nth power and nth roots of a complex number.

Let $z = r(\cos \theta + i \sin \theta)$ be the polar form of the complex number z. By taking $z_1 = z_2 = z$ in Eq. (21) we obtain the formula

$$z^2 = r^2(\cos 2\theta + i \sin 2\theta).$$

Since $z^3 = zz^2$, we can apply Eq. (21) a second time (with $z_1 = z, z_2 = z^2$) to deduce that

$$z^3 = r^3(\cos 3\theta + i \sin 3\theta).$$

Continuing in this manner we arrive inductively at the formula for the nth power of z:

(30) $\qquad z^n = r^n(\cos n\theta + i \sin n\theta) \qquad (n = 1, 2, 3, \ldots).$

When $|z| = 1$, Eq. (30) becomes

(31) $\qquad (\cos \theta + i \sin \theta)^n = \cos n\theta + i \sin n\theta,$

which is known as *De Moivre's formula*. This relation is useful in deriving trigonometric identities and in finding the roots of a complex number.

A complex number z is said to be an nth *root* of the complex number z_0 if

(32) $\qquad\qquad\qquad\qquad z^n = z_0 \, .$

Regarding $z_0(\neq 0)$ as a given number let us solve Eq. (32) for the unknown z. Substituting the polar forms

$$z = r(\cos \theta + i \sin \theta), \qquad z_0 = r_0(\cos \theta_0 + i \sin \theta_0)$$

into Eq. (32) and applying formula (30) we obtain

$$r^n(\cos n\theta + i \sin n\theta) = r_0(\cos \theta_0 + i \sin \theta_0).$$

Hence $r^n = r_0$ and $n\theta = \theta_0 + 2k\pi$ for some integer k. Since r and r_0 are positive real numbers, it follows that r is the positive nth root of r_0; i.e., $r = \sqrt[n]{r_0}$. Also $\theta = (\theta_0 + 2k\pi)/n$. Thus

(33) $\qquad z = \sqrt[n]{r_0} \left[\cos\left(\dfrac{\theta_0 + 2k\pi}{n} \right) + i \sin\left(\dfrac{\theta_0 + 2k\pi}{n} \right) \right]$

for some integer k. It may appear to the reader that by choosing $k = 0$, $\pm 1, \pm 2, \ldots$ we can generate infinitely many values for z. This, however, is not the case, for observe that as k takes on the integral values from $k = 0$ to $k = n - 1$ the roots given by (33) all have the same modulus but have different arguments. For other values of k we merely duplicate these n

roots because of the periodicity of the sine and cosine functions. Thus we have shown that *there are exactly n distinct nth roots of z_0*. These roots are denoted by $z_0^{1/n}$ and are given by

$$(34) \qquad z_0^{1/n} = \sqrt[n]{|z_0|}\left[\cos\left(\frac{\theta_0 + 2k\pi}{n}\right) + i \sin\left(\frac{\theta_0 + 2k\pi}{n}\right)\right]$$

$$(k = 0, 1, 2, \ldots, n - 1).$$

Notice that the arguments of the roots are $2\pi/n$ radians apart.

EXAMPLE 9   Find all the cube roots of $\sqrt{2} + i\sqrt{2}$.

SOLUTION   The polar form for $\sqrt{2} + i\sqrt{2}$ is

$$\sqrt{2} + i\sqrt{2} = 2\left(\cos\frac{\pi}{4} + i \sin\frac{\pi}{4}\right).$$

Putting $|z_0| = 2$, $\theta_0 = \pi/4$, and $n = 3$ in Eq. (34) we obtain

$$(\sqrt{2} + i\sqrt{2})^{1/3} = \sqrt[3]{2}\left[\cos\left(\frac{\pi}{12} + \frac{2k\pi}{3}\right) + i \sin\left(\frac{\pi}{12} + \frac{2k\pi}{3}\right)\right]$$

$$(k = 0, 1, 2).$$

Hence $\sqrt[3]{2}(\cos \pi/12 + i \sin \pi/12)$, $\sqrt[3]{2}(\cos 3\pi/4 + i \sin 3\pi/4)$,   and $\sqrt[3]{2}(\cos 17\pi/12 + i \sin 17\pi/12)$ are the cube roots of $\sqrt{2} + i\sqrt{2}$.   ∎

If $z_0 \neq 0$, we see from the above that there are two distinct values for $z_0^{1/2}$. Let w denote one of these values. Then $-w$ must be the other value since $(-w)^2 = w^2 = z_0$. This is consistent with Eq. (34), for the arguments of w and $-w$ differ by π. In particular, $4^{1/2} = \pm 2$, $(2i)^{1/2} = \pm(1 + i)$, and $(-r)^{1/2} = \pm i\sqrt{r}$, where r is any positive real number.

EXAMPLE 10   Let a, b, and c be complex constants and let $a \neq 0$. Prove that the solutions of the equation

$$(35) \qquad az^2 + bz + c = 0$$

are given by the quadratic formula

$$(36) \qquad z = \frac{-b \pm \sqrt{b^2 - 4ac}}{2a},$$

where $\sqrt{b^2 - 4ac}$ denotes one of the values of $(b^2 - 4ac)^{1/2}$.

SOLUTION   By completing the square, Eq. (35) can be written in the equivalent form

$$(2az + b)^2 = b^2 - 4ac.$$

Hence $2az + b = (b^2 - 4ac)^{1/2} = \pm\sqrt{b^2 - 4ac}$, which implies Eq. (36). ∎

Let us apply formula (34) to the number $z_0 = 1$. Since $1 = 1(\cos 0 + i \sin 0)$, we deduce that

$$(37) \qquad 1^{1/n} = \cos\left(\frac{2k\pi}{n}\right) + i \sin\left(\frac{2k\pi}{n}\right) \qquad (k = 0, 1, 2, \ldots, n-1).$$

Taking $k = 1$ we obtain the root

$$\omega_n \equiv \cos\left(\frac{2\pi}{n}\right) + i \sin\left(\frac{2\pi}{n}\right),$$

which is called a primitive nth root of unity.† From De Moivre's formula (31) it is easy to see that the roots (37) are given by

$$1, \omega_n, \omega_n^2, \ldots, \omega_n^{n-1}.$$

Geometrically, the n nth roots of unity form the vertices of a regular n-sided polygon inscribed in the circle of radius 1 about the origin (see Fig. 1.11). One of the vertices of the polygon is the point 1.

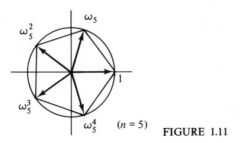

$(n = 5)$ FIGURE 1.11

EXAMPLE 11 Prove that

$$(38) \qquad 1 + \omega_n + \omega_n^2 + \cdots + \omega_n^{n-1} = 0.$$

SOLUTION This result is obvious from a physical point of view since, by symmetry, the center of mass $(1 + \omega_n + \cdots + \omega_n^{n-1})/n$ of the system of n unit masses located at the nth roots of unity must be at the origin.

To give an algebraic proof of Eq. (38) we simply note that

$$(\omega_n - 1)(1 + \omega_n + \omega_n^2 + \cdots + \omega_n^{n-1}) = \omega_n^n - 1 = 0.$$

Since $\omega_n \neq 1$, Eq. (38) follows. ∎

---

†A number w is said to be a *primitive nth root of unity* if $w^n = 1$ but $w^k \neq 1$ for $k = 1, 2, \ldots, n - 1$.

Observe that if z_1 is an nth root of the complex number $z_0 (\neq 0)$, then

$$z_1, \, z_1 \omega_n, \, z_1 \omega_n^2, \, \ldots, \, z_1 \omega_n^{n-1},$$

are *all* the nth roots of z_0, for these n numbers are distinct and satisfy

$$(z_1 \omega_n^k)^n = z_1^n (\omega_n^n)^k = z_1^n = z_0.$$

Exercises 1.4

1. Prove identity (30) by using induction.

2. Show that formula (30) also holds for negative integers n.

3. Let n be a positive integer. Prove that $\arg z^n = n \, \mathrm{Arg} \, z + 2k\pi$, $k = 0, \pm 1, \pm 2, \ldots$.

4. Use identity (30) to show that
   (a) $(\sqrt{3} - i)^7 = -64\sqrt{3} + i64$ (b) $(1 + i)^{95} = 2^{47}(1 - i)$

5. Find all the values of the following.
   (a) $(-16)^{1/4}$ (b) $1^{1/5}$
   (c) $i^{1/4}$ (d) $(1 - \sqrt{3}i)^{1/3}$
   (e) $(i - 1)^{1/2}$

6. Use De Moivre's formula to prove the following identities.
   (a) $\cos 3\theta = \cos^3 \theta - 3 \cos \theta \sin^2 \theta$
   (b) $\sin 3\theta = 3 \cos^2 \theta \sin \theta - \sin^3 \theta$

7. Let a, b, and c be real numbers and let $a \neq 0$. Show that the equation $az^2 + bz + c = 0$ has
   (a) two real solutions if $b^2 - 4ac > 0$.
   (b) two nonreal conjugate solutions if $b^2 - 4ac < 0$.

8. Solve each of the following equations.
   (a) $2z^2 + z + 3 = 0$
   (b) $z^2 - (3 - 2i)z + 1 - 3i = 0$
   (c) $z^2 - 2z + i = 0$

9. Solve the equation $z^3 - 3z^2 + 6z - 4 = 0$.

10. Solve the equation $(z + 1)^5 = z^5$.

11. Find all four roots of the equation $z^4 + 1 = 0$ and use them to deduce the factorization $z^4 + 1 = (z^2 - \sqrt{2}z + 1)(z^2 + \sqrt{2}z + 1)$.

12. Show that the n points $z_0^{1/n}$ form the vertices of a regular n-sided polygon inscribed in the circle of radius $\sqrt[n]{|z_0|}$ about the origin.

13. Show that $\omega_3 = (-1 + \sqrt{3}\,i)/2$ and that $\omega_4 = i$. Use these values to verify identity (38) for the special cases $n = 3$ and $n = 4$.

14. Figure 1.11 shows the root ω_5 raised to the first, second, third, fourth, and fifth powers. Using the same diagram, label the points which correspond to the first five powers of ω_5^2. Do the same for the powers of ω_5^3 and ω_5^4.

15. Let m and n be positive integers which have no common factor. Prove that the set of numbers $(z^{1/n})^m$ is the same as the set of numbers $(z^m)^{1/n}$. We denote this common set of numbers by $z^{m/n}$. Show that

$$z^{m/n} = \sqrt[n]{|z|^m}\left[\cos\frac{m}{n}(\theta + 2k\pi) + i\sin\frac{m}{n}(\theta + 2k\pi)\right]$$

$$(k = 0,\ 1,\ 2,\ \ldots,\ n-1).$$

16. Use the result of Prob. 15 to find all the values of $(1 - i)^{3/2}$.

17. Prove that if $z \neq 1$, then

$$1 + z + z^2 + \cdots + z^n = \frac{z^{n+1} - 1}{z - 1}.$$

Use this result and De Moivre's formula to establish the following identities.

(a) $1 + \cos\theta + \cos 2\theta + \cdots + \cos n\theta = \dfrac{1}{2} + \dfrac{\sin[(n + \frac{1}{2})\theta]}{2\sin\left(\dfrac{\theta}{2}\right)}$

(b) $\sin\theta + \sin 2\theta + \cdots + \sin n\theta = \dfrac{1}{2}\cot\left(\dfrac{\theta}{2}\right) - \dfrac{\cos[(n + \frac{1}{2})\theta]}{2\sin\left(\dfrac{\theta}{2}\right)},$

where $0 < \theta < 2\pi$.

18. Let n be a fixed positive integer and let l be an integer that is not divisible by n. Prove the following generalization of Eq. (38):

$$1 + \omega_n^l + \omega_n^{2l} + \cdots + \omega_n^{(n-1)l} = 0.$$

1.5 PLANAR SETS

In this section we shall give some basic definitions concerning point sets in the plane.

The set of all points z which satisfy the inequality

$$|z - z_0| < \rho,$$

where ρ is a positive real number, is called an *open disk* or *neighborhood of* z_0. This set consists of all the points which lie inside the circle of radius ρ about z_0. In particular, the solution sets of the inequalities

$$|z - 2| < 3, \qquad |z + i| < \frac{1}{2}, \qquad |z| < 8$$

are neighborhoods of the respective points 2, $-i$, and 0. We shall make frequent reference to the neighborhood $|z| < 1$, which is called the *open unit disk*.

A point z_0 which lies in a set S is called an *interior point of* S if there is some neighborhood of z_0 which is completely contained in S. For example, if S is the right half-plane Re $z > 0$ and $z_0 = .01$, then z_0 is an interior point of S because S contains the neighborhood $|z - z_0| < .01$ (see Fig. 1.12).

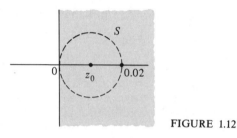

FIGURE 1.12

If every point of a set S is an interior point of S, we say that S is an *open set*. Any neighborhood is an open set (Prob. 1). Each of the following inequalities also describes an open set: (a) $\rho_1 < |z - z_0| < \rho_2$, (b) $|z - 3| > 2$, (c) Im $z > 0$, and (d) $1 < $ Re $z < 2$. These sets are sketched in Fig. 1.13. Note that the solution set T of the inequality $|z - 3| \geq 2$ is

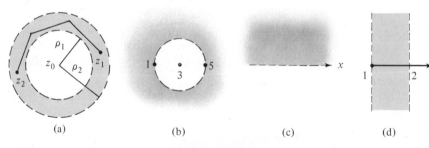

(a)　　　　　　(b)　　　　(c)　　　　(d)

FIGURE 1.13

not an open set since no point on the circle $|z - 3| = 2$ is an interior point of T. Note also that an open interval of the real axis is not an open set since it contains no open disk.

Let $w_1, w_2, \ldots, w_{n+1}$ be $n + 1$ points in the plane. For each $k = 1, 2, \ldots, n$, let l_k denote the line segment joining w_k to w_{k+1}. Then the successive line segments l_1, l_2, ..., l_n form a continuous chain known as a *polygonal line* that joins w_1 to w_{n+1}.

An open set S is said to be *connected* if every pair of points z_1, z_2 in S can be joined by a polygonal line that lies entirely in S [see Fig. 1.13(a)]. Roughly speaking, this means that S consists of a "single piece." Each of the sets in Fig. 1.13 is connected. The set consisting of all those points in the plane that do not lie on the circle $|z| = 1$ is an example of an open set which is not connected; indeed, if z_1 is a point inside the circle and z_2 is a point outside, then every polygonal line that joins z_1 and z_2 must intersect the circle.

We call an open connected set a *domain*. Therefore, all the sets in Fig. 1.13 are domains.

A point z_0 is said to be a *boundary point* of a set S if every neighborhood of z_0 contains at least one point of S and at least one point not in S. The set of all boundary points of S is called the *boundary* or *frontier* of S. The boundaries of the sets in Fig. 1.13 are as follows: (a) the two circles $|z - z_0| = \rho_1$ and $|z - z_0| = \rho_2$, (b) the circle $|z - 3| = 2$, (c) the real axis, and (d) the two lines Re $z = 1$ and Re $z = 2$. Since each point of a domain D is an interior point of D, it follows that a domain cannot contain any of its boundary points.

A *region* is a domain together with some, none, or all of its boundary points. In particular, every domain is a region. If a region contains all its boundary points, we say that the region is *closed*. The region described by the inequality $0 < |z| \le 1$ is not closed since it does not contain the boundary point 0. The set of points z which satisfy the inequality

$$|z - z_0| \le \rho \qquad (\rho > 0)$$

is a closed region, for it consists of the open disk $|z - z_0| < \rho$ together with its boundary $|z - z_0| = \rho$. Therefore we call this set a *closed disk*.

A set of points S is said to be *bounded* if there exists a positive real number R such that $|z| < R$ for every z in S. In other words, S is bounded if it is contained in some neighborhood of the origin. An *unbounded* set is one which is not bounded. Of the sets in Fig. 1.13 only (a) is bounded. A region which is both closed and bounded is said to be *compact*.

Exercises 1.5

1. Prove that the neighborhood $|z - z_0| < \rho$ is an open set.

Problems 2–8 refer to the sets described by the following inequalities:

(a) $|z - 1 + i| \le 3$ (b) $|\operatorname{Arg} z| < \dfrac{\pi}{4}$

(c) $0 < |z - 2| < 3$ (d) $-1 < \operatorname{Im} z \le 1$

(e) $|z| \ge 2$ (f) $(\operatorname{Re} z)^2 > 1$

2. Sketch each of the given sets.

3. Which of the given sets are open?

4. Which of the given sets are domains?

5. Which of the given sets are bounded?

6. Describe the boundary of each of the given sets.

7. Which of the given sets are regions?

8. Which of the given sets are closed regions?

9. Prove that any set consisting of finitely many points is bounded.

10. Prove that the closed disk $|z - z_0| \le \rho$ is bounded.

11. Let S be the set consisting of the points $1, \frac{1}{2}, \frac{1}{3}, \ldots$. What is the boundary of S?

12. Let z_0 be a point of the set S. Prove that if z_0 is not an interior point of S, then z_0 must be a boundary point of S.

13. Let R be a closed region (not the whole plane). Prove that the set of all points not in R (i.e., the *complement* of R) is an open set.

14. A point z_0 is said to be an *accumulation point* of a set S if every neighborhood of z_0 contains infinitely many points of the set S. Prove that a closed region contains all its accumulation points.

Problems 15–18 refer to the following definitions: Let S and T be sets. The set consisting of all points belonging to S or T or both S and T is called the *union* of S and T and is denoted by $S \cup T$. The set consisting of all points belonging to both S and T is called the *intersection* of S and T and is denoted by $S \cap T$.

15. Let S and T be the sets described by $|z + 1| < 2$ and $|z - i| < 1$, respectively. Sketch the sets $S \cup T$ and $S \cap T$.

16. If S and T are open sets, prove that $S \cup T$ is an open set.

17. Show that if S and T are domains which have at least one point in common, then $S \cup T$ is a domain.

18. If S and T are domains, is $S \cap T$ necessarily a domain?

SUMMARY

The complex number system is an extension of the real number system and consists of all expressions of the form $a + bi$, where a and b are real and $i^2 = -1$. The operations of addition, subtraction, multiplication, and division with complex numbers are performed in a manner analogous to "computing with radicals." Geometrically, complex numbers can be represented by points or vectors in the plane. Thus certain theorems from geometry, such as the triangle inequality, can be translated into the language of complex numbers. Associated with a complex number $z = a + bi$ are its absolute value, given by $|z| = \sqrt{a^2 + b^2}$, and its complex conjugate, given by $\bar{z} = a - bi$. The former is the distance from the point z to the origin, while the latter is the reflection of the point z in the x-axis. The numbers z, $\bar{z}$, and $|z|$ are related by $z\bar{z} = |z|^2$.

Every nonzero complex number z can be written in the polar form $z = r(\cos \theta + i \sin \theta)$, where $r = |z|$ and θ is the angle of inclination of the vector z. Any of the equivalent angles $\theta + 2k\pi$, $k = 0, \pm 1, \pm 2, \ldots$, is called the argument of z (arg z). The polar form is useful in finding powers and roots of z.

Special terminology is used in describing point sets in the plane. Important is the concept of a domain D. Such a set is characterized by two properties: (i) Each point z of D is the center of an open disk completely contained in D. (ii) Each pair of points z_1 and z_2 in D can be joined by a polygonal (broken) line that lies entirely in D. If some of the boundary points are adjoined to a domain, the resulting set is called a region.

SUGGESTED READING

Introductory Level

[1]   CHURCHILL, R. V. *Complex Variables and Applications*, 2nd ed. McGraw-Hill Book Company, New York, 1960.

[2]   FISHER, ROBERT C., and ZIEBUR, ALLEN D. *Integrated Algebra and Trigonometry*. Prentice-Hall, Inc., Englewood Cliffs, N.J., 1962.

[3]   LEVINSON, NORMAN, and REDHEFFER, RAYMOND M. *Complex Variables*. Holden-Day, Inc., San Francisco, 1970.

Advanced Level

[4]   AHLFORS, L. V. *Complex Analysis*, 2nd ed. McGraw-Hill Book Company, New York, 1966.

[5]   HILLE, E. *Analytic Function Theory*, Vol. I. Ginn/Blaisdell, Waltham, Mass., 1959.

[6]   NEHARI, Z. *Conformal Mapping*. McGraw-Hill Book Company, New York, 1952.

Extending Number Fields

[7]   BIRKHOFF, G., and MACLANE, S. *A Brief Survey of Modern Algebra*. The Macmillan Company, New York, 1953.

2

Analytic Functions

2.1 FUNCTIONS OF
A COMPLEX VARIABLE

The concept of a complex number was introduced in Chapter 1 in order
to solve certain algebraic equations. We shall now proceed to a study of
the functions defined on these complex variables. It turns out that certain
of these functions, called analytic functions, have remarkable properties
which make them very useful in the application of mathematics to physical problems.

Recall that a function f is a rule which assigns to each element in a set
A one and only one element in a set B. If f assigns the value b to the
element a in A, we write

$$b = f(a),$$

and call b the *image* of a under f. The set A is the *domain of definition* of f
(even if A is not a domain in the sense of Chapter 1), and the set of all

30

images $f(a)$ is the *range* of f. We sometimes refer to f as a *mapping* of A into B.

Here we shall be concerned with complex-valued functions of a complex variable, so that the domains of definition and ranges are subsets of the complex numbers. If w denotes the value of the function f at the point z, we then write $w = f(z)$. Just as z decomposes into real and imaginary parts as $z = x + iy$, the real and imaginary parts of w are each (real-valued) functions of z or, equivalently, of x and y, and so we customarily write

$$w = u(x, y) + iv(x, y),$$

with u and v denoting the real and imaginary parts, respectively, of w. Thus a complex-valued function of a complex variable is, in essence, a pair of real functions of two real variables.

EXAMPLE 1 Write the function $w = f(z) = z^2 + 2z$ in the form $w = u(x, y) + iv(x, y)$.

SOLUTION Setting $z = x + iy$ we obtain

$$w = f(z) = (x + iy)^2 + 2(x + iy) = x^2 - y^2 + i2xy + 2x + i2y.$$

Hence $w = (x^2 - y^2 + 2x) + i(2xy + 2y)$ is the desired form. ∎

Unfortunately, it is generally impossible to draw the graph of a complex function; to display two real functions of two real variables graphically would require four dimensions. We can, however, visualize some of the properties of a complex function $w = f(z)$ by sketching its domain of definition in the z-plane and its range in the w-plane, and depicting the relationship as in Fig. 2.1.

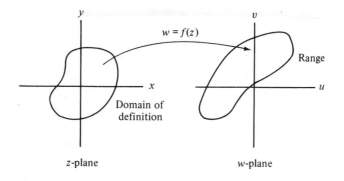

FIGURE 2.1

EXAMPLE 2   Describe the range of the function $f(z) = x^2 + 2i$ defined on the closed unit disk $|z| \leq 1$.

SOLUTION   We have $u(x, y) = x^2$ and $v(x, y) = 2$. Thus as z varies over the closed unit disk, u varies between 0 and 1, and v is constant. The range is therefore the line segment from $w = 2i$ to $w = 1 + 2i$.   ∎

EXAMPLE 3   Describe the function $f(z) = z^3$ for z in the semidisk given by $|z| \leq 2$, Im $z \geq 0$ [see Fig. 2.2(a)].

SOLUTION   From Sec. 1.4, we know that the points z in the sector of the semidisk from Arg $z = 0$ to Arg $z = 2\pi/3$, when cubed, cover the entire disk $|w| \leq 8$. The cubes of the remaining z-points also fall in this disk, overlapping it in the upper half-plane as depicted in Fig. 2.2(b).   ∎

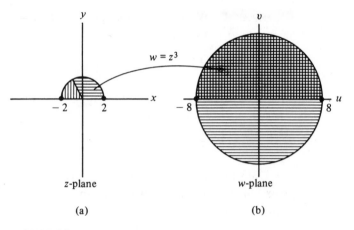

(a)   (b)

FIGURE 2.2

2.2 LIMITS AND CONTINUITY

As we observed in Chapter 1, the definition of absolute value can be used to designate the distance between two complex numbers. Having a concept of distance, we can proceed to introduce the notions of limit and continuity.

Informally, when we have an infinite sequence of complex numbers z_1, z_2, z_3, ..., we say that the number z_0 is the limit of the sequence if the z_n eventually (i.e., for large enough n) stay arbitrarily close to z_0. More precisely,

DEFINITION 1 *A sequence of complex numbers* $\{z_n\}_1^\infty$ *is said* **to have the limit** z_0 *or* **to converge to** z_0, *and we write*

$$\lim_{n \to \infty} z_n = z_0$$

or, equivalently,

$$z_n \to z_0 \quad as \quad n \to \infty,$$

if for any $\varepsilon > 0$ *there exists an integer* N *such that* $|z_n - z_0| < \varepsilon$ *for all* $n > N$.

Geometrically, this means that each term z_n, for $n > N$, lies in the open disk of radius ε about z_0 (see Fig. 2.3).

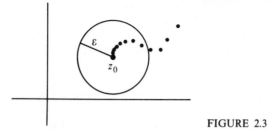

FIGURE 2.3

A related concept is the limit of a complex-valued function $f(z)$. Roughly speaking, we say that the number w_0 is the limit of the function $f(z)$ as z approaches z_0, if $f(z)$ stays close to w_0 whenever z is sufficiently near z_0. In precise terms we state

DEFINITION 2 *Let* $f(z)$ *be a function defined in some neighborhood of* z_0, *with the possible exception of the point* z_0 *itself. We say that the* **limit of** $f(z)$ **as** z **approaches** z_0 **is the number** w_0 *and write*

$$\lim_{z \to z_0} f(z) = w_0$$

or, equivalently,

$$f(z) \to w_0 \quad as \quad z \to z_0,$$

if for any $\varepsilon > 0$ *there exists a positive number* δ *such that*

$$|f(z) - w_0| < \varepsilon \quad whenever \quad 0 < |z - z_0| < \delta.$$

Geometrically, this says that *any* neighborhood of w_0 contains all the values assumed by f in some full neighborhood of z_0, except possibly the value $f(z_0)$; see Fig. 2.4.

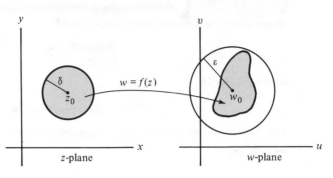

FIGURE 2.4

EXAMPLE 4   Prove that $\lim_{z \to i} z^2 = -1$.

SOLUTION   We must show that for given $\varepsilon > 0$ there is a positive number δ such that

$$|z^2 - (-1)| < \varepsilon \quad \text{whenever} \quad 0 < |z - i| < \delta.$$

Since

$$z^2 - (-1) = z^2 + 1 = (z - i)(z + i) = (z - i)(z - i + 2i),$$

it follows from the properties of absolute value derived in Chapter 1 [particularly relations (14) and (23)] that

(1) $|z^2 - (-1)| = |z - i|\,|z - i + 2i| \leq |z - i|(|z - i| + 2).$

Thus to ensure that the left-hand member of (1) is less than ε we merely have to insist that z lie in a neighborhood of i whose radius δ is less than each of the numbers 1 and $\varepsilon/3$, for then the right-hand member of (1) will satisfy

$$|z - i|(|z - i| + 2) < \frac{\varepsilon}{3}(1 + 2) = \varepsilon. \quad \blacksquare$$

There is an obvious relation between the limit of a function and the limit of a sequence; namely, if $\lim_{z \to z_0} f(z) = w_0$, then for every sequence $\{z_n\}_1^\infty$ converging to z_0 ($z_n \neq z_0$) the sequence $\{f(z_n)\}_1^\infty$ converges to w_0. The converse of this statement is also valid and is left as an exercise.

The condition of continuity is expressed in

DEFINITION 3   Let $f(z)$ be a function defined in a neighborhood of z_0. Then $f(z)$ is **continuous at** *z_0 if*

$$\lim_{z \to z_0} f(z) = f(z_0).$$

In other words, for $f(z)$ to be continuous at z_0, it must have a limiting value at z_0, and this limiting value must be $f(z_0)$.

A function $f(z)$ is said to be *continuous on a set S* if it is continuous at each point of S.

Clearly the definitions of this section are direct analogues of concepts introduced in elementary calculus. In fact, one can show that $f(z)$ approaches a limit precisely when its real and imaginary parts approach limits (see Prob. 19); similarly, the continuity of the latter functions is equivalent to the continuity of f. Because of the analogy, many of the familiar theorems on real sequences, limits, and continuity remain valid in the complex case. Two such theorems are stated below.

THEOREM 1 *If* $\lim_{z \to z_0} f(z) = A$ *and* $\lim_{z \to z_0} g(z) = B$, *then*

(i) $\lim_{z \to z_0} (f(z) \pm g(z)) = A \pm B$,

(ii) $\lim_{z \to z_0} f(z)g(z) = AB$,

(iii) $\lim_{z \to z_0} \dfrac{f(z)}{g(z)} = \dfrac{A}{B}$ *if* $B \neq 0$.

THEOREM 2 *If* $f(z)$ *and* $g(z)$ *are continuous at* z_0, *then so are* $f(z) \pm g(z)$ *and* $f(z)g(z)$. *The quotient* $f(z)/g(z)$ *is also continuous at* z_0 *provided* $g(z_0) \neq 0$.

(Theorem 2 is, in fact, an immediate consequence of Theorem 1.)

One can easily verify that the constant functions as well as the function $f(z) = z$ are continuous on the whole plane. Thus from Theorem 2 we deduce that the *polynomial functions* in z, i.e., functions of the form

$$a_0 + a_1 z + a_2 z^2 + \cdots + a_n z^n,$$

where the a_i are constants, are also continuous on the whole plane. *Rational functions* in z, which are defined as quotients of polynomials, i.e.,

$$\frac{a_0 + a_1 z + \cdots + a_n z^n}{b_0 + b_1 z + \cdots + b_m z^m},$$

are therefore continuous at each point where the denominator does not vanish. These considerations provide a much simpler solution for problems such as Example 4, as we illustrate below:

EXAMPLE 5 Find the limits, as $z \to 2i$, of the functions $f_1(z) = z^2 - 2z + 1$, $f_2(z) = (z + 2i)/z$, and $f_3(z) = (z^2 + 4)/z(z - 2i)$.

SOLUTION Since $f_1(z)$ and $f_2(z)$ are continuous at $z = 2i$, we simply evaluate them there, i.e.,

$$\lim_{z \to 2i} f_1(z) = f_1(2i) = (2i)^2 - 2(2i) + 1 = -3 - 4i,$$

$$\lim_{z \to 2i} f_2(z) = f_2(2i) = \frac{2i + 2i}{2i} = 2.$$

The function $f_3(z)$ is not continuous at $z = 2i$ because it is not defined there (the denominator vanishes). However, for $z \neq 2i$ and $z \neq 0$ we have

$$f_3(z) = \frac{(z + 2i)(z - 2i)}{z(z - 2i)} = \frac{z + 2i}{z} = f_2(z),$$

and so

$$\lim_{z \to 2i} f_3(z) = \lim_{z \to 2i} f_2(z) = 2. \quad \blacksquare$$

Note that in the above example the discontinuity of $f_3(z)$ at $z = 2i$ can be removed by suitably defining the function at this point [set $f_3(2i) = 2$]. In general, if a function can be defined or redefined at a single point z_0 so as to be continuous there, we say that this function has a *removable discontinuity* at z_0.

In closing this section, we wish to emphasize an important distinction between the concepts of limit in the (one-dimensional) real and complex cases. For the latter situation, observe that a sequence $\{z_n\}_1^\infty$ may approach a limit z_0 from *any* direction in the plane, or even along a spiral, etc. Thus the manner in which a sequence of numbers approaches its limit can be much more complicated in the complex case.

Exercises 2.2

1. Write each of the following functions in the form $w = u(x, y) + iv(x, y)$.

    (a) $f(z) = 3z^2 + 5z + i + 1$ (b) $g(z) = \dfrac{1}{z}$

    (c) $h(z) = \dfrac{z + i}{z^2 + 1}$ (d) $q(z) = \dfrac{2z^2 + 3}{|z - 1|}$

2. Find the domain of definition of each of the functions in Prob. 1.

3. Describe the range of each of the following functions.

    (a) $f(z) = z + 5$ for Re $z > 0$

    (b) $g(z) = z^2$ for z in the first quadrant, Re $z \geq 0$, Im $z \geq 0$

(c) $h(z) = \dfrac{1}{z}$ for $0 < |z| \le 1$

(d) $p(z) = -2z^3$ for z in the quarter-disk $|z| < 1$, $0 < \text{Arg } z < \dfrac{\pi}{2}$

4. A uniformly charged infinite rod, standing perpendicular to the z-plane at the point z_0, generates an electric field at every point in the plane. The intensity of this field varies inversely as the distance from z_0 to the point and is directed along the line from z_0 to the point.

   (a) Show that the (vector) field at the point z is given by the function $F(z) = 1/(\bar{z} - \bar{z}_0)$, in appropriate units.

   (b) If three such rods are located at the points $1 + i$, $-1 + i$, and 0, find the positions of equilibrium (i.e., the points where the vector sum of the fields is zero).

5. Using Definition 1, prove that $\lim_{n \to \infty} z_n = 0$ if and only if $\lim_{n \to \infty} |z_n| = 0$.

6. Sketch the first five terms of the sequence $(i/2)^n$, $n = 1, 2, 3, \ldots$, and then describe the convergence of this sequence.

7. Prove that if $|z_0| < 1$, then $z_0^n \to 0$ as $n \to \infty$.

8. Decide whether each of the following sequences converge, and if so, find its limit.

   (a) $z_n = \dfrac{i}{n}$ (b) $z_n = i(-1)^n$

   (c) $z_n = \dfrac{n(2 + i)}{n + 1}$ (d) $z_n = \left(\dfrac{1 - i}{4}\right)^n$

9. Use Definition 2 to prove that $\lim_{z \to 1 + i}(6z - 4) = 2 + 6i$.

10. Use Definition 2 to prove that $\lim_{z \to -i} 1/z = i$.

11. Prove Theorem 1.

12. Find each of the following limits.

    (a) $\lim_{z \to 2 + 3i} (z - 5i)^2$ (b) $\lim_{z \to 2} \dfrac{z^2 + 3}{iz}$

    (c) $\lim_{z \to 3i} \dfrac{z^2 + 9}{z - 3i}$ (d) $\lim_{z \to i} \dfrac{z^2 + 1}{z^4 - 1}$

    (e) $\lim_{\Delta z \to 0} \dfrac{(z_0 + \Delta z)^2 - z_0^2}{\Delta z}$

13. Let $f(z)$ be defined by

$$f(z) = \begin{cases} \dfrac{2z}{z+1} & \text{if } z \neq 0, \\ 1 & \text{if } z = 0. \end{cases}$$

At which points does $f(z)$ have a limit, and at which points is it continuous? Which of the discontinuities of $f(z)$ are removable?

14. Prove that the function $g(z) = \bar{z}$ is continuous on the whole plane.

15. Prove that if $f(z)$ is continuous at z_0, then so are the functions $\overline{f(z)}$, Re $f(z)$, Im $f(z)$, and $|f(z)|$. [HINT: To show that $|f(z)|$ is continuous at z_0 use inequality (15) of Chapter 1.]

16. Let g be a function defined in a neighborhood of z_0 and let f be a function defined in a neighborhood of the point $g(z_0)$. Show that if g is continuous at z_0 and f is continuous at $g(z_0)$, then the composite function $f(g(z))$ is continuous at z_0.

17. Let $f(z) = [x^2/(x^2 + y^2)] + 2i$. Does f have a limit at $z = 0$? [HINT: Investigate $\{f(z_n)\}$ for sequences $\{z_n\}$ approaching 0 along the real and imaginary axes separately.]

18. Show that the function Arg z is discontinuous at each point on the nonpositive real axis.

19. Let $f(z) = u(x, y) + iv(x, y)$, $z_0 = x_0 + iy_0$, and $w_0 = u_0 + iv_0$. Prove that

$$\lim_{z \to z_0} f(z) = w_0$$

if, and only if,

$$\lim_{\substack{x \to x_0 \\ y \to y_0}} u(x, y) = u_0 \quad \text{and} \quad \lim_{\substack{x \to x_0 \\ y \to y_0}} v(x, y) = v_0.$$

[HINT: Use the Triangle Inequality and Prob. 5 in Exercises 1.2.]

20. Use Prob. 19 to find $\lim_{z \to 1 - i}[x/(x^2 + 3y)] + ixy$.

21. Show that if $\lim_{n \to \infty} f(z_n) = w_0$ for every sequence $\{z_n\}_1^\infty$ converging to z_0 ($z_n \neq z_0$), then $\lim_{z \to z_0} f(z) = w_0$. [HINT: Show that if this were not true, then one could construct a sequence $\{z_n\}_1^\infty$ violating the hypothesis.]

2.3 ANALYTICITY

The derivative of a complex function is defined by a natural extension of the definition in the real case:

DEFINITION 4 *Let $f(z)$ be a complex-valued function defined in a neighborhood of z_0. Then the **derivative** of $f(z)$ at z_0 is given by*

$$f'(z_0) = \lim_{\Delta z \to 0} \frac{f(z_0 + \Delta z) - f(z_0)}{\Delta z},$$

provided this limit exists.

It is important to observe that Δz is a complex number, and so for $f(z)$ to be differentiable at z_0 the above difference quotient must tend to a unique complex number $f'(z_0)$ independent of the manner in which Δz approaches zero. Sometimes the symbol df/dz will be used instead of f'.

As in the real case, differentiability implies continuity (see Prob. 3). Furthermore, the usual laws of differentiation remain valid; i.e., if f and g are differentiable, then

(2) $$(f \pm g)'(z_0) = f'(z_0) \pm g'(z_0),$$

(3) $$(cf)'(z_0) = cf'(z_0) \text{ (for any constant } c),$$

(4) $$(fg)'(z_0) = f(z_0)g'(z_0) + f'(z_0)g(z_0),$$

(5) $$\left(\frac{f}{g}\right)'(z_0) = \frac{g(z_0)f'(z_0) - f(z_0)g'(z_0)}{g(z_0)^2} \quad \text{if } g(z_0) \neq 0,$$

and

(6) $$\frac{d}{dz} f(g(z)) = f'(g(z))g'(z),$$

which is the "chain rule." The proof of any of these laws can be established by referring to the proof for the real case and interpreting the quantities as complex numbers. As an example, we verify the usual rule for derivatives of powers of z.

EXAMPLE 6   Show that, for any positive integer n,

(7) $$\frac{d}{dz} z^n = nz^{n-1}.$$

SOLUTION   Using the Binomial Theorem (Prob. 19 in Exercises 1.1) we find

$$\frac{(z + \Delta z)^n - z^n}{\Delta z} = \frac{nz^{n-1}\,\Delta z + \frac{n(n-1)}{2} z^{n-2}(\Delta z)^2 + \cdots + (\Delta z)^n}{\Delta z}.$$

Thus

$$\frac{d}{dz} z^n = \lim_{\Delta z \to 0} \frac{(z + \Delta z)^n - z^n}{\Delta z} = nz^{n-1}. \quad \blacksquare$$

In particular, it follows from this example and rules (2) and (3) that any polynomial in z

$$P(z) = a_n z^n + a_{n-1} z^{n-1} + \cdots + a_1 z + a_0$$

is differentiable in the whole plane and that its derivative is given by

$$P'(z) = n a_n z^{n-1} + (n-1) a_{n-1} z^{n-2} + \cdots + a_1.$$

Consequently, from rule (5), any rational function of z is differentiable at every point in its domain of definition. We see then that for purposes of differentiation, polynomial and rational functions in z can be treated as if z were a real variable.

In contrast, the simple function $f(z) = \bar{z}$ is *not* differentiable at any point. The difference quotient of Definition 4 becomes, in this case,

$$\frac{\overline{(z_0 + \Delta z)} - \bar{z}_0}{\Delta z} = \frac{\overline{\Delta z}}{\Delta z}.$$

Now if $\Delta z \to 0$ through real values, then $\overline{\Delta z} = \Delta z$, and the difference quotient equals 1. On the other hand, if $\Delta z \to 0$ along the imaginary axis, then $\overline{\Delta z} = -\Delta z$, and the quotient equals -1. Consequently there is no way of assigning a unique value to the derivative of $\bar{z}$ at any point.

As we shall show in Prob. 8, it is possible for a complex function to be differentiable solely at isolated points. Of course, this also occurs in real analysis. Such functions are treated there as exceptional cases, while the general theorems usually apply only to functions differentiable over open intervals of the real line. By analogy, then, we distinguish a special class of complex functions in

DEFINITION 5   A complex-valued function $f(z)$ is said to be **analytic**†
on an open set G if it has a derivative at every point of G.

We emphasize that analyticity is a property defined over open sets, while differentiability could conceivably hold at one point only. Occasionally, however, we shall use the abbreviated phrase "$f(z)$ *is analytic at the point z_0*" to mean that $f(z)$ is analytic in some neighborhood of z_0. Thus we can say that a rational function of z is analytic at every point for which its denominator is nonzero. If $f(z)$ is analytic on the whole complex plane, then it is said to be *entire*. For example, all polynomial functions of z are entire.

When a function $f(z)$ is given in terms of x and y as $u(x, y) + iv(x, y)$, it may be very tedious to apply these definitions to determine if f is analytic. One has to establish the existence of the limit in Definition 4 for Δz approaching zero in any manner, and this must be done at all points of

†Some authors use the words *holomorphic* or *regular* instead of analytic.

a (two-dimensional) open set. It is thus desirable to establish other criteria for analyticity.

If the functions u and v are rational functions of x and y, it may be possible to establish the analyticity of $f(z)$ by making the substitutions

$$(8) \qquad x = \frac{z + \bar{z}}{2}, \qquad y = \frac{z - \bar{z}}{2i}$$

in the expression $u(x, y) + iv(x, y)$; if the appearances of $\bar{z}$ ultimately cancel, we are left with a rational function of z, which we know to be analytic (except at the zeros of the denominator). We illustrate this procedure in

EXAMPLE 7   Express the following functions in terms of z and $\bar{z}$:

$$f_1(z) = \frac{x - 1 - iy}{(x - 1)^2 + y^2}, \qquad f_2(z) = x^2 + y^2 + 3x + 1 + i3y.$$

SOLUTION   Using relations (8) we obtain

$$f_1(z) = \frac{\dfrac{z + \bar{z}}{2} - 1 - i\dfrac{z - \bar{z}}{2i}}{\left(\dfrac{z + \bar{z}}{2} - 1\right)^2 + \left(\dfrac{z - \bar{z}}{2i}\right)^2}$$

$$= \frac{\bar{z} - 1}{z\bar{z} - z - \bar{z} + 1} = \frac{1}{z - 1},$$

$$f_2(z) = \frac{(z + \bar{z})^2}{4} + \frac{(z - \bar{z})^2}{4i^2} + 3\left(\frac{z + \bar{z}}{2}\right) + 1 + i3\left(\frac{z - \bar{z}}{2i}\right)$$

$$= z\bar{z} + 3z + 1. \qquad \blacksquare$$

We see from the example that $f_1(z)$ is analytic except at $z = 1$. On the other hand, the presence of $\bar{z}$ in the final expression for $f_2(z)$ leads one to doubt its analyticity. In fact, since we can write

$$\bar{z} = \frac{f_2(z) - 3z - 1}{z},$$

analyticity for $f_2(z)$ would imply analyticity for $\bar{z}$ (when $z \neq 0$), which we know is false. Thus the fact that f_2 depends on $\bar{z}$ in an essential way negates its analyticity.

This leads one to conjecture that analytic functions can always be computed using x and y only in the combination $z = x + iy$, i.e., that they can be expressed in terms of sums, products, and quotients of constant multiples of z alone, without involving $\bar{z}$. Later in the book we shall prove that this conjecture is true if we interpret the word "sums" to include infinite series. For this reason one could say that the analytic functions are distinguished by the fact that they respect the complex structure of z, rather than depending on x and y in an arbitrary fashion.

As a method for determining analyticity the technique of substituting relations (8) has two drawbacks. First, it can only be applied to certain elementary functions. Ultimately we would like to consider such functions as e^x and $\sin y$, for which the substitutions lead to expressions which we don't yet know how to manipulate. Second, even for simple functions it may be difficult to detect whether or not $\bar{z}$ cancels in the final expression. For example, it is not readily apparent that the function

$$\frac{z^2\bar{z}^2 + z^2 + \bar{z}^2 - 2\bar{z}z^2 - 2\bar{z} + 1}{10\bar{z} + z\bar{z}^2 - 2z\bar{z} - 5\bar{z}^2 + z - 5}$$

has a common factor of $(\bar{z} - 1)^2$ in its numerator and denominator. In the next section we shall demonstrate a procedure for establishing analyticity which overcomes these difficulties.

Exercises 2.3

1. Let $f(z)$ be defined in a neighborhood of z_0. Show that finding $\lim_{\Delta z \to 0}[f(z_0 + \Delta z) - f(z_0)]/\Delta z$ is equivalent to finding $\lim_{z \to z_0}[f(z) - f(z_0)]/(z - z_0)$.

2. Prove that if $f(z)$ is differentiable at z_0, then $f(z) = f(z_0) + f'(z_0) \times (z - z_0) + \lambda(z)(z - z_0)$, where $\lambda(z) \to 0$ as $z \to z_0$.

3. Prove that if $f(z)$ is differentiable at z_0, then it is continuous at z_0.

4. Prove rules (2) and (4).

5. Prove that formula (7) is also valid for negative integers n.

6. Use rules (2)–(7) to find the derivatives of the following functions.

   (a) $f(z) = 6z^3 + 8z^2 + iz + 10$

   (b) $f(z) = (z^2 - 3i)^{-6}$

   (c) $f(z) = \dfrac{z^2 - 9}{iz^3 + 2z + \pi}$

   (d) $f(z) = \dfrac{(z + 2)^3}{(z^2 + iz + 1)^4}$

   (e) $f(z) = 6i(z^3 - 1)^4(z^2 + iz)^{100}$

7. Let $F(z) = f(z)g(z)h(z)$, where f, g, and h are each differentiable at z_0. Prove that

   $$F'(z_0) = f'(z_0)g(z_0)h(z_0) + f(z_0)g'(z_0)h(z_0) + f(z_0)g(z_0)h'(z_0).$$

8. Let $f(z) = |z|^2$. Use Definition 4 to show that f is differentiable at $z = 0$ but is not differentiable at any other point. [HINT: Write

$$\frac{|z_0 + \Delta z|^2 - |z_0|^2}{\Delta z} = \frac{(z_0 + \Delta z)(\bar{z}_0 + \overline{\Delta z}) - z_0 \bar{z}_0}{\Delta z}$$

$$= \bar{z}_0 + \overline{\Delta z} + z_0 \frac{\overline{\Delta z}}{\Delta z} .]$$

9. For each of the following determine the points at which the function is not analytic.

(a) $\dfrac{1}{z - 2 + 3i}$ (b) $\dfrac{iz^3 + 2z}{z^2 + 1}$

(c) $\dfrac{3z - 1}{z^2 + z + 4}$

10. Discuss the analyticity of each of the following functions.

(a) $8\bar{z} + i$

(b) $\dfrac{z}{\bar{z} + 2}$

(c) $\dfrac{z^3 + 2z + i}{z - 5}$

(d) $x^2 - y^2 + 2xyi$

(e) $x^2 + y^2 + y - 2 + ix$

(f) $\left(x + \dfrac{x}{x^2 + y^2} \right) + i\left(y - \dfrac{y}{x^2 + y^2} \right)$

(g) $|z|^2 + 2z$

(h) $\dfrac{|z| + z}{2}$

11. Let $f(z)$ and $g(z)$ be entire functions. Decide which of the following statements are always true.

(a) $f(z)^3$ is entire.   (b) $f(z)g(z)$ is entire.
(c) $f(z)/g(z)$ is entire.   (d) $5f(z) + ig(z)$ is entire.
(e) $f(1/z)$ is entire.   (f) $g(z^2 + 2)$ is entire.
(g) $f(g(z))$ is entire.

12. Let $P(z) = (z - z_1)(z - z_2) \cdots (z - z_n)$. Show that

$$\frac{P'(z)}{P(z)} = \frac{1}{z - z_1} + \frac{1}{z - z_2} + \cdots + \frac{1}{z - z_n} .$$

[NOTE: $P'(z)/P(z)$ is called the *logarithmic derivative* of $P(z)$.]

13. Prove L'Hospital's Rule: If $f(z)$ and $g(z)$ are analytic at z_0 and $f(z_0) = g(z_0) = 0$, but $g'(z_0) \neq 0$, then

$$\lim_{z \to z_0} \frac{f(z)}{g(z)} = \frac{f'(z_0)}{g'(z_0)}.$$

[HINT: Write

$$\frac{f(z)}{g(z)} = \frac{f(z) - f(z_0)}{z - z_0} \Bigg/ \frac{g(z) - g(z_0)}{z - z_0}.\Bigg]$$

14. Use L'Hospital's Rule to find $\lim_{z \to i}(1 + z^6)/(1 + z^{10})$.

15. Let $f(z) = z^3 + 1$, and let $z_1 = (-1 + \sqrt{3}\,i)/2, z_2 = (-1 - \sqrt{3}\,i)/2$. Show that there is no point w on the line segment from z_1 to z_2 such that

$$f(z_2) - f(z_1) = f'(w)(z_2 - z_1).$$

This shows that the Mean-Value Theorem of Calculus does not extend to complex functions.

2.4 THE CAUCHY-RIEMANN EQUATIONS

The property of analyticity of a function dictates a relationship between its real and imaginary parts. In this section we shall investigate the nature of this relationship.

If the function $f(z) = u(x, y) + iv(x, y)$ is differentiable at $z_0 = x_0 + iy_0$, then the limit

$$f'(z_0) = \lim_{\Delta z \to 0} \frac{f(z_0 + \Delta z) - f(z_0)}{\Delta z}$$

can be computed by allowing $\Delta z \, (= \Delta x + i\,\Delta y)$ to approach zero from any convenient direction in the complex plane. If it approaches horizontally, then $\Delta z = \Delta x$, and we obtain

$$f'(z_0) = \lim_{\Delta x \to 0} \frac{u(x_0 + \Delta x, y_0) + iv(x_0 + \Delta x, y_0) - u(x_0, y_0) - iv(x_0, y_0)}{\Delta x}$$

$$= \lim_{\Delta x \to 0} \left[\frac{u(x_0 + \Delta x, y_0) - u(x_0, y_0)}{\Delta x} \right]$$

$$+ i \lim_{\Delta x \to 0} \left[\frac{v(x_0 + \Delta x, y_0) - v(x_0, y_0)}{\Delta x} \right].$$

Since the limits of the bracketed expressions are just the first partial derivatives of u and v with respect to x, we deduce that

(9) $$f'(z_0) = \frac{\partial u}{\partial x}(x_0, y_0) + i\frac{\partial v}{\partial x}(x_0, y_0).$$

On the other hand, if Δz approaches zero vertically, then $\Delta z = i \, \Delta y$, and we have

$$f'(z_0) = \lim_{\Delta y \to 0} \left[\frac{u(x_0, y_0 + \Delta y) - u(x_0, y_0)}{i \, \Delta y} \right]$$

$$+ \, i \lim_{\Delta y \to 0} \left[\frac{v(x_0, y_0 + \Delta y) - v(x_0, y_0)}{i \, \Delta y} \right].$$

Hence

(10) $$f'(z_0) = -i \frac{\partial u}{\partial y}(x_0, y_0) + \frac{\partial v}{\partial y}(x_0, y_0).$$

But the right-hand members of Eqs. (9) and (10) are equal to the same complex number $f'(z_0)$, so by equating real and imaginary parts we see that the equations

(11) $$\frac{\partial u}{\partial x} = \frac{\partial v}{\partial y}, \qquad \frac{\partial u}{\partial y} = -\frac{\partial v}{\partial x}$$

must hold at $z_0 = x_0 + iy_0$. Equations (11) are called the *Cauchy-Riemann* equations. We have thus established

THEOREM 3 *A necessary condition for a function $f(z) = u(x, y) + iv(x, y)$ to be differentiable at a point z_0 is that the Cauchy-Riemann equations hold at z_0.*

Consequently, if $f(z)$ is analytic in an open set G, then the Cauchy-Riemann equations must hold at every point of G.

EXAMPLE 8 Show that the function $f(z) = (x^2 + y) + i(y^2 - x)$ is not analytic at any point.

SOLUTION Since $u(x, y) = x^2 + y$ and $v(x, y) = y^2 - x$, we have

$$\frac{\partial u}{\partial x} = 2x, \qquad \frac{\partial v}{\partial y} = 2y,$$

$$\frac{\partial u}{\partial y} = 1, \qquad \frac{\partial v}{\partial x} = -1.$$

Hence the Cauchy-Riemann equations are simultaneously satisfied only on the line $x = y$, and therefore in no open disk. Thus, by Theorem 3, $f(z)$ is nowhere analytic. ∎

The Cauchy-Riemann equations alone are *not* sufficient to prove differentiability (see Prob. 4). We need the additional hypothesis of the continuity of the first partial derivatives of u and v.

THEOREM 4 *Let $f(z) = u(x, y) + iv(x, y)$ be defined in some open set G containing the point z_0. If the first partial derivatives of u and v exist in G, are continuous at z_0, and satisfy the Cauchy-Riemann equations at z_0, then $f(z)$ is differentiable at z_0.*

Consequently, if the first partial derivatives are continuous and satisfy the Cauchy-Riemann equations at all points of G, then $f(z)$ is analytic in G.

Proof The difference quotient for f at z_0 can be written in the form

(12) $\dfrac{f(z_0 + \Delta z) - f(z_0)}{\Delta z}$

$$= \frac{[u(x_0 + \Delta x, y_0 + \Delta y) - u(x_0, y_0)] + i[v(x_0 + \Delta x, y_0 + \Delta y) - v(x_0, y_0)]}{\Delta x + i\,\Delta y},$$

where $z_0 = x_0 + iy_0$ and $\Delta z = \Delta x + i\,\Delta y$. The above expressions are well defined if $|\Delta z|$ is so small that the closed disk with center z_0 and radius $|\Delta z|$ lies entirely in G. Let us rewrite the difference $u(x_0 + \Delta x, y_0 + \Delta y) - u(x_0, y_0)$ as

(13) $[u(x_0 + \Delta x, y_0 + \Delta y) - u(x_0, y_0 + \Delta y)]$
$$+ [u(x_0, y_0 + \Delta y) - u(x_0, y_0)].$$

Because the partial derivatives exist in G, the Mean-Value Theorem says that there is a number x^* between x_0 and $x_0 + \Delta x$ such that

$$u(x_0 + \Delta x, y_0 + \Delta y) - u(x_0, y_0 + \Delta y) = \Delta x \frac{\partial u}{\partial x}(x^*, y_0 + \Delta y).$$

Furthermore, since the partial derivatives are continuous at (x_0, y_0), we can write

$$\frac{\partial u}{\partial x}(x^*, y_0 + \Delta y) = \frac{\partial u}{\partial x}(x_0, y_0) + \varepsilon_1,$$

where the function $\varepsilon_1 \to 0$ as $x^* \to x_0$ and $\Delta y \to 0$ (in particular, as $\Delta z \to 0$). Thus the first bracketed expression in (13) can be written as

$$u(x_0 + \Delta x, y_0 + \Delta y) - u(x_0, y_0 + \Delta y) = \Delta x \left[\frac{\partial u}{\partial x}(x_0, y_0) + \varepsilon_1 \right].$$

The second bracketed expression in (13) is treated similarly, introducing the function ε_2. Then working the same strategy for the v-difference in Eq. (12), we ultimately have

$$\frac{f(z_0 + \Delta z) - f(z_0)}{\Delta z}$$

$$= \frac{\Delta x \left[\dfrac{\partial u}{\partial x} + \varepsilon_1 + i\dfrac{\partial v}{\partial x} + i\varepsilon_3 \right] + \Delta y \left[\dfrac{\partial u}{\partial y} + \varepsilon_2 + i\dfrac{\partial v}{\partial y} + i\varepsilon_4 \right]}{\Delta x + i\,\Delta y},$$

where each partial derivative is evaluated at (x_0, y_0) and where each $\varepsilon_i \to 0$ as $\Delta z \to 0$. Now we use the Cauchy-Riemann equations to express the difference quotient as

(14)
$$\frac{\Delta x \left[\frac{\partial u}{\partial x} + i \frac{\partial v}{\partial x} \right] + i \, \Delta y \left[\frac{\partial u}{\partial x} + i \frac{\partial v}{\partial x} \right]}{\Delta x + i \, \Delta y} + \frac{\lambda}{\Delta x + i \, \Delta y},$$

where $\lambda \equiv \Delta x (\varepsilon_1 + i\varepsilon_3) + \Delta y(\varepsilon_2 + i\varepsilon_4)$. Since

$$\left| \frac{\lambda}{\Delta x + i \, \Delta y} \right| \leq \left| \frac{\Delta x}{\Delta x + i \, \Delta y} \right| |\varepsilon_1 + i\varepsilon_3| + \left| \frac{\Delta y}{\Delta x + i \, \Delta y} \right| |\varepsilon_2 + i\varepsilon_4|$$

$$\leq |\varepsilon_1 + i\varepsilon_3| + |\varepsilon_2 + i\varepsilon_4|,$$

we see that the last term in (14) approaches zero as $\Delta z \to 0$, and so

$$\lim_{\Delta z \to 0} \frac{f(z_0 + \Delta z) - f(z_0)}{\Delta z} = \frac{\partial u}{\partial x}(x_0, y_0) + i \frac{\partial v}{\partial x}(x_0, y_0);$$

i.e., $f'(z_0)$ exists. ∎

It follows from Theorem 4 that the nowhere analytic function $f(z)$ of Example 8 is, nonetheless, differentiable at each point on the line $x = y$. Another application of this theorem is given in

EXAMPLE 9 Prove that the function $f(z) = e^x \cos y + ie^x \sin y$ is entire, and find its derivative.

SOLUTION Since $\partial u / \partial x = e^x \cos y$, $\partial v / \partial y = e^x \cos y$, $\partial u / \partial y = -e^x \sin y$, and $\partial v / \partial x = e^x \sin y$, the first partial derivatives are continuous and satisfy the Cauchy-Riemann equations at every point in the plane. Hence $f(z)$ is entire. From Eq. (9) we see that

$$f'(z) = \frac{\partial u}{\partial x} + i \frac{\partial v}{\partial x} = e^x \cos y + ie^x \sin y.$$

Surprisingly, $f'(z) = f(z)$. ∎

As a further application of these techniques, let us prove the following theorem whose analogue in the real case is well known.

THEOREM 5 *If $f(z)$ is analytic in a domain D and if $f'(z) = 0$ everywhere in D, then $f(z)$ is constant in D.*

Before we proceed with the proof, we observe that the *connectedness* property of the domain is essential. Indeed, if $f(z)$ is defined by

$$f(z) = \begin{cases} 0 & \text{if } |z| < 1, \\ 1 & \text{if } |z| > 2, \end{cases}$$

then f is analytic and $f'(z) = 0$ on its *domain of definition* (which is not a *domain!*), yet f is not constant.

Proof of Theorem 5 Since any two points in a domain can be joined by a polygonal line in the domain, it suffices to show that $f(z)$ is constant along any straight-line segment lying in D. Such a segment can always be represented by the parametric equations

$$x = at + b, \qquad y = ct + d,$$

where a, b, c, and d are real constants and t ranges between 0 and 1. The values of $f(z) = u(x, y) + iv(x, y)$ on the segment can thus be represented by two real functions, $U(t)$ and $V(t)$, according to

$$U(t) + iV(t) = u(at + b, ct + d) + iv(at + b, ct + d).$$

The derivatives of U and V can be computed by the chain rule of elementary calculus,

$$\frac{dU}{dt} = \frac{\partial u}{\partial x}\frac{dx}{dt} + \frac{\partial u}{\partial y}\frac{dy}{dt} = \frac{\partial u}{\partial x}a + \frac{\partial u}{\partial y}c,$$

and, similarly,

$$\frac{dV}{dt} = \frac{\partial v}{\partial x}a + \frac{\partial v}{\partial y}c.$$

Since $f'(z) = 0$ in D, we see from Eqs. (9) and (11) that all the first partial derivatives of u and v vanish. Thus $U(t)$ and $V(t)$ are two real functions of a real variable whose derivatives are zero, and therefore they are constant. Since $f = U + iV$ along the segment, we are done. ∎

Using the above theorem and the Cauchy-Riemann equations, one can show that an analytic function $f(z)$ must be constant when any one of the following conditions hold in the domain D:

(15)   Re $f(z)$ is constant.

(16)   Im $f(z)$ is constant
         [this follows from (15) by considering $if(z)$].

(17) $|f(z)|$ is constant.

The proofs are left as problems.

Exercises 2.4

1. Use the Cauchy-Riemann equations to show that the following functions are nowhere differentiable.

    (a) $w = \bar{z}$ (b) $w = $ Re z (c) $w = 2y - ix$

2. Show that $h(z) = x^3 + 3xy^2 - 3x + i(y^3 + 3x^2y - 3y)$ is differentiable on the coordinate axes but is nowhere analytic.

3. Use Theorem 4 to show that $g(z) = 3x^2 + 2x - 3y^2 - 1 + i(6xy + 2y)$ is entire. Write this function in terms of z.

4. Let

$$f(z) = \begin{cases} \dfrac{x^{4/3}y^{5/3} + ix^{5/3}y^{4/3}}{x^2 + y^2} & \text{if } z \neq 0, \\ 0 & \text{if } z = 0. \end{cases}$$

Show that the Cauchy-Riemann equations hold at $z = 0$ but that f is not differentiable at this point. [HINT: Consider the difference quotient $f(\Delta z)/\Delta z$ for $\Delta z \to 0$ along the real axis and along the line $y = x$.]

5. Prove that the function $f(z) = e^{x^2 - y^2}[\cos(2xy) + i \sin(2xy)]$ is entire, and find its derivative.

6. If u and v are expressed in terms of polar coordinates (r, θ), show that the Cauchy-Riemann equations can be written in the form

$$\frac{\partial u}{\partial r} = \frac{1}{r}\frac{\partial v}{\partial \theta}, \qquad \frac{\partial v}{\partial r} = -\frac{1}{r}\frac{\partial u}{\partial \theta}.$$

7. Prove that if two analytic functions have the same derivative throughout a domain D, then they differ only by an additive constant.

8. Prove that if condition (15) holds in a domain D, then the analytic function $f(z)$ must be constant in D.

9. Show, by contradiction, that the function $f(z) = |z^2 - z|$ is nowhere analytic because of condition (16).

10. Prove that if $f(z)$ is analytic and real-valued in a domain D, then $f(z)$ is constant in D.

11. Suppose that $f(z)$ and $\overline{f(z)}$ are analytic in a domain D. Show that $f(z)$ is constant in D.

12. Prove that if condition (17) holds in a domain D, then the analytic function $f(z)$ must be constant in D. [HINT: $|f|^2$ is constant, so $\partial|f|^2/\partial x = \partial|f|^2/\partial y = 0$ throughout D. Using these two relations and the Cauchy-Riemann equations, deduce that $f'(z) = 0$.]

13. Given that $f(z)$ and $|f(z)|$ are each analytic in a domain D, prove that $f(z)$ is constant in D.

Solutions of the two-dimensional *Laplace equation*

(18)
$$\frac{\partial^2 \phi}{\partial x^2} + \frac{\partial^2 \phi}{\partial y^2} = 0$$

are among the most important functions in mathematical physics. The electrostatic potential solves Eq. (18) in two-dimensional free space, as does the scalar magnetostatic potential; the corresponding field in any direction is given by the directional derivative of $\phi(x, y)$. Two-dimensional fluid flow problems are described by such functions under certain idealized conditions, and ϕ can also be interpreted as the displacement of a membrane stretched across a loop of wire, if the loop is nearly flat. One of the most important applications of analytic function theory to applied mathematics is the abundance of solutions of Eq. (18) that it supplies. We shall adopt the following standard terminology for these solutions:

DEFINITION 6   *A real-valued function $\phi(x, y)$ is said to be* **harmonic** *in a domain D if all its second-order partial derivatives are continuous in D and if at each point of D, ϕ satisfies Eq. (18).*

The sources of these harmonic functions are the real and imaginary parts of analytic functions, as we prove in

THEOREM 6   *If $f(z) = u(x, y) + iv(x, y)$ is analytic in a domain D, then each of the functions $u(x, y)$ and $v(x, y)$ is harmonic in D.*

Proof In a later chapter we shall show that the real and imaginary parts of any analytic function have continuous partial derivatives of all orders. Assuming this fact, we recall from elementary calculus that under such conditions mixed partial derivatives can be taken in any order; i.e.,

(19)
$$\frac{\partial}{\partial y}\frac{\partial u}{\partial x} = \frac{\partial}{\partial x}\frac{\partial u}{\partial y}.$$

Using the Cauchy-Riemann equations for the first derivatives, we transform Eq. (19) into

$$\frac{\partial^2 v}{\partial y^2} = -\frac{\partial^2 v}{\partial x^2},$$

which is equivalent to Eq. (18). Thus v is harmonic in D, and a similar computation proves that u is also. ∎

Conversely, if we are given a function $u(x, y)$ harmonic in, say, an open disk, then we can find another harmonic function $v(x, y)$ so that $u + iv$ is an analytic function of z in the disk. Such a function v is called a *harmonic conjugate* of u. The procedure is illustrated in

EXAMPLE 10 Construct an analytic function whose real part is $u(x, y) = x^3 - 3xy^2 + y$.

SOLUTION   First we verify that

$$\frac{\partial^2 u}{\partial x^2} + \frac{\partial^2 u}{\partial y^2} = 6x - 6x = 0,$$

and so u is harmonic in the whole plane. Now we have to find a mate, $v(x, y)$, for u such that the Cauchy-Riemann equations are satisfied. Thus we must have

(20)
$$\frac{\partial v}{\partial y} = \frac{\partial u}{\partial x} = 3x^2 - 3y^2$$

and

(21)
$$\frac{\partial v}{\partial x} = -\frac{\partial u}{\partial y} = 6xy - 1.$$

If we hold x constant and integrate Eq. (20) with respect to y, we get

$$v(x, y) = 3x^2 y - y^3 + \text{constant},$$

but the "constant" could conceivably be any differentiable function of x; it need only be independent of y. Therefore we write

$$v(x, y) = 3x^2 y - y^3 + \psi(x).$$

We can find $\psi(x)$ by plugging this last expression into Eq. (21);

(22)
$$\frac{\partial v}{\partial x} = 6xy + \psi'(x) = 6xy - 1.$$

This yields $\psi'(x) \equiv -1$, and so $\psi(x) = -x + a$, where a is some (genuine) constant. It follows that a harmonic conjugate of $u(x, y)$ is given by

$$v(x, y) = 3x^2 y - y^3 - x + a,$$

and the analytic function

$$f(z) = x^3 - 3xy^2 + y + i(3x^2 y - y^3 - x + a),$$

which we recognize as $z^3 - i(z - a)$, solves the problem.   ∎

This procedure will always work for any $u(x, y)$ harmonic in a disk, as is shown in the exercises. Thus we can learn a great deal about analytic functions by studying harmonic functions, and vice versa.

The harmonic functions forming the real and imaginary parts of an analytic function $f(z)$ each generate a family of curves in the xy-plane, namely, the "level curves"

(23) $$u(x, y) = \text{constant}$$

and

(24) $$v(x, y) = \text{constant}.$$

If u is interpreted as an electrostatic potential, then the curves (23) are the "equipotentials." For the function $f(z) = z^2 = x^2 - y^2 + 2ixy$, the two sets of level curves are shown in Fig. 2.5.

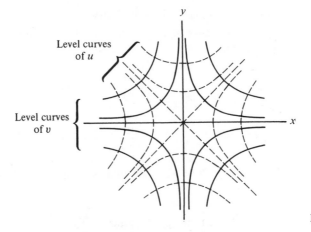

FIGURE 2.5

Now it is quite possible that some of the level curves defined by Eqs. (23) or (24) are irregular, in the sense that they do not possess well-defined tangents at each point. For the present discussion we shall ignore these anomalies and deal only with smooth level curves. Then, as is shown in elementary calculus, the vector with components

$$\left(\frac{\partial u}{\partial x}(x_0, y_0), \frac{\partial u}{\partial y}(x_0, y_0) \right),$$

which is known as the *gradient* of u, is perpendicular to the level curve $u(x, y) = u(x_0, y_0)$ at the point (x_0, y_0). Similarly, the gradient of v is perpendicular to the level curve for v. Observe that, because of the Cauchy-Riemann equations, the scalar (dot) product of these two gradients (evaluated at the same point) vanishes:

$$\frac{\partial u}{\partial x}\frac{\partial v}{\partial x} + \frac{\partial u}{\partial y}\frac{\partial v}{\partial y} = \frac{\partial v}{\partial y}\frac{\partial v}{\partial x} - \frac{\partial v}{\partial x}\frac{\partial v}{\partial y} = 0.$$

This implies that these two gradients are orthogonal (unless one of them is the zero vector, which is the anomalous case we are ignoring). Therefore the level curves themselves are orthogonal; i.e., we have shown that if $v(x, y)$ is a harmonic conjugate of $u(x, y)$, then the level curves of v intersect those of u at right angles. This is also illustrated in Fig. 2.5.

The curves orthogonal to a family of equipotentials are known as "lines of force." Analytic functions therefore provide both equipotentials and lines of force for two-dimensional electrostatic problems.

Exercises 2.5

1. Verify directly that the real and imaginary parts of the following analytic functions satisfy Laplace's equation.

   (a) $f(z) = z^2 + 2z + 1$ (b) $g(z) = \dfrac{1}{z}$

   (c) $h(z) = e^x \cos y + i e^x \sin y$

2. Find the most general harmonic polynomial of the form $ax^2 + bxy + cy^2$.

3. Prove that if r and θ are polar coordinates, then the functions $r^n \cos n\theta$ and $r^n \sin n\theta$ are harmonic as functions of x and y. [HINT: Recall De Moivre's formula.]

4. Find a function harmonic inside the wedge bounded by the non-negative x-axis and the half-line $y = x$ ($x \geq 0$) which goes to zero on these sides but is not identically zero. [HINT: See Prob. 3.] The level curves for this function can be interpreted as streamlines for a fluid flowing inside this wedge, under certain idealized conditions.

5. Find a harmonic function, not identically zero, which vanishes on the hyperbola $y = 1/x$. [HINT: Begin by considering z^2.] As in Prob. 4, the level curves can be interpreted as streamlines.

6. Find a function $\phi(x, y)$ harmonic in the upper half-plane Im $z > 0$ and continuous on Im $z \geq 0$ such that

   (a) $\phi(x, 0) = x^2 + 5x + 1$ for all x.

   (b) $\phi(x, 0) = \dfrac{2x^3}{(x^2 + 4)}$ for all x.

   $\left[\text{HINT: Write } \dfrac{2x^3}{x^2 + 4} = \dfrac{x^2}{x - 2i} + \dfrac{x^2}{x + 2i}.\right]$

7. Verify that each given function u is harmonic, and then find a harmonic conjugate of u.

    (a) $u = y$ (b) $u = e^x \sin y$

    (c) $u = xy - x + y$ (d) $u = \sin x \cosh y$

    (e) $u = \ln |z|$

8. Show that if $v(x, y)$ is a harmonic conjugate of $u(x, y)$ in a domain D, then *every* harmonic conjugate of $u(x, y)$ in D must be of the form $v(x, y) + a$, where a is a real constant.

9. Show that if v is a harmonic conjugate for u, then $-u$ is a harmonic conjugate for v.

10. Show that if v is a harmonic conjugate of u in a domain D, then uv is harmonic in D.

11. Suppose that $f(z)$ is analytic and nonzero in a domain D. Prove that $\ln |f(z)|$ is harmonic in D.

12. Illustrate the orthogonality property mentioned in the text by sketching the level curves for the real and imaginary parts of the following analytic functions.

    (a) $f(z) = z$,   (b) $g(z) = \dfrac{1}{z}$

    (c) $g(z) = z^3$ (use polar form)

13. Let $f(z) = z + 1/z$. Show that the level curve Im $f(z) = 0$ consists of the real axis and the circle $|z| = 1$. [The level curves Im $f(z) = $ constant can be interpreted as streamlines for fluid flow around a cylindrical obstacle.]

14. By tracing the steps in Example 10, show that every function $u(x, y)$ harmonic in a disk has a harmonic conjugate $v(x, y)$. [HINT: The only difficulty which could occur is in the step corresponding to Eq. (22), where in order to find $\psi'(x)$ we must be certain that all appearances of the variable y cancel. Show that this is guaranteed because u is harmonic.]

*2.6 VISUALIZATION OF HARMONIC FUNCTIONS

We shall now present an informal physical interpretation of harmonic functions in order to help the reader visualize some of their properties. Let us imagine that we are given a smooth closed curve, i.e., a loop, in three-dimensional space and that we stretch a membrane across the loop, like a drumhead; we do not, however, assume that the loop is planar. Furthermore, we adjust the membrane so that the tension is uniform and isotropic over the surface.

It is intuitively clear that if the loop is reasonably simple, the position occupied by the membrane is a minimal surface; i.e., it has the least area of all the surfaces having the given loop as boundary. Indeed, if the area were not minimal, some slack would result, and the stretching would not be uniform. As a matter of fact, a detailed mechanical analysis would show that the potential energy due to the stretching is proportional to the area, in the case of uniform tension; the position of minimal area thus corresponds to mechanical equilibrium.

To see the connection of this problem with harmonic functions, we coordinatize the model.

The surface is described in three-dimensional space by specifying the height, z, as a function of x and y, say,

$$z = \phi(x, y).$$

(Notice that z is not a complex variable here; it is the vertical coordinate in a Cartesian axis system.) For simplicity, we are assuming that the loop is aligned so that any vertical line cuts the surface at most once, implying that ϕ is a single-valued function (see Fig. 2.6).

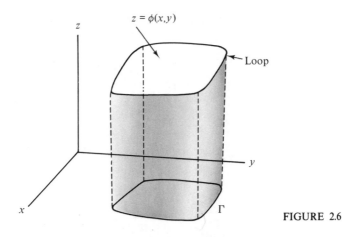

FIGURE 2.6

To tie this in with harmonic functions, we must make two more assumptions. First, we suppose that $\phi(x, y)$ has continuous second-order derivatives (actually this can be proved, by advanced methods, if the curve is smooth enough). Second, we restrict ourselves to the case where the loop is approximately horizontal; we do not assume that it is planar, but it should be sufficiently flat so that we can consider fourth-order terms, such as $(\partial\phi/\partial x)^4$, $(\partial\phi/\partial y)^4$, and $(\partial\phi/\partial x)^2(\partial\phi/\partial y)^2$, as negligible.

Under these conditions, we claim that the function $\phi(x, y)$ is harmonic.

The verification is a classic argument from the calculus of variations. We shall compare the area of the minimal surface, $z = \phi(x, y)$, with the areas of nearby surfaces having the same loop as boundary. To implement a parametrization of these surfaces, we first project the minimal surface (including the bounding loop) into the xy-plane, yielding a region bounded by the curve Γ, which is the projection of the loop. This is depicted in Fig. 2.6. Now let $\eta(x, y)$ be any twice-differentiable function defined in the region bounded by Γ and vanishing on Γ itself. Then the equation

$$(25) \qquad z = \phi(x, y) + \varepsilon\eta(x, y),$$

for any real ε, will describe a surface with the given loop as its boundary; for $\varepsilon = 0$ it will be the minimal surface, and for small ε it will be "nearly flat" in the sense described earlier.

The area of the surface described by Eq. (25) can be considered as a function of ε; it is computed from the following formula given in calculus:

$$(26) \qquad A(\varepsilon) = \iint \sqrt{1 + \left(\frac{\partial z}{\partial x}\right)^2 + \left(\frac{\partial z}{\partial y}\right)^2}\ dx\ dy,$$

where the integration is taken over the region in the xy-plane bounded by Γ.

For the "nearly horizontal" surfaces we are considering, we can make the approximation

$$1 + \left(\frac{\partial z}{\partial x}\right)^2 + \left(\frac{\partial z}{\partial y}\right)^2 \approx 1 + \left(\frac{\partial z}{\partial x}\right)^2 + \left(\frac{\partial z}{\partial y}\right)^2 + \frac{1}{4}\left[\left(\frac{\partial z}{\partial x}\right)^2 + \left(\frac{\partial z}{\partial y}\right)^2\right]^2$$

$$= \left\{1 + \frac{1}{2}\left[\left(\frac{\partial z}{\partial x}\right)^2 + \left(\frac{\partial z}{\partial y}\right)^2\right]\right\}^2,$$

where we have completed the square by adding a "negligible" term. Thus from Eq. (26)

$$(27) \qquad A(\varepsilon) \approx \iint \left[1 + \frac{1}{2}\left(\frac{\partial z}{\partial x}\right)^2 + \frac{1}{2}\left(\frac{\partial z}{\partial y}\right)^2\right]\ dx\ dy$$

$$= \iint \left[1 + \frac{1}{2}\left(\frac{\partial\phi}{\partial x} + \varepsilon\frac{\partial\eta}{\partial x}\right)^2 + \frac{1}{2}\left(\frac{\partial\phi}{\partial y} + \varepsilon\frac{\partial\eta}{\partial y}\right)^2\right]\ dx\ dy.$$

Since the surface with $\varepsilon = 0$ in Eq. (25) has minimal area, we know that

$$\frac{dA(\varepsilon)}{d\varepsilon} = 0 \qquad \text{at } \varepsilon = 0.$$

Accepting the approximation in (27), we also have

$$(28) \qquad \frac{dA}{d\varepsilon}\bigg|_{\varepsilon=0} = \iint \left(\frac{\partial \phi}{\partial x} \frac{\partial \eta}{\partial x} + \frac{\partial \phi}{\partial y} \frac{\partial \eta}{\partial y} \right) dx\, dy.$$

Let us look at the first term in this integral; it appears in the identity

$$(29) \qquad \iint \frac{\partial}{\partial x}\left[\eta \frac{\partial \phi}{\partial x} \right] dx\, dy = \iint \frac{\partial \phi}{\partial x} \frac{\partial \eta}{\partial x} dx\, dy + \iint \eta \frac{\partial^2 \phi}{\partial x^2} dx\, dy.$$

We shall show that the left-hand side of Eq. (29) is zero. To see this, we refer to Fig. 2.7, illustrating the region of integration bounded by Γ in the xy-plane. (For simplicity, we have drawn the region as convex; our results will hold for more complicated regions, but the reasoning involves a theorem of Green[†] with which the reader may not be familiar.) The

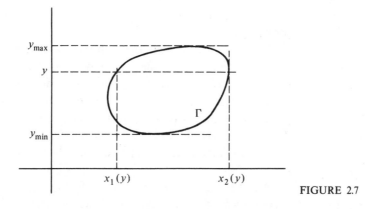

FIGURE 2.7

integration of the left-hand member of Eq. (29) is performed iteratively. Holding y constant, we integrate with respect to x across a horizontal line cutting Γ at the points with x-coordinates x_1 and x_2 (which are functions of y); then we integrate the result with respect to y between the extreme values y_{min} and y_{max}. The x-integration is immediate because of the Fundamental Theorem of Calculus:

$$\int_{x_1(y)}^{x_2(y)} \frac{\partial}{\partial x}\left[\eta \frac{\partial \phi}{\partial x} \right] dx = \eta \frac{\partial \phi}{\partial x}\bigg|_{(x_1(y),\, y)}^{(x_2(y),\, y)}.$$

Since the end points lie on Γ, where η vanishes, the last expression is zero, which implies that the double integral is zero. Thus, from Eq. (29), we have

$$\iint \frac{\partial \phi}{\partial x} \frac{\partial \eta}{\partial x} dx\, dy = - \iint \eta \frac{\partial^2 \phi}{\partial x^2} dx\, dy.$$

†See, e.g., Ref. [1] at the end of this chapter.

A similar technique can be applied to the second term in the integral of Eq. (28), resulting in

(30)
$$\frac{dA}{d\varepsilon}\bigg|_{\varepsilon=0} = -\iint \eta\left(\frac{\partial^2\phi}{\partial x^2} + \frac{\partial^2\phi}{\partial y^2}\right) dx\, dy.$$

Since this integral is zero for any function $\eta(x, y)$ vanishing on Γ, the reader should be able to convince himself that $\partial^2\phi/\partial x^2 + \partial^2\phi/\partial y^2$ must be identically zero; in other words, $\phi(x, y)$ is harmonic (to the order of accuracy we are considering).

Observe that if the loop were *exactly* planar, then the stretched membrane would naturally be planar also, so that the function $z = \phi(x, y)$ would have the form $ax + by + c$; such a function is obviously harmonic.

Thus the stretched membrane provides us with a physical model of harmonic functions, and we can use it to visualize some of their properties.

First we observe that there should be no "bulges" in a uniformly stretched membrane; if there were, we could "gather up slack" and produce a surface with less area. Mathematically, this means that the harmonic function $\phi(x, y)$ can have no local maxima or minima at interior points, and the extremal values must occur on the boundary of the domain. This is the *Maximum Principle* for harmonic functions, and we shall show later in the book that it is related to a similar statement for analytic functions.

Another fact that we can predict from the physical model is that the specification of the values of any harmonic function $\psi(x, y)$ on a closed curve Γ in the xy-plane completely determines its values inside Γ. After all, the specification of the values on Γ suffices to determine a loop, through the equation $z = \psi(x, y)$ for (x, y) lying on Γ. This loop (if sufficiently regular) then has a minimal surface. We need merely to verify, therefore, that $z = \psi(x, y)$ *is* this minimal surface. Again working under the assumption that the surface $z = \psi(x, y)$ has the "nearly flat" property, we argue as follows:

Reversing the steps in our earlier calculations (with ϕ replaced by ψ) we can see, conversely, that the harmonicity of ψ implies that $dA(\varepsilon)/d\varepsilon = 0$ at $\varepsilon = 0$; here $A(\varepsilon)$ is the area of the surface $z = \psi(x, y) + \varepsilon\eta(x, y)$. Thus the area of the surface $z = \psi(x, y)$ is at least stationary, if not minimal. But since the area is proportional to the potential energy, the latter is also stationary in this configuration. From mechanics we know that the surface must therefore be in equilibrium. Now, our intuition tells us that there is only one position of equilibrium for membranes stretched over reasonably shaped loops, so we conclude that $\psi(x, y)$ must correspond to this minimal surface. Thus ψ is uniquely determined by its values on Γ.

The analogue of this statement for analytic functions is Cauchy's Integral Formula, which we shall derive in a later chapter.

SUMMARY

A complex-valued function f of a complex variable $z = x + iy$ can be considered as a pair of real functions of two real variables in accordance with $f(z) = u(x, y) + iv(x, y)$. The definitions of limit, continuity, and derivative for such functions are direct analogues of the corresponding concepts introduced in calculus, but the greater freedom of z to vary in two dimensions lends added strength to these conditions. In particular, the existence of a derivative, defined as the limit of $[f(z + \Delta z) - f(z)]/\Delta z$ as $\Delta z \to 0$, implies a strong relationship between the functions u and v, namely, the Cauchy-Riemann equations

$$\frac{\partial u}{\partial x} = \frac{\partial v}{\partial y}, \qquad \frac{\partial u}{\partial y} = -\frac{\partial v}{\partial x}.$$

If the function f is differentiable in an open set, it is said to be analytic. This property can be established by showing that the first-order partial derivatives of u and v are continuous and satisfy the Cauchy-Riemann equations on the open set. Analyticity of a function f is the mathematical expression of the intuitive condition that f respects the complex structure of z; i.e., f can be computed using x and y only in the combination $(x + iy)$. If f is given in terms of z alone, the basic techniques of calculus can be used to find its derivative.

The real and imaginary parts of an analytic function are harmonic; i.e., they satisfy Laplace's equation

$$\frac{\partial^2 \phi}{\partial x^2} + \frac{\partial^2 \phi}{\partial y^2} = 0,$$

and their second-order partial derivatives are continuous. Furthermore, the level curves of the real part intersect those of the imaginary part orthogonally. Given a harmonic function $u(x, y)$ it is possible to construct another harmonic function $v(x, y)$ so that $u(x, y) + iv(x, y)$ is analytic in some domain; such a function v is called a harmonic conjugate of u. Harmonic functions can be physically interpreted as describing small displacements of stretched membranes.

SUGGESTED READING

In addition to the references following Chapter 1, the following texts may be helpful for special topics:

Green's Theorem

[1]   DAVIS, H. F., and SNIDER, A. D. *Introduction to Vector Analysis*, 3rd ed. Allyn and Bacon, Inc., Boston, 1975.

Stretched Membranes

[2]   COURANT, R., and HILBERT, D. *Methods of Mathematical Physics*, Vol. 1. John Wiley & Sons, Inc. (Interscience Division), New York, 1953.

[3]   WYLIE, C. R. *Advanced Engineering Mathematics*. McGraw-Hill Book Company, New York, 1966.

3

Elementary Functions

3.1 THE EXPONENTIAL, TRIGONOMETRIC, AND HYPERBOLIC FUNCTIONS

The complex exponential function e^z plays a prominent role in analytic function theory, not only because of its own important properties but because it is used to define the complex trigonometric and hyperbolic functions.

In defining e^z, we seek a function which agrees with the exponential function of calculus when z is real and which has, by analogy, the following properties:

(1) $$e^{z_1}e^{z_2} = e^{z_1+z_2},$$

(2) $$\frac{e^{z_1}}{e^{z_2}} = e^{z_1-z_2},$$

(3) $$\frac{d}{dz}e^z = e^z.$$

Let us, then, turn to the task of finding such a function.

If property (1) is to hold for all complex numbers z_1 and z_2, we must have

$$e^{x+iy} = e^x e^{iy} \tag{4}$$

for real x and y. We have stipulated that the value of e^x must agree with its definition in the real case, so Eq. (4) demonstrates that e^z will be defined once we specify the function e^{iy}, for real y.

Since property (3) is to hold, $f(y) = e^{iy}$ must satisfy the equation

$$\frac{df}{dy} = ie^{iy} = if.$$

Differentiating a second time we obtain

$$\frac{d^2f}{dy^2} = i\frac{df}{dy} = i(if) = -f;$$

i.e.,

$$\frac{d^2f}{dy^2} + f = 0. \tag{5}$$

Observe that any function of the form

$$A \cos y + B \sin y \qquad (A, B \text{ complex constants})$$

satisfies Eq. (5). In fact from the theory of differential equations it is known that every solution of Eq. (5) must have this form. Hence we can write

$$f(y) = A \cos y + B \sin y. \tag{6}$$

To evaluate A and B we use the conditions that

$$f(0) = e^{i0} = e^0 = 1,$$

and

$$\frac{df}{dy}(0) = if(0) = i.$$

Thus $A = 1$ and $B = i$, leading us to the identification

$$e^{iy} = \cos y + i \sin y. \tag{7}$$

Combining this with Eq. (4) we adopt

DEFINITION 1 *If $z = x + iy$, then e^z is defined to be the complex number*

$$e^z = e^x(\cos y + i \sin y). \tag{8}$$

Observe that since Definition 1 validates Eq. (7), the polar form (Sec. 1.3) for a nonzero complex number z can now be conveniently written as

$$z = |z|e^{i \arg z}.$$

As a consequence of Example 9 in Sec. 2.4, e^z is *an entire function* satisfying property (3). Also properties (1) and (2) are readily verified. It is useful to notice that the right-hand member of Eq. (8) displays e^z in polar form. Therefore we have

(9) $$|e^z| = |e^{x+iy}| = e^x$$

and

(10) $$\arg e^z = \arg e^{x+iy} = y + 2k\pi \qquad (k = 0, \pm 1, \pm 2, \ldots).$$

Since $e^x > 0$ for all real numbers x, we see from Eq. (9) that the function e^z is *never zero*. However, it assumes every other complex value (see Prob. 5).

Recall that a function f is *one-to-one* on a set S if the equation $f(z_1) = f(z_2)$, where z_1 and z_2 are in S, implies that $z_1 = z_2$. As is shown in calculus, the exponential function is one-to-one on the real axis. However it is *not* one-to-one on the whole plane. In fact, we have

THEOREM 1

(i) *A necessary and sufficient condition that*

$$e^z = 1$$

is that $z = 2k\pi i$, where k is an integer.

(ii) *A necessary and sufficient condition that*

$$e^{z_1} = e^{z_2}$$

is that $z_1 = z_2 + 2k\pi i$, where k is an integer.

Proof of (i) First suppose that $e^z = 1$, with $z = x + iy$. Then we must have

$$|e^z| = |e^{x+iy}| = e^x = 1,$$

and so $x = 0$. This implies that

$$e^z = e^{iy} = \cos y + i \sin y = 1,$$

or, equivalently,

$$\cos y = 1, \qquad \sin y = 0.$$

These two simultaneous equations are clearly satisfied only when $y = 2k\pi$ for some integer k; i.e., $z = 2k\pi i$.

Conversely, if $z = 2k\pi i$, where k is an integer, then

$$e^z = e^{2k\pi i} = e^0(\cos 2k\pi + i \sin 2k\pi) = 1. \quad \blacksquare$$

Proof of (ii) It follows from property (1) that

$$e^{z_1} = e^{z_2} \quad \text{if, and only if,} \quad e^{z_1 - z_2} = e^0 = 1.$$

But, by part (i), the last equation holds precisely when $z_1 - z_2 = 2k\pi i$, where k is an integer. $\quad \blacksquare$

One important consequence of Theorem 1 is the fact that e^z is periodic. In general, a function $f(z)$ is said to be *periodic* in a domain D if there exists a constant λ such that the equation $f(z + \lambda) = f(z)$ holds for every z in D. Any constant λ with this property is called a *period* of $f(z)$. Since, for all z,

$$e^{z + 2\pi i} = e^z,$$

we see that e^z is periodic with complex period $2\pi i$. Consequently, if we divide up the z-plane into the infinite horizontal strips

$$S_n = \{x + iy \mid -\infty < x < \infty, (2n-1)\pi < y \le (2n+1)\pi\}$$

$$(n = 0, \pm 1, \pm 2, \ldots),$$

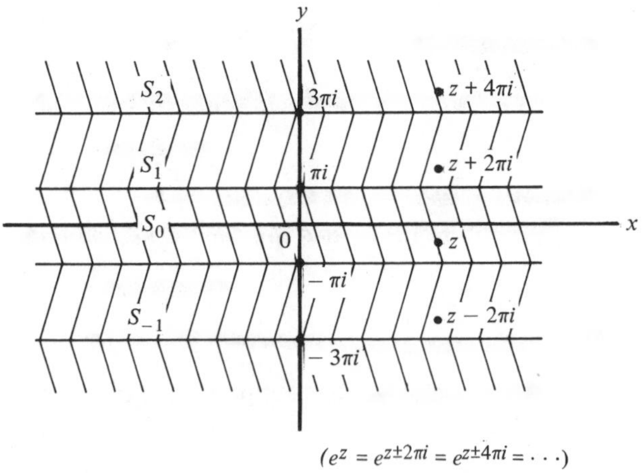

$$(e^z = e^{z \pm 2\pi i} = e^{z \pm 4\pi i} = \cdots)$$

FIGURE 3.1

as shown in Fig. 3.1, then e^z will behave in the same manner on each strip. Furthermore, from part (ii) of Theorem 1, it follows that e^z is one-to-one on each strip S_n. For these reasons any one of these strips is called a *fundamental region* for e^z.

From the identity

$$e^{iy} = \cos y + i \sin y,$$

and its obvious consequence

$$e^{-iy} = \cos y - i \sin y,$$

we deduce, by subtracting and adding these equations, that

$$\sin y = \frac{e^{iy} - e^{-iy}}{2i}, \qquad \cos y = \frac{e^{iy} + e^{-iy}}{2}.$$

These real variable formulas suggest

DEFINITION 2 *Given any complex number z, we define*

$$\sin z = \frac{e^{iz} - e^{-iz}}{2i}, \qquad \cos z = \frac{e^{iz} + e^{-iz}}{2}.$$

Since e^{iz} and e^{-iz} are entire functions, so are $\sin z$ and $\cos z$. In fact,

(11) $$\frac{d}{dz} \sin z = \frac{d}{dz}\left(\frac{e^{iz} - e^{-iz}}{2i}\right) = \frac{1}{2i}(ie^{iz} - (-i)e^{-iz}) = \cos z,$$

and, similarly,

(12) $$\frac{d}{dz} \cos z = -\sin z.$$

We recognize that Eqs. (11) and (12) agree with the familiar formulas derived in calculus. Some further identities which remain valid in the complex case are listed below:

(13) $$\sin(z + 2\pi) = \sin z, \qquad \cos(z + 2\pi) = \cos z.$$

(14) $$\sin(-z) = -\sin z, \qquad \cos(-z) = \cos z.$$

(15) $$\sin^2 z + \cos^2 z = 1.$$

(16) $$\sin(z_1 \pm z_2) = \sin z_1 \cos z_2 \pm \sin z_2 \cos z_1.$$

(17) $$\cos(z_1 \pm z_2) = \cos z_1 \cos z_2 \mp \sin z_1 \sin z_2.$$

(18) $$\sin 2z = 2 \sin z \cos z, \qquad \cos 2z = \cos^2 z - \sin^2 z.$$

The proofs of these identities follow directly from the properties of the exponential function and are left to the exercises. Notice that Eqs. (13) imply that $\sin z$ and $\cos z$ are both periodic with period 2π.

EXAMPLE 1 Prove that $\sin z = 0$ if, and only if, $z = k\pi$, where k is an integer.

SOLUTION   If $z = k\pi$, then clearly $\sin z = 0$. Now suppose, conversely, that $\sin z = 0$. Then we have

$$\frac{e^{iz} - e^{-iz}}{2i} = 0,$$

or, equivalently,

$$e^{iz} = e^{-iz}.$$

By Theorem 1(ii) there follows

$$iz = -iz + 2k\pi i,$$

which implies that $z = k\pi$ for some integer k.   ∎

Thus the only zeros of $\sin z$ are its real zeros. The same is true of the function $\cos z$; i.e.,

$$\cos z = 0 \quad \text{if, and only if,} \quad z = \frac{\pi}{2} + k\pi,$$

where k is an integer.

The other four complex trigonometric functions are defined by

$$\tan z = \frac{\sin z}{\cos z}, \qquad \cot z = \frac{\cos z}{\sin z}, \qquad \sec z = \frac{1}{\cos z},$$

$$\csc z = \frac{1}{\sin z}.$$

Notice that the functions $\cot z$ and $\csc z$ are analytic except at the zeros of $\sin z$, i.e., the points $z = k\pi$, while the functions $\tan z$ and $\sec z$ are analytic except at the points $z = \pi/2 + k\pi$, where k is any integer. Furthermore, the usual rules for differentiation remain valid for these functions:

$$\frac{d}{dz} \tan z = \sec^2 z, \qquad \frac{d}{dz} \sec z = \sec z \tan z,$$

$$\frac{d}{dz} \cot z = -\csc^2 z, \qquad \frac{d}{dz} \csc z = -\csc z \cot z.$$

The above discussion has emphasized the similarity between the real trigonometric functions and their complex extensions. However, this analogy should not be carried too far. For example, the real sine function is bounded by 1, i.e.,

$$|\sin x| \le 1, \qquad \text{for all real } x,$$

but

$$|\sin iy| = \left| \frac{e^{-y} - e^y}{2i} \right| = |\sinh y|,$$

which becomes arbitrarily large as $y \to \pm\infty$.

The complex hyperbolic functions are defined by a natural extension of their definitions in the real case:

DEFINITION 3 *For any complex number z we define*

(19)
$$\sinh z = \frac{e^z - e^{-z}}{2}, \qquad \cosh z = \frac{e^z + e^{-z}}{2}.$$

Notice that the functions (19) are entire and satisfy

(20)
$$\frac{d}{dz}\sinh z = \cosh z, \qquad \frac{d}{dz}\cosh z = \sinh z.$$

By using the trigonometric identities and the readily verified formulas

(21)
$$\sinh z = -i \sin iz, \qquad \cosh z = \cos iz,$$

one can easily show that the familiar hyperbolic identities are valid in the complex case. The four remaining complex hyperbolic functions are given by

$$\tanh z = \frac{\sinh z}{\cosh z}, \qquad \coth z = \frac{\cosh z}{\sinh z}, \qquad \operatorname{sech} z = \frac{1}{\cosh z},$$

$$\operatorname{csch} z = \frac{1}{\sinh z}.$$

Exercises 3.1

1. Verify properties (1) and (2) using Definition 1.

2. Show that
   (a) $e^{z+\pi i} = -e^z$
   (b) $(e^z)^n = e^{nz}$ for any integer n

3. Prove that if $|e^z| = 1$, then z is pure imaginary.

4. Show that $\overline{e^z} = e^{\bar{z}}$.

5. Let $\omega \ (\neq 0)$ have the polar representation $\omega = re^{i\theta}$. Show that $\exp(\ln r + i\theta) = \omega.$†

6. Write each of the following numbers in the form $a + bi$.

   (a) $\exp\!\left(2 + \dfrac{i\pi}{4}\right)$ (b) $\exp(1 + i3\pi)/\exp\!\left(-1 + \dfrac{i\pi}{2}\right)$

   (c) $\sin(2i)$ (d) $\cos(1 - i)$

   (e) $\sinh(1 + \pi i)$ (f) $\cosh\!\left(\dfrac{i\pi}{2}\right)$

†As a convenience in printing we sometimes use the notation $\exp(z)$ instead of e^z.

7. Establish the trigonometric identities (15) and (16).

8. Show that the formula $e^{iz} = \cos z + i \sin z$ holds for all complex numbers z.

9. Verify the differentiation formulas (20).

10. Find dw/dz for each of the following.

   (a) $w = \exp(\pi z^2)$

   (b) $w = \cos(2z) + i \sin\left(\dfrac{1}{z}\right)$

   (c) $w = \exp[\sin(2z)]$

   (d) $w = \tan^3 z$

   (e) $w = [\sinh z + 1]^2$

   (f) $w = \tanh z$

11. Explain why the function $f(z) = \sin(z^2) + e^{-z} + iz$ is entire.

12. Explain why the function $\operatorname{Re}(\cos z/e^z)$ is harmonic in the whole plane.

13. Establish the following hyperbolic identities by using the relations (21) and the corresponding trigonometric identities.
   (a) $\cosh^2 z - \sinh^2 z = 1$
   (b) $\sinh(z_1 + z_2) = \sinh z_1 \cosh z_2 + \cosh z_1 \sinh z_2$
   (c) $\cosh(z_1 + z_2) = \cosh z_1 \cosh z_2 + \sinh z_1 \sinh z_2$

14. Show the following.
   (a) $\sin(x + iy) = \sin x \cosh y + i \cos x \sinh y$
   (b) $\cos(x + iy) = \cos x \cosh y - i \sin x \sinh y$

15. Prove the following.
   (a) e^{iz} is periodic with period 2π.
   (b) $\tan z$ is periodic with period π.
   (c) $\sinh z$ and $\cosh z$ are both periodic with period $2\pi i$.
   (d) $\tanh z$ is periodic with period πi.

16. Prove that $\cos z = 0$ if and only if $z = \pi/2 + k\pi$, where k is an integer.

17. Find all numbers z such that
   (a) $e^{4z} = 1$.   (b) $e^{iz} = 3$.   (c) $\cos z = i \sin z$.

18. Verify the identity

$$\cos z_2 - \cos z_1 = 2 \sin \frac{z_1 + z_2}{2} \sin \frac{z_1 - z_2}{2},$$

and use it to show that $\cos z_1 = \cos z_2$ if and only if $z_2 = \pm z_1 + 2k\pi$, where k is an integer.

19. Prove the following.

(a) $\displaystyle \lim_{z \to 0} \frac{\sin z}{z} = 1$ (b) $\displaystyle \lim_{z \to 0} \frac{\cos z - 1}{z} = 0$

[HINT: Use the fact that $f'(0) = \lim_{z \to 0} [f(z) - f(0)]/z$.]

20. Prove that the function e^z is one-to-one on any open disk of radius π.

21. Show that the function $w = e^z$ maps the rectangle in Fig. 3.2(a) one-to-one onto the semiannulus in Fig. 3.2(b).

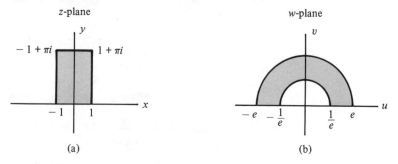

(a)   (b)

FIGURE 3.2

22. Below is an outline of an alternative proof that $\sin^2 z + \cos^2 z = 1$ for all z. Justify each step in the proof.
    (a) The function $f(z) = \sin^2 z + \cos^2 z$ is entire.
    (b) $f'(z) = 0$ for all z.
    (c) $f(z)$ is a constant function.
    (d) $f(0) = 1$.
    (e) $f(z) = 1$ for all z.

3.2 THE LOGARITHMIC FUNCTION

In discussing a correspondence $w = f(z)$ we have used the word "function" to mean that f assigns a single value w to each permissible value of z. Sometimes this fact is emphasized by saying "f is a single-valued function." Of course there are equations which do not define single-valued functions, for example, $w = \arg z$ and $w = z^{1/2}$. Indeed, for each nonzero z there are an infinite number of distinct values of $\arg z$ and two distinct values of $z^{1/2}$. In general, if for some values of z there corresponds more than one value of $w = f(z)$, then we say that $w = f(z)$ is a *multiple-valued function*. We commonly obtain multiple-valued functions by taking the inverse of single-valued functions which are not one-to-one. This is, in fact, how we obtain the complex logarithmic function $\log z$.

We want to define $\log z$ as the inverse of the exponential function; i.e.,

$$w = \log z \quad \text{if} \quad z = e^w.$$

Since e^w is never zero, we must assume that $z \neq 0$. To find $\log z$ explicitly, let us write z in polar form as $z = re^{i\theta}$, and write w in standard form as $w = u + iv$. Then the equation $z = e^w$ becomes

(22)
$$re^{i\theta} = e^{u+iv} = e^u e^{iv}.$$

Because the end members of Eq. (22) are both expressed in polar form, we deduce that

$$v = \theta + 2k\pi,$$

where k is an integer, and that

$$e^u = r, \quad \text{i.e., } u = \text{Log } r,$$

where the capitalization ("Log") is used to designate the natural logarithmic function of real variables. Thus

$$w = u + iv = \text{Log } r + i(\theta + 2k\pi),$$

and since k can be any integer, we see that $w = \log z$ is a multiple-valued function which can be explicitly defined as follows:

DEFINITION 4 *If $z \neq 0$, then we define $\log z$ to be any of the infinitely many values*

$$\log z = \text{Log}|z| + i(\theta + 2k\pi) \quad (k = 0, \pm 1, \pm 2, \ldots),$$

where θ denotes a particular value of $\arg z$.

Notice that the values of $\log z$ differ by integer multiples of $2\pi i$. This reflects the fact that $2\pi i$ is a period of the exponential function. Using the notation of Sec. 1.3, we can write $\log z$ in the equivalent forms

(23) $\log z = \text{Log}|z| + i \arg z,$

(24) $\log z = \text{Log}|z| + i \text{ Arg } z + i2k\pi \quad (k = 0, \pm 1, \pm 2, \ldots).$

In particular,

$$\log 3 = \text{Log } 3 + i \arg 3 = (1.098\ldots) + i2k\pi,$$

$$\log(-1) = \text{Log } 1 + i \arg(-1) = i(2k+1)\pi,$$

$$\log(1+i) = \text{Log}|1+i| + i \arg(1+i)$$

$$= \frac{\text{Log } 2}{2} + i\left(\frac{\pi}{4} + 2k\pi\right),$$

where $k = 0, \pm 1, \pm 2, \ldots$.

The familiar properties of the real logarithmic function extend to the complex case, but the precise statements of these extensions are complicated by the fact that $\log z$ is multiple-valued. For example, if $z \neq 0$, then we have

$$z = e^{\log z},$$

but

$$\log e^z = z + 2k\pi i \qquad (k = 0, \pm 1, \pm 2, \ldots).$$

Furthermore, using the representation (23) and the equations

$$\arg z_1 z_2 = \arg z_1 + \arg z_2,$$

(25)

$$\arg\left(\frac{z_1}{z_2}\right) = \arg z_1 - \arg z_2,$$

one can readily verify that

(26) $$\log z_1 z_2 = \log z_1 + \log z_2,$$

and that

(27) $$\log\left(\frac{z_1}{z_2}\right) = \log z_1 - \log z_2.$$

As with Eqs. (25), we must interpret Eqs. (26) and (27) to mean that if particular values are assigned to any two of their terms, then one can find a value of the third term so that the equation is satisfied. For example, if $z_1 = z_2 = -1$ and we select πi to be the value of $\log z_1$ and $\log z_2$, then Eq. (26) will be satisfied if we use the particular value $2\pi i$ for $\log(z_1 z_2) = \log 1$.

Since the concept of analyticity was defined only for single-valued functions, we cannot directly discuss this property for multiple-valued functions such as $\log z$. We shall, however, investigate the analyticity of certain single-valued functions which can be derived from a multiple-valued function $w = f(z)$. We construct these single-valued functions by focusing our attention on a domain D and selecting a single value of $f(z)$

for each z in D. If it turns out that the single-valued function obtained in this manner is analytic in D, then we call it a *branch* of $f(z)$ in D.

To obtain a branch of the logarithmic function we consider the single-valued function derived from log z by taking $k = 0$ in the representation (24). We denote this function by Log z, i.e.,

(28) $$\text{Log } z = \text{Log}|z| + i \text{ Arg } z,$$

noticing that if z is a positive real number, Arg $z = 0$, and so the left- and right-hand members of Eq. (28) are notationally consistent. The number given by Eq. (28) is called the *principal value* of log z. The reader should verify that the function Log z is actually the inverse of the function obtained by restricting e^z to the horizontal strip S_0 discussed in Sec. 3.1.

Before investigating the analyticity of Log z we can make some immediate remarks concerning its continuity. Observe that Log$|z|$ is not continuous at $z = 0$. Also, since $-\pi < \text{Arg } z \leq \pi$, the function Arg z is not continuous at any point x_0 on the negative real axis, for if we let z approach x_0 from the upper half-plane, then Arg z tends to π, while if we approach x_0 from the lower half-plane, Arg z tends to $-\pi$. This means that the function Log z cannot be continuous at any point on the nonpositive real axis, and so it is not analytic there. However, at all points off the nonpositive real axis Log z *is* continuous and this fact enables us to prove

THEOREM 2 *The function* Log z *is analytic in the domain D^* consisting of all points of the complex plane except those lying on the nonpositive real axis (see Fig. 3.3). Furthermore,*

(29) $$\frac{d}{dz} \text{Log } z = \frac{1}{z}, \qquad \text{for } z \text{ in } D^*.$$

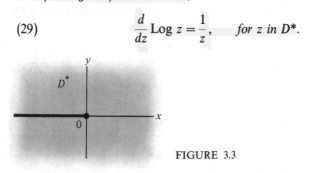

FIGURE 3.3

Proof Let us set $w = \text{Log } z$. Our goal is to prove that, for z_0 in D^* and $w_0 = \text{Log } z_0$, the limit of the difference quotient†

(30) $$\lim_{z \to z_0} \frac{w - w_0}{z - z_0}$$

†It is convenient here to use form (30) instead of the usual form $\lim_{\Delta z \to 0}[\text{Log}(z_0 + \Delta z) - \text{Log } z_0]/\Delta z$. The equivalence of these limits can be seen by putting $\Delta z = z - z_0$.

exists and equals $1/z_0$. We are guided in this endeavor by the knowledge that $z = e^w$ and that the exponential function *is* analytic so that

$$\lim_{w \to w_0} \frac{z - z_0}{w - w_0} = \left. \frac{dz}{dw} \right|_{w = w_0} = e^{w_0} = z_0 .$$

Observe that we will have accomplished our goal if we can show that

$$(31) \qquad \lim_{z \to z_0} \frac{w - w_0}{z - z_0} = \lim_{w \to w_0} \frac{1}{\dfrac{z - z_0}{w - w_0}} ,$$

because the limit on the right exists and equals $1/z_0$. But (31) will follow from the trivial identity

$$(32) \qquad \frac{w - w_0}{z - z_0} = \frac{1}{\dfrac{z - z_0}{w - w_0}} ,$$

provided we show that

(*i*) As z approaches z_0, w must approach w_0, and

(*ii*) For $z \neq z_0$, w will not coincide with w_0 [so that the terms in Eq. (31) are meaningful].

Condition (*i*) follows from the continuity, in D^*, of $w = \text{Log } z$. Condition (*ii*) is even more immediate; if w coincided with w_0, z would have to equal z_0, since $z = e^w$. Thus $w = \text{Log } z$ is differentiable at every point in D^*, and hence analytic there. ∎

Theorem 2 shows that $\text{Log } z$ is a branch of the logarithm in the domain D^*, and we refer to it as the *principal branch of* log z. To obtain a domain of analyticity for $\text{Log } z$ it was necessary to slit the complex plane along the nonpositive real axis. For this reason the nonpositive real axis is called a *branch cut* for $\text{Log } z$.

There are infinitely many other branches of log z in D^*, namely the single-valued functions

$$(33) \qquad \text{Log}_n z \equiv \text{Log } z + 2n\pi i,$$

where n is an integer. Although none of these functions is analytic on the negative real axis, we can easily construct branches of log z which are analytic there, for observe that the proof of Theorem 2 essentially depends on the fact that $\text{Log } z$ is continuous in D^*, a property which followed from the continuity, in D^*, of $\text{Arg } z$. In other words, if we had originally defined the principal value of arg z to lie, say, between 0 and 2π, then we would have concluded that $\text{Arg } z$ was continuous, and hence

Log z was analytic, in the domain D_0 consisting of all points in the plane except those on the *nonnegative* real axis. Thus the single-valued function

$$(34) \qquad \mathscr{L}_0(z) = \text{Log}\,|z| + i\theta,$$

where θ is the particular value of arg z which satisfies $0 < \theta \leq 2\pi$, is a branch of log z analytic in D_0. More generally, the function

$$(35) \qquad \mathscr{L}_\tau(re^{i\theta}) = \text{Log}\,r + i\theta, \qquad \text{where } \tau < \theta \leq \tau + 2\pi,$$

is a branch of log z in the domain D_τ consisting of all points in the plane except those on the ray R_τ emanating from the origin with angle of inclination equal to τ. The ray R_τ is therefore the branch cut for the function given in Eq. (35).

Hence we see that for each nonzero point we can find a branch of log z which is analytic at this point and that, regardless of the branch chosen, its derivative will always be given by $1/z$. The exceptional point zero, at which no branch of log z is analytic, is called a *branch point* of log z.

EXAMPLE 2   Determine a branch of $\log(z^3 - 2)$ which is analytic at $z = 0$, and find its derivative there.

   SOLUTION   The given multiple-valued function is the composition of the logarithm with the analytic function $f(z) = z^3 - 2$. Thus, by the chain rule, it suffices to choose any branch of the logarithm which is analytic at $f(0) = -2$. In particular, $w = \mathscr{L}_0(f(z))$, where $\mathscr{L}_0$ is given by (34), solves the problem. Furthermore,

$$\left.\frac{dw}{dz}\right|_{z=0} = \mathscr{L}_0'(f(0))f'(0) = \frac{1}{f(0)}f'(0) = 0. \qquad \blacksquare$$

Exercises 3.2

1. Evaluate each of the following.

   (a) log i   (b) log$(1 - i)$

   (c) Log$(-i)$ (d) Log$(\sqrt{3} + i)$

2. Verify formulas (26) and (27).

3. Show that if $z_1 = i$ and $z_2 = i - 1$, then

$$\text{Log}\,z_1 z_2 \neq \text{Log}\,z_1 + \text{Log}\,z_2.$$

4. Prove that Log $e^z = z$ if, and only if, $-\pi < \text{Im}\,z \leq \pi$.

5. Find the error in the following "proof" that $z = -z$: Since $z^2 = (-z)^2$, $2 \text{ Log } z = 2 \text{ Log}(-z)$, and hence $\text{Log } z = \text{Log}(-z)$, which implies that $z = e^{\text{Log } z} = e^{\text{Log}(-z)} = -z$.

6. Solve the following equations.

$$\text{(a) } e^z = 2i \qquad \text{(b) } \text{Log}(z^2 - 1) = \frac{i\pi}{2}, \qquad \text{(c) } e^{2z} + e^z + 1 = 0$$

7. Use the polar form of the Cauchy-Riemann equations (Prob. 6 in Exercises 2.4) to give another proof of Theorem 2.

8. Without directly verifying Laplace's equation, explain why the function $\text{Log}|z|$ is harmonic in every domain which does not contain the origin.

9. Show that the function $\text{Log}(-z) + i\pi$ is a branch of $\log z$ analytic in the domain D_0 consisting of all points in the plane except those on the nonnegative real axis.

10. Prove that the branch $\text{Log}_n z$ defined by Eq. (33) is the inverse of the function obtained by restricting e^z to the horizontal strip S_n discussed in Sec. 3.1.

11. Determine a branch of $\log(z^2 + 2z + 3)$ which is analytic at $z = -1$, and find its derivative there.

12. Find a branch of $\log(z^2 + 1)$ which is analytic at $z = 0$ and takes the value $2\pi i$ there.

13. Find a branch of $\log(2z - 1)$ which is analytic at all points in the plane except those on the following rays.
    (a) $\{x + iy \,|\, x \le \frac{1}{2}, y = 0\}$
    (b) $\{x + iy \,|\, x \ge \frac{1}{2}, y = 0\}$
    (c) $\{x + iy \,|\, x = \frac{1}{2}, y \ge 0\}$

14. Find a one-to-one analytic mapping of the upper half-plane $\text{Im } z > 0$ onto the infinite horizontal strip

$$\mathscr{H} : \{u + iv \,|\, -\infty < u < \infty, 0 < v < 1\}.$$

    [HINT: Start by considering $w = \text{Log } z$.]

15. Prove that there exists no function $F(z)$ analytic in the annulus D: $1 < |z| < 2$ such that $F'(z) = 1/z$ for all z in D. [HINT: Assume that F exists and show that for z in D, z not a negative real number, $F(z) = \text{Log } z + c$, where c is a constant.]

3.3 COMPLEX POWERS AND INVERSE TRIGONOMETRIC FUNCTIONS

One important use of the logarithmic function is to define complex powers of z. The definition is motivated by the identity

$$z^n = \left(e^{\log z}\right)^n = e^{n \log z},$$

which holds for any integer n.

DEFINITION 5 *If α is a complex constant and $z \neq 0$, then we define z^α by*

$$z^\alpha = e^{\alpha \log z}.$$

This means that a particular value of $\log z$ leads to a particular value of z^α.

EXAMPLE 3 Find all the values of $(-2)^i$.

SOLUTION Since $\log(-2) = \text{Log } 2 + (\pi + 2k\pi)i$, we have

$$(-2)^i = e^{i \log(-2)} = e^{i \text{ Log } 2} e^{-\pi - 2k\pi} \qquad (k = 0, \pm 1, \pm 2, \ldots).$$

Thus $(-2)^i$ has infinitely many different values. ∎

Using the representation (24) we can write

(36) $$z^\alpha = e^{\alpha(\text{Log}|z| + i \text{ Arg } z + 2k\pi i)} = e^{\alpha(\text{Log}|z| + i \text{ Arg } z)} e^{\alpha 2k\pi i},$$

where $k = 0, \pm 1, \pm 2, \ldots$. The values of z^α obtained by taking $k = k_1$ and $k = k_2$ ($\neq k_1$) in Eq. (36) will therefore be the same when

$$e^{\alpha 2k_1\pi i} = e^{\alpha 2k_2\pi i}.$$

But by Theorem 1 this occurs only if

$$\alpha 2k_1 \pi i = \alpha 2k_2 \pi i + 2m\pi i,$$

where m is an integer. Solving the equation we get $\alpha = m/(k_1 - k_2)$; i.e., formula (36) yields some identical values of z^α only when α is a real rational number. Consequently, if α is not a real rational number, we obtain infinitely many different values for z^α, one for each choice of the integer k in Eq. (36). On the other hand, if $\alpha = m/n$, where m and $n > 0$ are integers having no common factor, then one can verify that there are exactly n distinct values of $z^{m/n}$, namely

(37) $$z^{m/n} = e^{(m/n)\text{Log}|z|} e^{(m/n)i(\text{Arg } z + 2k\pi)} \qquad (k = 0, 1, \ldots, n-1).$$

Notice that when $m = 1$, formula (37) agrees with the results of Chapter 1. Also, if $n = 1$ and m is any integer, only one value of $z^{m/n}$ results. Thus we see that unless α is an integer, $w = z^\alpha$ is a multiple-valued function.

It is clear from Definition 5 that each branch of log z yields a branch of z^α. For example, using the principal branch of log z we obtain the *principal branch of z^α*, namely $e^{\alpha \, \text{Log} \, z}$. Since e^z is entire and Log z is analytic in the slit domain D^* of Theorem 2, the chain rule implies that *the principal branch of z^α is also analytic in D^**. Furthermore, for z in D^*, we have

$$\frac{d}{dz}(e^{\alpha \, \text{Log} \, z}) = e^{\alpha \, \text{Log} \, z}\frac{d}{dz}(\alpha \, \text{Log} \, z) = e^{\alpha \, \text{Log} \, z}\frac{\alpha}{z}.$$

Other branches of z^α can be constructed by using other branches of log z, and since each branch of the latter has derivative $1/z$, the formula

$$(38) \qquad \frac{d}{dz}(z^\alpha) = \alpha z^\alpha \frac{1}{z}$$

is valid for each corresponding branch of z^α (provided the same branch is used on both sides of the equation). Observe that if z_0 is any given nonzero point, then by selecting a branch cut for log z which avoids z_0 we get a branch of z^α which is analytic in a neighborhood of z_0.

As we shall show in the following example the inverse trigonometric functions can also be described in terms of logarithms.

EXAMPLE 4   The inverse sine function $w = \sin^{-1} z$ is defined by the equation $z = \sin w$. Show that $\sin^{-1} z$ is a multiple-valued function given by

$$(39) \qquad \sin^{-1} z = -i \log[iz + (1 - z^2)^{1/2}].$$

SOLUTION   From the equation

$$z = \sin w = \frac{e^{iw} - e^{-iw}}{2i},$$

we deduce that

$$(40) \qquad e^{2iw} - 2ize^{iw} - 1 = 0.$$

Using the quadratic formula we can solve Eq. (40) for e^{iw}:

$$e^{iw} = iz + (1 - z^2)^{1/2},$$

where, of course, the square root is two-valued. Formula (39) now follows by taking logarithms.† ∎

We can obtain a branch of the multiple-valued function $\sin^{-1} z$ by first choosing a branch of the square root and then selecting a suitable

---

†To ensure that our method has not introduced extraneous solutions one should verify that every value w given by Eq. (39) satisfies $z = \sin w$.

branch of the logarithm. Using the chain rule and formula (38) one can show that any such branch of $\sin^{-1} z$ satisfies

(41) $$\frac{d}{dz}(\sin^{-1} z) = \frac{1}{(1-z^2)^{1/2}} \qquad (z \neq \pm 1),$$

where the choice of the square root on the right must be the same as that used in the branch of $\sin^{-1} z$.

For the inverse cosine and inverse tangent functions a similar procedure leads to the expressions

(42) $$\cos^{-1} z = -i \log[z + (z^2 - 1)^{1/2}],$$

(43) $$\tan^{-1} z = \frac{i}{2} \log \frac{i+z}{i-z} = \frac{i}{2} \log \frac{1-iz}{1+iz} \qquad (z \neq \pm i)$$

and to the formulas

(44) $$\frac{d}{dz}(\cos^{-1} z) = \frac{-1}{(1-z^2)^{1/2}} \qquad (z \neq \pm 1),$$

(45) $$\frac{d}{dz}(\tan^{-1} z) = \frac{1}{1+z^2} \qquad (z \neq \pm i).$$

Notice that the derivative in Eq. (45) is independent of the branch chosen for $\tan^{-1} z$, while the derivative in Eq. (44) depends on the choice of the square root used in the branch of $\cos^{-1} z$.

The same methods can be applied to the inverse hyperbolic functions. The results are

(46) $$\sinh^{-1} z = \log[z + (z^2 + 1)^{1/2}],$$

(47) $$\cosh^{-1} z = \log[z + (z^2 - 1)^{1/2}],$$

(48) $$\tanh^{-1} z = \frac{1}{2} \log \frac{1+z}{1-z} \qquad (z \neq \pm 1).$$

Exercises 3.3

1. Find all the values of the following.
   (a) i^i (b) $(-1)^{2/3}$ (c) $2^{\pi i}$
   (d) $(1+i)^{1-i}$ (e) $(1+i)^3$

2. Find the principal value (i.e., the value given by the principal branch) of each of the following.
   (a) $4^{1/2}$ (b) i^{2i} (c) $(1+i)^{1+i}$

3. Show that if $z \neq 0$, then $z^0 = 1$.

4. Is 1 raised to any power always equal to 1?

5. Let α and β be complex constants and let $z \neq 0$. Show that the following identities hold when each power function is given by its principal branch.

(a) $z^{-\alpha} = \dfrac{1}{z^{\alpha}}$ (b) $z^{\alpha} z^{\beta} = z^{\alpha+\beta}$ (c) $\dfrac{z^{\alpha}}{z^{\beta}} = z^{\alpha-\beta}$

6. Give an example to show that the principal value of $(z_1 z_2)^{\alpha}$ need not be equal to the product of principal values $z_1^{\alpha} z_2^{\alpha}$.

7. Find the derivative of the principal branch of z^{1+i} at $z = i$.

8. Show that all the solutions of the equation $\sin z = 2$ are given by $\pi/2 + 2k\pi \pm i \operatorname{Log}(2 + \sqrt{3})$, where $k = 0, \pm 1, \pm 2, \dots$.

9. Derive formulas (42) and (44) concerning $\cos^{-1} z$.

10. Show that all the solutions of the equation $\cos z = 2i$ are given by $\pi/2 + 2k\pi - i \operatorname{Log}(\sqrt{5} + 2)$, $\quad -\pi/2 + 2k\pi + i \operatorname{Log}(\sqrt{5} + 2)$, for $k = 0, \pm 1, \pm 2, \dots$.

11. Derive formulas (43) and (45) concerning $\tan^{-1} z$.

12. Find all solutions of the equation $\sin z = \cos z$.

13. Derive formulas (46) and (47) for $\sinh^{-1} z$ and $\cosh^{-1} z$.

14. Derive the formula $d(\sinh^{-1} z)/dz = 1/(1 + z^2)^{1/2}$ and explain the conditions under which it is valid.

15. Show that the function

$$f(z) = e^{(1/2) \operatorname{Log}(z-1) + (1/2) \operatorname{Log}(z+1)}$$

is a branch of $(z^2 - 1)^{1/2}$ analytic at all points in the plane except those on the real line segment $[-1, 1]$. [HINT: From Theorem 2 it is easy to see that $f(z)$ is analytic except possibly at the points on the ray $\{x + iy \,|\, x \leq 1, y = 0\}$ (draw a diagram). Therefore the essence of the problem is to establish that $f(z)$ is actually analytic on the portion $\mathscr{R}$: $\{x + iy \,|\, x < -1, y = 0\}$. To see this, verify that for points z on, above, or below $\mathscr{R}$ we have

$$f(z) = -e^{(1/2) \operatorname{Log}(1-z) + (1/2) \operatorname{Log}(-z-1)}$$

and that the function on the right is analytic on $\mathscr{R}$.]

16. According to Definition 5 the multiple-valued function c^z, where c is a nonzero constant, is given by $c^z = e^{z \log c}$. Show that by selecting a particular value of $\log c$ we obtain a branch of c^z which is entire. Find the derivative of such a branch.

17. The following problem arises in the theory of fluid mechanics. Find a function, harmonic inside the wedge bounded by the positive x-axis and the ray emanating from the origin at an angle of inclination ϕ, which goes to zero on these sides but is not identically zero. [HINT: Recall Prob. 4 in Exercises 2.5.]

*3.4 APPLICATION TO OSCILLATING SYSTEMS

Many engineering problems are ultimately based upon the response of a system to a sinusoidal input. Naturally, all the parameters in such a situation are real, and the models can be analyzed using the techniques of real variables. However, the utilization of complex variables can greatly simplify the computations and lend some insight into the roles played by the various parameters. In this section we shall illustrate the technique for the analysis of a simple *RLC* (resistor-inductor-capacitor) circuit.

The electric circuit is shown in Fig. 3.4. We suppose that the power supply is driving the system with a sinusoidal voltage V_s oscillating at a frequency of v cycles per second. To be precise, let us say that at time t

$$(49) \qquad V_s = A \sin \omega t,$$

where A is the amplitude of the signal and $\omega = 2\pi v$. We shall try to find the current output I_s of the power supply.

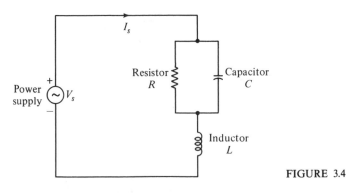

FIGURE 3.4

Across each element of the circuit there is a voltage drop V which is related to the current I flowing through the element. The relationships between these two quantities are as follows: For the resistor

$$(50) \qquad V_r = I_r R,$$

where R is a constant known as the resistance; for the capacitor

$$(51) \qquad C \frac{dV_c}{dt} = I_c,$$

where C is the capacitance; and for the inductor

(52)
$$V_l = L\frac{dI_l}{dt},$$

in terms of the inductance L.

One can incorporate formulas (49)–(52) together with Kirchhoff's laws,† which express conservation of charge and energy, to produce a system of differential equations involving all the currents and voltages. The solution of this system is, however, a laborious process, and a simpler technique is clearly desirable.

Before we demonstrate the utilization of complex variables in this problem, we shall make some observations based upon physical considerations.

First, notice that if the capacitor and inductor were replaced by resistors, as we illustrate in Fig. 3.5, the solution would be simple. The pair

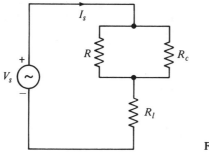

FIGURE 3.5

of resistances R and R_c are wired in parallel, so they can be replaced by an equivalent resistance R_1 given by the familiar law

$$\frac{1}{R_1} = \frac{1}{R} + \frac{1}{R_c},$$

or

(53)
$$R_1 = \frac{RR_c}{R + R_c}.$$

This resistance then appears in series with R_l, yielding an effective total resistance R_{eff} given by

(54)
$$R_{\text{eff}} = R_1 + R_l,$$

and the circuit is equivalently (from the point of view of the power supply) represented by Fig. 3.6.

†See the references at the end of this chapter.

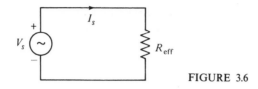

FIGURE 3.6

Equations (49) and (50) now yield the current output I_s:

$$I_s = \frac{A \sin \omega t}{R_{\text{eff}}} = \frac{A \sin \omega t}{\dfrac{RR_c}{R + R_c} + R_l}.$$

Since this model is so easy to solve, it would clearly be advantageous to replace, if possible, the capacitor and inductor by equivalent resistors.

The second point we wish to make involves the nature of the solution I_s for the original circuit. If the power supply is turned on at time $t = 0$, there will be a fairly complicated initial current response; this so-called "transient," however, eventually dies out, and the system enters a steady state in which all the currents and voltages oscillate sinusoidally at the same frequency as the driving voltage. This behavior is common to all damped linear systems, and it will be familiar to anyone who has ridden a bicycle over railroad ties or mastered the art of dribbling a basketball. For many applications it is this steady-state response, which is independent of the initial state of the system, that is of interest. The complex-variable technique which we shall describe is ideally suited for finding this response.

The technique is based upon the equation

(55) $$e^{i\omega t} = \cos \omega t + i \sin \omega t$$

for real t. Observe first that the simple sinusoids can be written, from Eq. (55), as

(56) $$\cos \omega t = \text{Re } e^{i\omega t}, \qquad \sin \omega t = \text{Im } e^{i\omega t}.$$

Second, the most general sinusoid with this frequency, $a \cos \omega t + b \sin \omega t$, can be written in the form

(57) $$a \cos \omega t + b \sin \omega t = \text{Re } \alpha e^{i\omega t}$$

$$= \text{Im } \alpha' e^{i\omega t},$$

where $\alpha = a - ib$ and $\alpha' = i\alpha$. The advantage of the exponential representation for these functions lies in its compactness and in the simplicity of the expression for the derivative; differentiating $e^{i\omega t}$ (with respect to t) merely amounts to multiplying by $i\omega$. Consequently, the derivative of the general sinusoid in Eq. (57) becomes

$$\frac{d}{dt}(a \cos \omega t + b \sin \omega t) = \frac{d}{dt} \operatorname{Re} \alpha e^{i\omega t} = \operatorname{Re} \frac{d}{dt}(\alpha e^{i\omega t})$$

$$= \operatorname{Re} i\omega\alpha e^{i\omega t},$$

or, alternatively,

$$\frac{d}{dt}(a \cos \omega t + b \sin \omega t) = \operatorname{Im} i\omega\alpha' e^{i\omega t}.$$

To take advantage of these notions, we must make one more observation about the circuit in Fig. 3.4. Each of the elements is a *linear* device in the sense that the superposition principle holds; that is, if an element responds to the excitation voltages $V_1(t)$ and $V_2(t)$ by producing the currents $I_1(t)$ and $I_2(t)$, respectively, then the response to the voltage $V_1(t) + \beta V_2(t)$ will be the current $I_1(t) + \beta I_2(t)$, for any constant β. This is a mathematical consequence of the linearity of Eqs. (50)–(52), and it will hold true even if the functions $V_1(t)$ and $V_2(t)$ take complex values, although this corresponds to no physically meaningful situation. In particular, the response to the voltage $V_1(t) + iV_2(t)$ will be the current $I_1(t) + iI_2(t)$. And since we know that the circuit responds to real voltages with real currents, it follows that if the response to the complex voltage $V(t)$ is $I(t)$, then the responses to the (real) voltages $\operatorname{Re} V(t)$ and $\operatorname{Im} V(t)$ will be $\operatorname{Re} I(t)$ and $\operatorname{Im} I(t)$, respectively.

With all these mathematical tools at hand, let us return to the solution of the problem depicted in Fig. 3.4. The supply voltage, given in Eq. (49), can be represented by

$$(58) \qquad V_s(t) = \operatorname{Im} A e^{i\omega t}.$$

Our strategy will be to find the steady-state response of the circuit to the (fictional) voltage $Ae^{i\omega t}$, and to take the imaginary part of the answer as our solution, in accordance with the previous paragraph.

From the earlier observations about the nature of the steady-state response and Eq. (57), we are led to postulate that the current or voltage in any part of the circuit can be written as a (possibly complex) constant times $e^{i\omega t}$; furthermore, differentiation of such a function is equivalent to multiplication by the factor $i\omega$. Thus, in the situation at hand, Eq. (51) for the behavior of the capacitor becomes

$$(59) \qquad i\omega C V_c = I_c,$$

and Eq. (52) for the inductor becomes

$$(60) \qquad V_l = i\omega L I_l.$$

But the voltage-current relationships (59) and (60) now have the same form as Eq. (50) for a resistor; in other words, operated at the frequency $v = \omega/2\pi$, a capacitor behaves mathematically like a resistor with resistance

$$(61) \qquad R_c = \frac{1}{i\omega C},$$

and an inductor behaves like a resistor with resistance

$$(62) \qquad R_l = i\omega L.$$

These are pure imaginary numbers, but we have already abandoned physical reality in assuming a complex power-supply voltage. As a concession to realists, however, engineers have adopted the term "impedance" to describe a complex resistance.

Having replaced, formally, the capacitor and inductor by resistors, we are back in the situation of the simple circuit of Fig. 3.5. The effective impedance of the series-parallel arrangement is displayed in Eqs. (53) and (54), and substitution of the relations (61) and (62) yields the expression

$$R_{\text{eff}} = \frac{\dfrac{R}{i\omega C}}{R + \dfrac{1}{i\omega C}} + i\omega L.$$

The current output of the power supply is therefore

$$I_s = \operatorname{Im} \frac{A e^{i\omega t}}{R_{\text{eff}}},$$

which after some manipulation can be written as

$$(63) \qquad I_s = \operatorname{Im} \frac{A e^{i\omega t}\left\{\dfrac{1}{R} + i\left[\omega C(1 - \omega^2 LC) - \dfrac{\omega L}{R^2}\right]\right\}}{(1 - \omega^2 LC)^2 + \dfrac{\omega^2 L^2}{R^2}}.$$

The expression for I_s as a linear combination of $\sin \omega t$ and $\cos \omega t$ may be found by expanding Eq. (63), but complex variables can once more be used in providing a more meaningful interpretation of the answer. If we define ϕ_0 and R_0 by

$$\phi_0 = \text{Arg } R_{\text{eff}}, \qquad R_0 = |R_{\text{eff}}|,$$

then I_s can be expressed as

(64) $\qquad I_s = \text{Im} \frac{Ae^{i\omega t}}{R_0 e^{i\phi_0}} = \text{Im} \frac{A}{R_0} e^{i(\omega t - \phi_0)} = \frac{A}{R_0} \sin(\omega t - \phi_0).$

This displays some easily visualized properties of the output current in relation to the input voltage (49). The current is, as we indicated earlier, a sinusoid with the same frequency, but its amplitude differs from the voltage amplitude by the factor $1/R_0$. Furthermore, the two sinusoids are out of phase, with ϕ_0 measuring the phase difference in radians. The numbers R_0 and ϕ_0 can be computed in terms of the circuit parameters and the frequency, and the circuit problem is solved.

In conclusion we wish to reiterate the two advantages of using complex notation in solving linear sinusoidal problems. First, the representation of general sinusoids (57) is compact and leads naturally to a reinterpretation of the sinusoid in terms of amplitudes and phases, as in Eqs. (63) and (64). And second, the process of differentiation is replaced by simple multiplication. It is important to keep in mind, however, that only the real (or imaginary) part of the solution corresponds to physical reality and that the condition of linearity is of the utmost importance.

Exercises 3.4

1. Verify the expression in Eq. (63).

2. In the limit of very low frequencies, a capacitor behaves like an open circuit (infinite resistance), while an inductor behaves like a short circuit (zero resistance). Draw and analyze the low-frequency limit of the circuit in Fig. 3.4, and verify expression (63) in this limit.

3. Repeat Prob. 2 for the high-frequency limit, where the capacitor behaves like a short circuit and the inductor like an open circuit.

4. Using the techniques of this section, find the steady-state current output I_s of the circuits in Fig. 3.7.

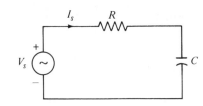

$V_s = A \cos \omega t$

(a)

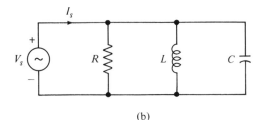

$V_s = A \cos \omega(t + \beta)$

(b)

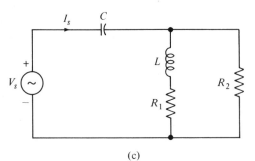

$V_s = A \cos \omega_1 t + B \sin \omega_2 t$

$C = L = R_1 = R_2 = 1$

(c)

FIGURE 3.7

SUMMARY

The complex exponential function e^z is defined by $e^z = e^{x+iy} = e^x(\cos y + i \sin y)$. This definition was motivated by the desire to preserve the fundamental properties of the real exponential function. In particular, e^z is entire and satisfies $d(e^z)/dz = e^z$ for all z. However, e^z is not one-to-one on the whole plane; in fact, it is periodic with complex period $2\pi i$; i.e., $e^{z+2\pi i} = e^z$ for all z.

The complex sine and cosine functions are defined in terms of the exponential function: $\sin z = (e^{iz} - e^{-iz})/2i$, $\cos z = (e^{iz} + e^{-iz})/2$. When z is real these definitions agree with those given in calculus. Furthermore, the usual differentiation formulas and trigonometric identities remain valid for these functions. Complex hyperbolic functions are also defined in terms of the exponential function, and, again, they retain many of their familiar properties.

The complex logarithmic function log z is the inverse of the exponential function and is given by log $z =$ Log $|z| + i$ arg z, where the capitalization signifies the natural logarithmic function of real variables. Since for each $z \neq 0$, arg z has infinitely many values, the same is true of log z. Thus log z is an example of a multiple-valued function.

The concept of analyticity for a multiple-valued function f is discussed in terms of branches of f. A single-valued function F is said to be a branch of f if it is analytic in some domain, at each point of which $F(z)$ coincides with one of the values of $f(z)$.

Branches of log z can be obtained by restricting arg z so that it is single-valued and continuous. For example, the function Log $z =$ Log $|z| + i$ Arg z, which is the principal branch of log z, is analytic in the domain D^* consisting of all points in the plane except those on the nonpositive real axis. The formula $d(\log z)/dz = 1/z$ is valid in the sense that it holds for every branch of log z.

Complex powers of z are defined by means of logarithms. Specifically, $z^\alpha = e^{\alpha \log z}$ for $z \neq 0$. Unless α is an integer, $w = z^\alpha$ is a multiple-valued function whose branches can be obtained by selecting branches of log z. The inverse trigonometric and hyperbolic functions can also be expressed in terms of logarithms, and they too are multiple-valued.

SUGGESTED READING

The following texts may be helpful for further study of the concepts used in Sec. 3.4:

Electrical Circuits

[1]   SCOTT, R. E. *Elements of Linear Circuits.* Addison-Wesley Publishing Company, Inc., Reading, Mass., 1965.

[2]   SKILLING, H. H. *Electrical Engineering Circuits.* John Wiley & Sons, Inc., New York, 1965.

Sinusoidal Analysis

[3]   CHEN, W. H. *The Analysis of Linear Systems.* McGraw-Hill Book Company, New York, 1963.

[4]   GUILLEMIN, E. A. *Theory of Linear Physical Systems.* John Wiley & Sons,
      Inc., New York, 1963.

[5]   KAPLAN, W. *Operational Methods for Linear Systems.* Addison-Wesley Pub-
      lishing Company, Inc., Reading, Mass., 1962.

4

Complex Integration

In Chapter 2 we saw that the *two*-dimensional nature of the complex plane required us to generalize our notion of a derivative because of the freedom of the variable to approach its limit along any of an infinite number of directions. This two-dimensional aspect will have an effect on the theory of integration as well, necessitating the consideration of integrals along general curves in the plane and not merely segments of the *x*-axis. Fortuitously, such well-known techniques as using antiderivatives to evaluate integrals carry over to the complex case.

When the function under consideration is analytic the theory of integration becomes an instrument of profound significance in studying its behavior. The main result is the theorem of *Cauchy*, which roughly says that the integral of a function around a closed loop is zero if the function is analytic "inside and on" the loop. Using this result we shall derive the *Cauchy Integral Formula*, which explicitly displays many of the important properties of analytic functions.

4.1 CONTOURS

Let us turn to the problem of finding a mathematical explication of our intuitive concept of a *curve* in the *xy*-plane. It is helpful in this regard to visualize an artist actually tracing the curve γ on graph paper. At any particular instant of time t, he is drawing a dot at, say, the point $z = x + iy$; the locus of dots generated over an interval of time $a \leq t \leq b$ constitutes the curve. Clearly we can interpret his actions as generating z as a function of t, and then the curve γ is the range of $z(t)$ as t varies between a and b. In such a case $z(t)$ is called a *parametrization* of γ.

To keep things simple, we shall set down certain ground rules for the artist as he draws γ. First, we restrain him from lifting his pen from the paper during the sketch; mathematically, we are requiring that $z(t)$ be continuous. Second, we restrict him to curves that can be drawn with an even, steady stroke; specifically, the pen point must move with a well-defined (finite) velocity which, also, must vary continuously. Now the velocity of the point tracing out the trajectory $(x(t), y(t))$ is the vector $(dx/dt, dy/dt) = x'(t) + iy'(t)$. It makes sense to call this vector $z'(t)$, since $x(t) + iy(t) = z(t)$ and therefore

$$x'(t) + iy'(t) = \lim_{\Delta t \to 0} \frac{z(t + \Delta t) - z(t)}{\Delta t}.\dagger$$

Thus we insist that $z'(t)$ exist and be continuous on $[a, b]$.

Finally, it is sometimes convenient to stipulate that no point be drawn twice; in other words, $z(t)$ must be one-to-one. Occasionally, however, we allow the possibility that the initial and terminal points coincide, as in the case of a circle.

Putting this all together, we now specify the class of *smooth curves*. These fall into two separate categories: *smooth arcs*, which have distinct endpoints, and *smooth closed curves*, whose endpoints coincide.

DEFINITION 1 A point set γ in the complex plane is said to be a **smooth‡ arc** *if it is the range of some continuous complex-valued function* $z = z(t)$, $a \leq t \leq b$, *which satisfies the following conditions:*

(i) $z(t)$ has a continuous derivative on $[a, b]$,§

(ii) $z'(t)$ never vanishes on $[a, b]$,

(iii) $z(t)$ is one-to-one on $[a, b]$.

---

†Observe, however, that since t is a real variable in this context, the existence of the derivative $z'(t)$ is not nearly so profound as it was in Chapter 2, where the independent variable was complex.

‡The term *regular* is sometimes used instead of smooth.

§At the endpoint $t = a$, $z'(t)$ denotes the right-hand derivative, while at $t = b$, $z'(t)$ is the left-hand derivative.

A point set γ is called a **smooth closed curve** *if it is the range of some continuous function $z = z(t)$, $a \le t \le b$, satisfying conditions (i) and (ii) above and the following:*

(iii)' $z(t)$ is one-to-one on the half-open interval $[a, b)$, but $z(b) = z(a)$ and $z'(b) = z'(a)$.

The phrase "γ is a smooth curve" means that γ is either a smooth arc or a smooth closed curve.

In elementary calculus it is shown that the vector $(x'(t), y'(t))$, if it exists and is nonzero, can be interpreted geometrically as being tangent to the curve at the point $(x(t), y(t))$. Hence the conditions of Definition 1 imply that a smooth curve possesses a unique tangent at every point and that the tangent direction varies continuously along the curve. Consequently a smooth curve has no corners or cusps; see Fig. 4.1.

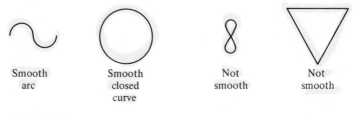

Smooth   Smooth   Not   Not
arc   closed   smooth   smooth
      curve

FIGURE 4.1

The functional relationship $z = z(t)$, $a \le t \le b$, which describes the curve γ is, of course, somewhat arbitrary. Another artist could draw the same curve faster, or use a different starting point, resulting in a different description $z = z_1(t)$, $a_1 \le t \le b_1$. For example, the straight-line segment joining 0 to $1 + i$ is generated by each of the following functions:

$$z_1(t) = t + it \qquad\qquad (0 \le t \le 1),$$

$$z_2(t) = (1 - 2t) + i(1 - 2t) \qquad \left(0 \le t \le \frac{1}{2}\right),$$

$$z_3(t) = \tan t + i \tan t \qquad \left(0 \le t \le \frac{\pi}{4}\right).$$

Any function $z(t)$ satisfying the conditions of Definition 1 for a given smooth curve γ is called an *admissible parametrization* of γ, and the variable t is called the *parameter*; the corresponding interval $[a, b]$ is known as the *parametric interval*. So we can describe our situation in the following manner: A given smooth curve γ will have many different admissible parametrizations, but we need produce only *one* admissible parametrization in order to demonstrate that a given curve is smooth.

EXAMPLE 1 Find an admissible parametrization for each of the following smooth curves: (a) the straight-line segment joining $-2 - 3i$ and
$5 + 6i$, (b) the circle of radius 2 centered at $1 - i$, and (c) the graph of the
function $y = x^3$ for $0 \leq x \leq 1$.

SOLUTION   (a) Given any two distinct points z_1 and z_2, every point
on the line segment joining z_1 and z_2 is of the form $z_1 + t(z_2 - z_1)$, where
$0 \leq t \leq 1$ (see Prob. 3 in Exercises 1.3). Therefore the given curve constitutes the range of

$$z(t) = -2 - 3i + t(7 + 9i) \qquad (0 \leq t \leq 1).$$

The verification of conditions (i), (ii), and (iii) is immediate; thus this is an
admissible parametrization.

(b) In Chapter 3 it was shown that any point on the unit circle
centered at the origin can be written in the form $e^{it} = \cos t + i \sin t$ for
$0 \leq t < 2\pi$; therefore an admissible parametrization for this smooth
closed curve is $z_0(t) = e^{it}$, $0 \leq t \leq 2\pi$ (notice that the endpoints are
joined properly). To parametrize the given circle (b) we simply shift the
center and double the radius:

$$z(t) = 1 - i + 2e^{it} \qquad (0 \leq t \leq 2\pi).$$

(c) The parametrization of the graph of any function $y = f(x)$ is quite
trivial; simply let $x = t$ and $y = f(t)$. Of course one must verify that this is
an admissible parametrization. For the graph (c) we have

$$z(t) = t + it^3 \qquad (0 \leq t \leq 1). \qquad \blacksquare$$

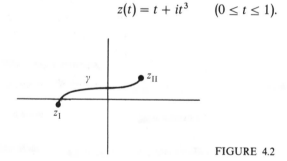

FIGURE 4.2

Let us carry our analysis of curve sketching a little further. Suppose
that the artist is to draw a smooth *arc* like that in Fig. 4.2 and that he is to
abide by our ground rules (in particular he is not permitted to retrace
points). Then it is intuitively clear that he must either start at z_1 and work
toward z_{II}, or start at z_{II} and terminate at z_1. Either mode produces an
ordering of the points along the curve (Fig. 4.3).

Thus we see that there are exactly two such "natural" orderings of the
points of a smooth arc γ, and either one can be specified by declaring
which endpoint of γ is the initial point. A smooth arc, together with a
specific ordering of its points, is called a *directed smooth arc*. The ordering
can be indicated by an arrow, as in Fig. 4.3.

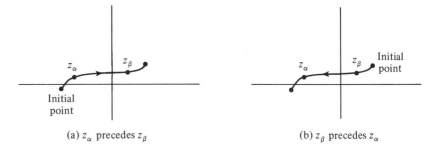

(a) z_α precedes z_β (b) z_β precedes z_α

FIGURE 4.3

The ordering which the artist generates as he draws γ is reflected in the parametrization function $z(t)$ which describes his pen's trajectory; specifically, the point $z(t_1)$ will precede the point $z(t_2)$ whenever $t_1 < t_2$. Since there are only two possible (natural) orderings, *any* admissible parametrization must fall into one of two categories, according to the particular ordering it respects. (The proof of this fact forms the basis of the rigorous treatment of directed curves; see Ref. [1].) For example, consider the ordering indicated on the line segment in Fig. 4.4. The (admissible) parametrization functions $z_1(t) = (1 + i)t$, $0 \le t \le 1$, and $z_2(t) = (1 + i)\tan t$, $0 \le t \le \pi/4$, are both consistent with the given ordering, but $z_3(t) = -(1 + i)t$, $-1 \le t \le 0$, corresponds to the opposite ordering.

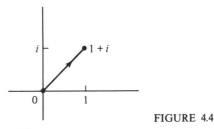

FIGURE 4.4

In general, if $z = z(t)$, $a \le t \le b$, is an admissible parametrization consistent with one of the orderings, then $z = z(-t)$, $-b \le t \le -a$, always corresponds to the opposite ordering.

The situation is slightly more complicated if the artist is to draw a smooth *closed* curve. First he must select an initial point; then he must choose in which of two directions to move as he begins to trace the curve (see Fig. 4.5). Having made these decisions, he has established the ordering of the points of γ. Now, however, there is one anomaly; the initial point both precedes and is preceded by every other point, since it also serves as the terminal point. Ignoring this schizophrenic pest, we shall say that the points of a smooth closed curve have been ordered when (*i*) a designation of the initial point is made and (*ii*) one of the two "directions

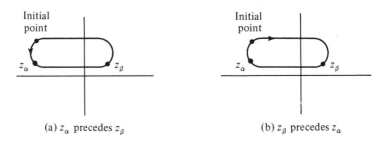

(a) z_α precedes z_β (b) z_β precedes z_α

FIGURE 4.5

of transit" from this point is selected. A smooth closed curve whose points have been ordered is called a *directed smooth closed curve*.

As in the case of smooth arcs, the parametrization of the trajectory of the artist's pen reflects the ordering which he generates as he sketches a smooth closed curve. If this parametrization is given by $z = z(t)$, $a \le t \le b$, then (*i*) the initial point must be $w = z(a)$ and (*ii*) the point $z(t_1)$ precedes the point $z(t_2)$ whenever $a < t_1 < t_2 < b$. Any other admissible parametrization having the same initial point w must reflect either the same or the opposite ordering. Thus in Fig. 4.6 the parametrizations $z = z_1(t) = e^{2\pi it}$, $0 \le t \le 1$, and $z = z_2(t) = e^{it}$, $0 \le t \le 2\pi$, are consistent with the ordering indicated. The parametrization functions $z_3(t) = e^{-2\pi it}$, $0 \le t \le 1$, and $z_4(t) = -e^{2\pi it}$, $0 \le t \le 1$, are *not* consistent with this ordering; indeed $z_3(t)$ reflects the opposite ordering, and $z_4(t)$ has the wrong initial point.

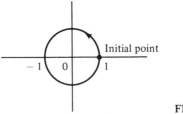

FIGURE 4.6

The phrase "directed smooth curve" will be used to mean either a directed smooth arc or a directed smooth closed curve.

Now we are ready to specify the kinds of curves which will be used in the theory of integration. They are formed by joining directed smooth curves together end-to-end; this allows self-intersections, cusps, and corners. In addition, it will be convenient to include single isolated points as members of this class. Let us explore the possibilities uncovered by

DEFINITION 2   A **contour** Γ *is either a single point z_0 or a finite sequence of directed smooth curves $(\gamma_1, \gamma_2, \ldots, \gamma_n)$ such that the terminal point of γ_k coincides with the initial point of γ_{k+1} for each $k = 1, 2, \ldots, n - 1$.*

Notice that a single directed smooth curve is a contour with $n = 1$.

Speaking loosely, we can say that the contour Γ inherits a direction from its components γ_k: If z_1 and z_2 lie on the same directed smooth curve γ_k, they are ordered by the direction on γ_k, and if z_1 lies on γ_i while z_2 lies on γ_j, we say that z_1 precedes z_2 if $i < j$. This is ambiguous because of the possibility that z_1, say, may lie on two different smooth curves, and therefore we must indicate which "occurrence" of z_1 is meant when we say z_1 precedes z_2.

Figure 4.7 illustrates four elementary examples of contours formed by joining directed smooth curves. In Fig. 4.7(d), z_α regarded as a point of γ_1 precedes z_β, but regarded as a point of γ_3, it is preceded by z_β.

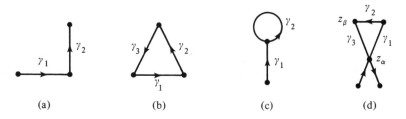

(a) (b) (c) (d)

FIGURE 4.7

In Fig. 4.8 we explore some of the more exotic possibilities allowed in Definition 2. The contour Γ in Fig. 4.8(a) is traced by moving along the segment from left to right and then halfway back from right to left. The circle in Fig. 4.8(b) is traversed twice in the counterclockwise direction. And in Fig. 4.8(c) the first segment of Γ is traced left to right, then we make one and one-half clockwise revolutions around the circle, and finally trace out the second segment.

We shall say that $z = z(t)$, $a \le t \le b$, is a parametrization of the contour $\Gamma = (\gamma_1, \gamma_2, \ldots, \gamma_n)$ if there is a subdivision of $[a, b]$ into n subintervals $[\tau_0, \tau_1], [\tau_1, \tau_2], \ldots, [\tau_{n-1}, \tau_n]$, where

$$a = \tau_0 < \tau_1 < \cdots < \tau_{n-1} < \tau_n = b,$$

such that on each subinterval $[\tau_{k-1}, \tau_k]$ the function $z(t)$ is an admissible parametrization of the smooth curve γ_k, consistent with the direction on γ_k. Since the endpoints of consecutive γ_k's are properly connected, $z(t)$ must be continuous on $[a, b]$. However, $z'(t)$ may have jump discontinuities at the points τ_k. The contour parametrization of a point is simply a constant function.

When we have admissible parametrizations of the components γ_k of a contour Γ, it is a simple matter to piece these together to get a contour parametrization for Γ; in fact, we can always use the "standard" parametric interval $[0, 1]$. We employ the techniques of *rescaling*, which are amply illustrated by the following example. (The general case is discussed in Prob. 6.)

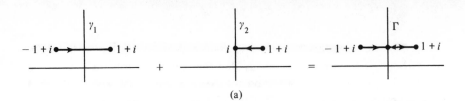

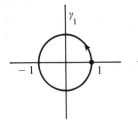 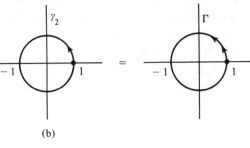

(a)

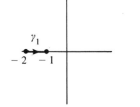 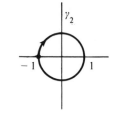 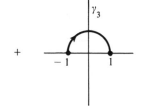

(b)

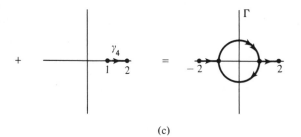

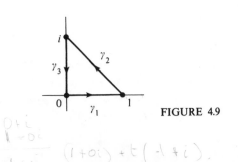

(c)

FIGURE 4.8

EXAMPLE 2 Parametrize the contour in Fig. 4.9, over the interval $[0, 1]$.

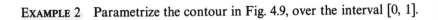

FIGURE 4.9

$(1 + 0i) + t(-1 + i).$

SOLUTION We have already seen how to parametrize straight lines. The following functions are admissible parametrizations for γ_1, γ_2, and γ_3, consistent with their directions:

$$\gamma_1: z_1(t) = t \qquad (0 \le t \le 1),$$
$$\gamma_2: z_2(t) = 1 + t(i - 1) \qquad (0 \le t \le 1),$$
$$\gamma_3: z_3(t) = i - ti \qquad (0 \le t \le 1).$$

Now we rescale so that γ_1 is traced as t varies between 0 and $\frac{1}{3}$, γ_2 is traced for $\frac{1}{3} \le t \le \frac{2}{3}$, and γ_3 is traced for $\frac{2}{3} \le t \le 1$. This is simply a matter of shifting and stretching the variable t.

For γ_1, observe that the range of the function $z_1(t) = t$, $0 \le t \le 1$, is the same as the range of $z_I(t) = 3t$, $0 \le t \le \frac{1}{3}$, and that $z_I(t)$ is an admissible parametrization corresponding to the same ordering. The curve γ_2 is the range of $z_2(t) = 1 + t(i - 1)$, $0 \le t \le 1$, and this is the same as the range of $z_{II}(t) = 1 + 3(t - \frac{1}{3})(i - 1)$, $\frac{1}{3} \le t \le \frac{2}{3}$, again preserving admissibility and ordering. Handling $z_3(t)$ similarly, we find

$$z(t) = \begin{cases} 3t & (0 \le t \le \frac{1}{3}), \\ 1 + 3(t - \frac{1}{3})(i - 1) & (\frac{1}{3} \le t \le \frac{2}{3}), \\ i - 3(t - \frac{2}{3})i & (\frac{2}{3} \le t \le 1). \end{cases} \quad \blacksquare$$

For obvious reasons, we call the (undirected) point set underlying a contour a *piecewise smooth curve*. We shall use the symbol Γ ambiguously to refer to both the contour and its underlying curve, allowing the context to provide the proper interpretation.

Much of the terminology of directed smooth curves is readily applied to contours. The initial point of Γ is the initial point of γ_1, and its terminal point is the terminal point of γ_n; therefore Γ can be regarded as a path connecting these points. If the directions on all the components of Γ are reversed and the components are taken in the opposite order, the resulting contour is called the *opposite contour* and is denoted by $-\Gamma$ (see Fig. 4.10). Notice that if $z = z(t)$, $a \le t \le b$, is a parametrization of Γ, then $z = z(-t)$, $-b \le t \le -a$, parametrizes $-\Gamma$.

Γ is said to be a *closed contour* or a *loop* if its initial and terminal points coincide. A *simple closed contour* is a closed contour with no multiple points other than its initial-terminal point; in other words, if

FIGURE 4.10

$z = z(t)$, $a \le t \le b$, is a parametrization of the closed contour, then $z(t)$ is one-to-one on the half-open interval $[a, b)$. Figure 4.9 illustrates a simple closed contour.

A useful tool in specifying the direction along a simple closed contour is the *Jordan Curve Theorem*. A special case of this result is stated in

THEOREM 1 *Any simple closed contour separates the plane into two domains, each having the contour as its boundary. One of these domains, called the "interior," is bounded; the other, called the "exterior," is unbounded.*

This result actually holds for more general curves, but the proof is quite involved, even for contours; hence we shall omit it. However, the meaning of the theorem is quite evident. See Fig. 4.11.

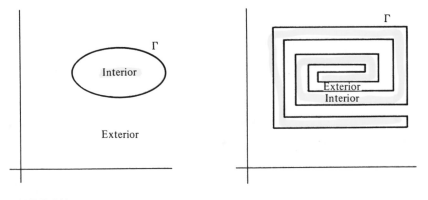

FIGURE 4.11

Now given a simple closed contour Γ we can imagine a dwarf bicycling around the curve and tracing out its points in the order specified by its direction. If the bicycle has training wheels and if it is small enough, then one of the training wheels will always remain in the interior domain of the contour, while the other remains in the exterior (otherwise we would have a path connecting these domains without crossing Γ, in contradiction to the Jordan Curve Theorem). Consequently the direction along Γ can be completely specified by declaring its initial-terminal point and stating which domain (interior or exterior) lies to the *left* of an observer tracing out the points in order. When the interior domain lies to the left, we say that Γ is *positively* oriented. Otherwise Γ is said to be oriented *negatively*. A positive orientation generalizes the concept of counterclockwise motion; see Fig. 4.9.

The final topic we want to discuss in this section is the length of a contour. We begin by considering a smooth curve γ, with any admissible parametrization $z = z(t)$, $a \leq t \leq b$. Let $s(t)$ denote the length of the arc of γ traversed in going from the point $z(a)$ to the point $z(t)$. We wish to find an expression for ds/dt. Accordingly, we must consider $s(t + \Delta t) - s(t)$, which is the length of the curve between $z(t)$ and $z(t + \Delta t)$. See Fig. 4.12.

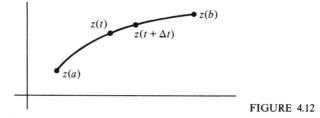

FIGURE 4.12

We appeal to the reader's geometric intuition and his experience with differential calculus to establish the following claim: Because the curve is smooth, the arc between $z(t)$ and $z(t + \Delta t)$ lies very close to the chord joining these points, and in fact the limit of the ratio $(s(t + \Delta t) - s(t))/|z(t + \Delta t) - z(t)|$ is 1 as Δt decreases to zero. Accepting this, we have

$$\frac{ds(t)}{dt} = \lim_{\Delta t \to 0+} \frac{s(t + \Delta t) - s(t)}{\Delta t}$$

$$= \lim_{\Delta t \to 0+} \frac{s(t + \Delta t) - s(t)}{|z(t + \Delta t) - z(t)|} \frac{|z(t + \Delta t) - z(t)|}{\Delta t}$$

$$= 1 \cdot \left| \frac{dz(t)}{dt} \right|,$$

i.e., $ds/dt = |dz/dt|$.

Now we can compute the length of the smooth curve by integrating ds/dt, and we have the important formula

$$l(\gamma) = \text{length of } \gamma = \int_a^b \frac{ds}{dt} dt = \int_a^b \left| \frac{dz}{dt} \right| dt;$$

i.e.,

$$(1) \qquad l(\gamma) = \int_a^b \left| \frac{dz}{dt} \right| dt = \int_a^b \sqrt{\left(\frac{dx}{dt} \right)^2 + \left(\frac{dy}{dt} \right)^2} \, dt.$$

This formula is established rigorously in the references. Sometimes the shorthand $\int_\gamma |dz|$ is used to indicate $\int_a^b |dz/dt| \, dt$; it emphasizes the intuitively evident fact that $l(\gamma)$ is a geometric quantity which depends only on the point set γ, and is independent of the particular admissible parametrization used in the computation.

The *length of a contour* is simply defined to be the sum of the lengths of its component curves.

EXAMPLE 3 Compute the lengths of the contours in Fig. 4.8.

SOLUTION Simply sum the lengths of the components γ_i.

In Fig. 4.8(a), $l(\gamma_1) = 2$ and $l(\gamma_2) = 1$. Hence $l(\Gamma) = 2 + 1 = 3$.
In Fig. 4.8(b), $l(\Gamma) = 2\pi + 2\pi = 4\pi$.
In Fig. 4.8(c), $l(\Gamma) = 1 + 2\pi + \pi + 1 = 2 + 3\pi$.   ∎

The lengths of all the components in this example were determined by inspection. In Prob. 10 we invite the reader to verify formula (1) for these curves.

Exercises 4.1

1. Show that the ellipse $x^2/a^2 + y^2/b^2 = 1$ is a smooth curve.

2. Show why the condition that $z'(t)$ never vanishes is necessary to ensure that smooth curves have no cusps. [HINT: Consider the curve traced by $z(t) = t^2 + it^3$, $-1 \le t \le 1$.]

3. For each of the following smooth curves give an admissible parametrization which is consistent with the indicated direction.
    (a) the line segment from $z = 1 + i$ to $z = -2 - 3i$
    (b) the circle $|z - 2i| = 4$ traversed once in the clockwise direction starting from the point $z = 4 + 2i$
    (c) the arc of the circle $|z| = R$ lying in the second quadrant, from $z = Ri$ to $z = -R$
    (d) the segment of the parabola $y = x^2$ from the point $(1, 1)$ to the point $(3, 9)$

4. Show that the range of the function $z(t) = t^3 + it^6$, $-1 \le t \le 1$, is a smooth curve.

5. Identify the interior of the simple closed contour Γ in Fig. 4.13. Is Γ positively oriented?

FIGURE 4.13

6. Let γ be a directed smooth curve. Show that if $z = z(t), a \le t \le b$, is an admissible parametrization of γ consistent with the ordering on γ, then the same is true of

$$z_1(t) = z\left(\frac{b-a}{d-c}t + \frac{ad-bc}{d-c}\right) \qquad (c \le t \le d).$$

7. Parametrize the contour consisting of the perimeter of the square with vertices $-1 - i, 1 - i, 1 + i$, and $-1 + i$ traversed once in that order. What is the length of this contour?

8. Parametrize the contour Γ indicated in Fig. 4.14. Also give a parametrization for the opposite contour $-\Gamma$.

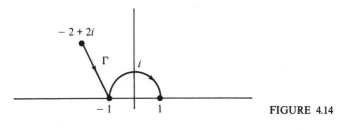

FIGURE 4.14

9. Parametrize the barbell-shaped contour in Fig. 4.15.

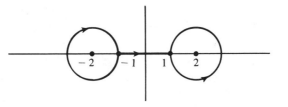

FIGURE 4.15

10. Using an admissible parametrization verify from formula (1) that
     (a) the length of the line segment from z_1 to z_2 is $|z_2 - z_1|$;
     (b) the length of the circle $|z - z_0| = r$ is $2\pi r$.

11. Does formula (1) remain valid when γ is a *contour* parametrized by $z = z(t), a \le t \le b$?

12. Find the length of the contour Γ parametrized by $z = z(t) = 5e^{3it}$, $0 \le t \le \pi$.

13. Interpreting t as time and the admissible parametrization $z = z(t)$, $a \le t \le b$, as the position function of a moving particle, give the physical meaning of the following quantities.
     (a) $z'(t)$ (b) $|z'(t)|$

     (c) $|z'(t)\, dt|$ (d) $\displaystyle\int_a^b |z'(t)|\, dt$

14. Let $z = z_1(t)$ be an admissible parametrization of the smooth curve γ. If $\phi(s)$, $c \leq s \leq d$, is a strictly increasing function such that (i) $\phi'(s)$ is positive and continuous on $[c, d]$ and (ii) $\phi(c) = a$, $\phi(d) = b$, then the function $z_2(s) = z_1(\phi(s))$, $c \leq s \leq d$, is also an admissible parametrization of γ. Verify that

$$\int_a^b |z_1'(t)| \; dt = \int_c^d |z_2'(s)| \; ds,$$

which demonstrates invariance of formula (1).

4.2 CONTOUR INTEGRALS

In calculus the definite integral of a real-valued function f over an interval $[a, b]$ is defined as the limit of certain sums $\sum_{k=1}^n f(c_k) \Delta x_k$ which are called Riemann sums. The aim of this section is to use the notion of Riemann sums to define the definite integral of a complex-valued function f along a contour Γ in the plane. This will be accomplished in two steps: first we define the integral along a single directed smooth curve, and second, we define the integral along a contour in terms of the integrals along its components.

Consider then a function $f(z)$ which is defined over a directed smooth curve γ with initial point α and terminal point β (possibly coinciding with α). As in the previous section, the points on γ are ordered in accordance with the direction.

For any positive integer n, we define a *partition* $\mathscr{P}_n$ of γ to be a finite number of points z_0, z_1, ..., z_n on γ such that $z_0 = \alpha$, $z_n = \beta$, and z_{k-1} precedes z_k on γ for $k = 1, 2, \ldots, n$ (see Fig. 4.16). If we compute the arc

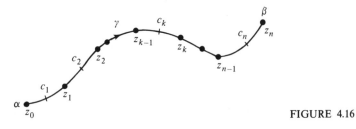

FIGURE 4.16

length along γ between every consecutive pair of points (z_{k-1}, z_k), the largest of these lengths provides a measure of the "fineness" of the subdivision; this maximum length is called the *mesh* of the partition and is denoted by $\mu(\mathscr{P}_n)$. It follows that if a given partition $\mathscr{P}_n$ has a "small" mesh, then n must be large and the successive points of the partition must be close to one another.

Now let $c_1, c_2, \ldots, c_n$ be any points on γ such that c_1 lies on the arc from z_0 to z_1, c_2 lies on the arc from z_1 to z_2, etc. Under these circumstances the sum $S(\mathcal{P}_n)$ defined by

$$S(\mathcal{P}_n) = f(c_1)(z_1 - z_0) + f(c_2)(z_2 - z_1) + \cdots + f(c_n)(z_n - z_{n-1})$$

is called a *Riemann sum* for the function f corresponding to the partition $\mathcal{P}_n$. On writing $z_k - z_{k-1} = \Delta z_k$, this becomes

$$S(\mathcal{P}_n) = \sum_{k=1}^{n} f(c_k)(z_k - z_{k-1}) = \sum_{k=1}^{n} f(c_k)\, \Delta z_k .$$

Now we can generalize the definition of definite integral given in calculus.

DEFINITION 3 *Let $f(z)$ be a complex-valued function defined on the directed smooth curve γ. We say that $f(z)$ is integrable along γ if there exists a complex number L which is the limit of every sequence of Riemann sums $S(\mathcal{P}_1), S(\mathcal{P}_2), \ldots, S(\mathcal{P}_n), \ldots$ corresponding to any sequence of partitions of γ satisfying $\lim_{n \to \infty} \mu(\mathcal{P}_n) = 0$; i.e.,*

$$\lim_{n \to \infty} S(\mathcal{P}_n) = L \quad \text{whenever} \quad \lim_{n \to \infty} \mu(\mathcal{P}_n) = 0.$$

The constant L is called the *integral of $f(z)$ along γ*, and we write

$$L = \int_{\gamma} f(z)\, dz \quad \text{or} \quad L = \int_{\gamma} f.$$

Because Definition 3 is analogous to the definition of the integral given in calculus, certain familiar properties of the latter integrals carry over to the complex case. For example, if $f(z)$ and $g(z)$ are integrable along γ, then

(2)
$$\int_{\gamma} [f(z) \pm g(z)]\, dz = \int_{\gamma} f(z)\, dz \pm \int_{\gamma} g(z)\, dz,$$

(3)
$$\int_{\gamma} cf(z)\, dz = c \int_{\gamma} f(z)\, dz$$

$$(c \text{ any complex constant}),$$

and

(4)
$$\int_{-\gamma} f(z)\, dz = - \int_{\gamma} f(z)\, dz,$$

where $-\gamma$ denotes the curve directed opposite to γ.

As we know from calculus, not all functions f are integrable. However, if we require that f be continuous, then its integral must exist.

THEOREM 2   If $f(z)$ is continuous† on the directed smooth curve γ, then $f(z)$ is integrable along γ.

For a proof of this theorem the reader is referred to Ref. [2] at the end of the chapter.

While Theorem 2 is of great theoretical importance, it gives us no information on how to compute the integral $\int_\gamma f(z)\, dz$. However, since we are already skilled in evaluating the definite integrals of calculus, it would certainly be advantageous if we could somehow express the complex integral in terms of real integrals.

For this purpose we first consider the special case when γ is the real line segment $[a, b]$ directed from left to right. Notice that if f happened to be a real-valued function defined on $[a, b]$, then Definition 3 would agree precisely with the definition of the integral $\int_a^b f(t)\, dt$ given in calculus. Hence, even when f is complex-valued, we shall use the symbol

$$\int_a^b f(t)\, dt$$

to denote the integral of f along the directed real line segment. In this case, when $f(t)$ is a complex-valued function continuous on $[a, b]$, we can write $f(t) = u(t) + iv(t)$, where u and v are each real-valued and continuous on $[a, b]$. Then from properties (2) and (3) we have

(5)
$$\int_a^b f(t)\, dt = \int_a^b [u(t) + iv(t)]\, dt$$

$$= \int_a^b u(t)\, dt + i \int_a^b v(t)\, dt;$$

this expresses the complex integral in terms of two real integrals.

If $f(t)$ has the antiderivative $F(t) = U(t) + iV(t)$, then $U' = u$, $V' = v$, and Eq. (5) leads immediately to the following generalization of the Fundamental Theorem of Calculus:

THEOREM 3   If the complex-valued function $f(t)$ is continuous on $[a, b]$ and $F'(t) = f(t)$ for all t in $[a, b]$, then

$$\int_a^b f(t)\, dt = F(b) - F(a).$$

This result is illustrated in

---

†The meaning of continuity for a function f having an arbitrary set S as its domain of definition is as follows: f is continuous on the set S if for any point z_0 in S and for every $\varepsilon > 0$, there exists a $\delta > 0$ such that $|f(z) - f(z_0)| < \varepsilon$ whenever z belongs to S and $|z - z_0| < \delta$.

EXAMPLE 4 Compute $\int_0^\pi e^{it} \, dt$.

SOLUTION Since $F(t) = e^{it}/i$ is an antiderivative of $f(t) = e^{it}$, we have by Theorem 3

$$\int_0^\pi e^{it} \, dt = \frac{e^{it}}{i} \Big|_0^\pi = \frac{e^{i\pi}}{i} - \frac{e^{i0}}{i} = \frac{-2}{i} = 2i. \quad \blacksquare$$

Now we move on to the general case where γ is any directed smooth curve along which $f(z)$ is continuous. We can obtain a formula for the integral $\int_\gamma f(z) \, dz$ by considering an admissible parametrization $z = z(t)$, $a \le t \le b$, for γ (consistent with its direction). Indeed, if $\mathscr{P}_n = \{z_0, z_1, \dots, z_n\}$ is a partition of γ, then we can write

$$z_0 = z(t_0), \ z_1 = z(t_1), \ \dots, \ z_n = z(t_n),$$

where

$$a = t_0 < t_1 < \cdots < t_n = b.$$

Furthermore, since the function $z(t)$ has a continuous derivative on $[a, b]$, the difference $\Delta z_k = z(t_k) - z(t_{k-1})$ is approximately equal to $z'(t_k) \times (t_k - t_{k-1}) = z'(t_k) \, \Delta t_k$, the error going to zero faster than Δt_k. Hence we see that the sum

$$\sum_{k=1}^n f(z_k) \, \Delta z_k = \sum_{k=1}^n f(z(t_k)) \, \Delta z_k,$$

which is a Riemann sum for $f(z)$ along γ, can be approximated by the sum

$$\sum_{k=1}^n f(z(t_k)) z'(t_k) \, \Delta t_k,$$

which is a Riemann sum for the continuous function $f(z(t))z'(t)$ over the interval $[a, b]$. These considerations suggest the following theorem (and provide the essential ingredients for its justification):

THEOREM 4 *Let $f(z)$ be a function continuous on the directed smooth curve γ. Then if $z = z(t)$, $a \le t \le b$, is any admissible parametrization of γ consistent with its direction, we have*

(6) $$\int_\gamma f(z) \, dz = \int_a^b f(z(t))z'(t) \, dt \left(= \int_a^b f(z(t)) \frac{dz}{dt}(t) \, dt \right).$$

The precise details of the proof of Theorem 4, though not difficult, are quite laborious and not particularly illuminating for our subject matter. Hence we shall omit them. A rigorous treatment of this theorem can be found in Ref. [2] at the end of the chapter.

Since Eq. (6) is valid for all suitable parametrizations of γ, and since the integral of f along γ was defined independently of any parametrization, we immediately deduce

COROLLARY 1 If $f(z)$ is continuous on the directed smooth curve γ and if $z = z_1(t), a \leq t \leq b$, and $z = z_2(t), c \leq t \leq d$, are any two admissible parametrizations of γ consistent with its direction, then

$$\int_a^b f(z_1(t))z_1'(t) \, dt = \int_c^d f(z_2(t))z_2'(t) \, dt.$$

EXAMPLE 5   Compute the integral $\int_{C_r} (z - z_0)^n \, dz$,† where n is an integer and C_r is the circle $|z - z_0| = r$ traversed once in the counterclockwise direction as indicated in Fig. 4.17.

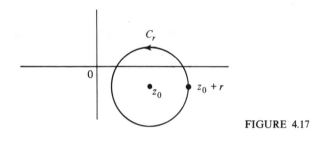

FIGURE 4.17

SOLUTION   A suitable parametrization for C_r is given by $z(t) = z_0 + re^{it}, 0 \leq t \leq 2\pi$. Setting $f(z) = (z - z_0)^n$, we have

$$f(z(t)) = (z_0 + re^{it} - z_0)^n = r^n e^{int}$$

and

$$z'(t) = ire^{it}.$$

Hence by formula (6)

$$\int_{C_r} (z - z_0)^n \, dz = \int_0^{2\pi} (r^n e^{int})(ire^{it}) \, dt = ir^{n+1} \int_0^{2\pi} e^{i(n+1)t} \, dt.$$

The evaluation of the last integral requires two separate computations. If $n \neq -1$, we obtain

$$ir^{n+1} \int_0^{2\pi} e^{i(n+1)t} \, dt = ir^{n+1} \frac{e^{i(n+1)t}}{i(n+1)} \bigg|_0^{2\pi} = ir^{n+1} \left[\frac{1}{i(n+1)} - \frac{1}{i(n+1)} \right] = 0,$$

†Occasionally we write

$$\oint_{C_r} f(z) \, dz$$

to emphasize the fact that the integration is taken in the positive direction.

while if $n = -1$, then

$$ir^{n+1} \int_0^{2\pi} e^{i(n+1)t}\, dt = i \int_0^{2\pi} dt = 2\pi i.$$

Thus (regardless of the value of r)

$$(7) \qquad \int_{C_r} (z - z_0)^n\, dz = \begin{cases} 0 & \text{for } n \neq -1, \\ 2\pi i & \text{for } n = -1. \end{cases} \qquad \blacksquare$$

Integrals along a contour are computed according to

DEFINITION 4 *Suppose that Γ is a contour consisting of the directed smooth curves $(\gamma_1, \gamma_2, \ldots, \gamma_n)$, and let $f(z)$ be a function continuous on Γ. Then the* **contour integral of $f(z)$ along** *Γ is denoted by the symbol $\int_\Gamma f(z)\, dz$ and is defined by the equation*

$$(8) \qquad \int_\Gamma f(z)\, dz = \int_{\gamma_1} f(z)\, dz + \int_{\gamma_2} f(z)\, dz + \cdots + \int_{\gamma_n} f(z)\, dz.\dagger$$

If Γ consists of a single point, then for obvious reasons we set

$$\int_\Gamma f(z)\, dz = 0.$$

EXAMPLE 6 Compute $\int_\Gamma 1/(z - z_0)\, dz$, where Γ is the circle $|z - z_0| = r$ traversed twice in the counterclockwise direction starting from the point $z_0 + r$.

SOLUTION Letting C_r denote the circle traversed once in the counterclockwise direction, we have $\Gamma = (C_r, C_r)$. Hence from formula (7), obtained in the solution of Example 5, there follows

$$\int_\Gamma \frac{dz}{z - z_0} = \int_{C_r} \frac{dz}{z - z_0} + \int_{C_r} \frac{dz}{z - z_0} = 2\pi i + 2\pi i = 4\pi i. \qquad \blacksquare$$

EXAMPLE 7 Compute $\int_\Gamma \bar{z}^2\, dz$ along the simple closed contour Γ indicated in Fig. 4.18.

SOLUTION According to Definition 4 we have

$$\int_\Gamma \bar{z}^2\, dz = \int_{\gamma_1} \bar{z}^2\, dz + \int_{\gamma_2} \bar{z}^2\, dz + \int_{\gamma_3} \bar{z}^2\, dz.$$

†For this reason we sometimes write $\Gamma = \gamma_1 + \gamma_2 + \cdots + \gamma_n$. More generally, if $\Gamma_1, \Gamma_2, \ldots,$ Γ_m are contours, we use the symbol $\int_{\Gamma_1 + \Gamma_2 + \cdots + \Gamma_m} f(z)\, dz$ to denote the sum $\sum_{k=1}^m \int_{\Gamma_k} f(z)\, dz$.

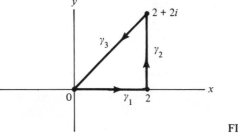

FIGURE 4.18

Suitable parametrizations for the line segments γ_k are

$\dfrac{dz}{dr} = 1$

$$\gamma_1: \quad z_1(t) = t \qquad\qquad (0 \le t \le 2),$$

$\dfrac{dr}{dt}$

$$\gamma_2: \quad z_2(t) = 2 + ti \qquad (0 \le t \le 2),$$

$$\gamma_3: \quad z_3(t) = -t(1 + i) \qquad (-2 \le t \le 0),$$

and so by Theorem 4 we have

$$\int_{\gamma_1} \bar{z}^2 \, dz = \int_0^2 \overline{z_1(t)}^2 z_1'(t) \, dt = \int_0^2 t^2 \, dt = \frac{t^3}{3}\bigg|_0^2 = \frac{8}{3},$$

$$\int_{\gamma_2} \bar{z}^2 \, dz = \int_0^2 \overline{z_2(t)}^2 z_2'(t) \, dt = \int_0^2 (2 - ti)^2 i \, dt$$

$$= \frac{i(2 - ti)^3}{-3i}\bigg|_0^2 = \frac{-(2 - 2i)^3}{3} + \frac{8}{3},$$

and

$$\int_{\gamma_3} \bar{z}^2 \, dz = \int_{-2}^0 \overline{z_3(t)}^2 z_3'(t) \, dt = \int_{-2}^0 [-t(1 - i)]^2 [-(1 + i)] \, dt$$

$$= -(1 + i)(1 - i)^2 \int_{-2}^0 t^2 \, dt = -(1 + i)(1 - i)^2 \frac{8}{3}.$$

Therefore

$$\int_{\Gamma} \bar{z}^2 \, dz = \frac{8}{3} + \left[\frac{-(2 - 2i)^3}{3} + \frac{8}{3}\right] + \left[-(1 + i)(1 - i)^2 \frac{8}{3}\right],$$

which after some computations turns out to equal $16/3 + 32i/3$. ∎

Using Definition 4 it is easy to see that the results discussed above for integrals along a directed smooth curve carry over to integrals along a contour. In particular, we have

(9)
$$\int_{\Gamma} [f(z) \pm g(z)] \, dz = \int_{\Gamma} f(z) \, dz \pm \int_{\Gamma} g(z) \, dz,$$

(10)
$$\int_{\Gamma} cf(z) \, dz = c \int_{\Gamma} f(z) \, dz$$

(*c* any complex constant),

and

(11)
$$\int_{-\Gamma} f(z)\, dz = -\int_{\Gamma} f(z)\, dz,$$

where $f(z)$ and $g(z)$ are both continuous on the contour Γ.

Furthermore, if we have a parametrization $z = z(t)$, $a \le t \le b$, for the whole contour $\Gamma = (\gamma_1, \gamma_2, \ldots, \gamma_n)$, then we know that there is a subdivision

$$a = \tau_0 < \tau_1 < \cdots < \tau_{n-1} < \tau_n = b$$

such that the function $z(t)$ restricted to the kth subinterval $[\tau_{k-1}, \tau_k]$ constitutes a suitable parametrization of γ_k. Hence by formula (6)

$$\int_{\gamma_k} f(z)\, dz = \int_{\tau_{k-1}}^{\tau_k} f(z(t))z'(t)\, dt \qquad (k = 1, 2, \ldots, n),$$

and so

$$\int_{\Gamma} f(z)\, dz = \sum_{k=1}^{n} \int_{\tau_{k-1}}^{\tau_k} f(z(t))z'(t)\, dt,$$

which we can write as

$$\int_{\Gamma} f(z)\, dz = \int_{a}^{b} f(z(t))z'(t)\, dt.$$

Using this formula it is not difficult to prove that integration around simple closed contours is independent of the choice of the initial-terminal point (see Prob. 17). Consequently, in problems dealing with integrals along such contours, we need only specify the direction of transit, not the starting point.

Many times in theory and in practice, it is not necessary to actually evaluate a contour integral. What may be required is simply a good upper bound on its magnitude. We therefore turn to the problem of estimating contour integrals.

Suppose that the function $f(z)$ is continuous on the directed smooth curve γ and that $f(z)$ is bounded by the constant M on γ; i.e., $|f(z)| \le M$ for all z on γ. If we consider a Riemann sum $\sum_{k=1}^{n} f(c_k)\, \Delta z_k$ corresponding to a partition $\mathscr{P}_n$ of γ, then we have by the Generalized Triangle Inequality

$$\left| \sum_{k=1}^{n} f(c_k)\, \Delta z_k \right| \le \sum_{k=1}^{n} |f(c_k)|\, |\Delta z_k| \le M \sum_{k=1}^{n} |\Delta z_k|.$$

Furthermore, notice that the sum of the chordal lengths $\sum_{k=1}^{n} |\Delta z_k|$ cannot be greater than the length of γ. Hence

(12)
$$\left| \sum_{k=1}^{n} f(c_k)\, \Delta z_k \right| \le Ml(\gamma).$$

Since inequality (12) is valid for all Riemann sums of $f(z)$, it follows by taking the limit [as $\mu(\mathscr{P}_n) \to 0$] that

(13)
$$\left| \int_\gamma f(z)\, dz \right| \leq Ml(\gamma).$$

Applying this fact and the triangle inequality to the Eq. (8) defining a contour integral, we deduce

THEOREM 5   *If $f(z)$ is continuous on the contour Γ and if $|f(z)| \leq M$ for all z on Γ, then*

(14)
$$\left| \int_\Gamma f(z)\, dz \right| \leq Ml(\Gamma),$$

where $l(\Gamma)$ denotes the length of Γ. In particular, we have

(15)
$$\left| \int_\Gamma f(z)\, dz \right| \leq \max_{z \text{ on } \Gamma} |f(z)| \cdot l(\Gamma).$$

EXAMPLE 8   Find an upper bound for $\left| \int_\Gamma e^z/(z^2 + 1)\, dz \right|$, where Γ is the circle $|z| = 2$ traversed once in the counterclockwise direction.

SOLUTION   First observe that the path of integration has length $l = 4\pi$. Next we seek an upper bound M for the function $e^z/(z^2 + 1)$ when $|z| = 2$. Writing $z = x + iy$ we have

$$|e^z| = |e^{x+iy}| = e^x \leq e^2, \quad \text{for } |z| = \sqrt{x^2 + y^2} = 2,$$

and by the triangle inequality

$$|z^2 + 1| \geq |z|^2 - 1 = 4 - 1 = 3, \quad \text{for } |z| = 2.$$

Hence $|e^z/(z^2 + 1)| \leq e^2/3$ for $|z| = 2$, and so by the theorem

$$\left| \int_\Gamma \frac{e^z}{z^2 + 1}\, dz \right| \leq \frac{e^2}{3} \cdot 4\pi. \quad \blacksquare$$

In concluding this section we remark that although the real definite integral can be interpreted, among other things, as an area, no corresponding geometric visualization is available for contour integrals. Nevertheless, the latter integrals are extremely useful in applied problems, as we shall see in subsequent chapters.

Exercises 4.2

1.   Let γ be a directed smooth curve with initial point α and terminal point β. Show directly from Definition 3 that $\int_\gamma c\, dz = c(\beta - \alpha)$,

where c is any complex constant. Does the same formula hold for integration along an arbitrary contour joining α to β?

2. Using Definition 3, prove properties (2), (3), and (4).

3. Furnish the details of the proof of Theorem 3.

4. Evaluate each of the following integrals.

(a) $\int_0^1 (2t + it^2)\, dt$ (b) $\int_{-2}^0 (1 + i)\cos(it)\, dt$

(c) $\int_0^1 (1 + 2it)^5\, dt$ (d) $\int_0^2 \dfrac{t}{(t^2 + i)^2}\, dt$

5. Utilize Example 5 to evaluate

$$\int_C \left[\frac{6}{(z - i)^2} + \frac{2}{z - i} + 1 - 3(z - i)^2 \right] dz,$$

where C is the circle $|z - i| = 4$ traversed once counterclockwise.

6. Compute $\int_\Gamma \bar{z}\, dz$, where
   (a) Γ is the circle $|z| = 2$ traversed once counterclockwise.
   (b) Γ is the circle $|z| = 2$ traversed once clockwise.
   (c) Γ is the circle $|z| = 2$ traversed three times clockwise.

7. Compute $\int_\gamma \mathrm{Re}\, z\, dz$ along the directed line segment from $z = 0$ to $z = 1 + 2i$.

8. Evaluate $\int (x - 2xyi)\, dz$ over the contour $\Gamma: z = t + it^2, 0 \le t \le 1$, where $x = \mathrm{Re}\, z,\, y = \mathrm{Im}\, z$.

9. Let C be the perimeter of the square with vertices at the points $z = 0, z = 1, z = 1 + i$, and $z = i$ traversed once in that order. Show that

$$\int_C e^z\, dz = 0.$$

10. Compute $\int_C \bar{z}^2\, dz$ along the perimeter of the square in Prob. 9.

11. Evaluate $\int_\Gamma (2z + 1)\, dz$, where Γ is the following contour from $z = -i$ to $z = 1$.
    (a) the simple line segment
    (b) two simple line segments, the first from $z = -i$ to $z = 0$ and the second from $z = 0$ to $z = 1$
    (c) the circular arc $z = e^{it}, -\pi/2 \le t \le 0$

12. Compute $\int_\Gamma (|z - 1 + i|^2 - z)\, dz$ along the semicircle $z = 1 - i + e^{it}, 0 \le t \le \pi$.

13. For each of the following use Theorem 5 to establish the indicated estimate.

(a) If C is the circle $|z| = 3$ traversed once, then

$$\left| \int_C \frac{dz}{z^2 - i} \right| \le \frac{3\pi}{4}.$$

(b) If γ is the vertical line segment from $z = R \, (> 0)$ to $z = R + 2\pi i$, then

$$\left| \int_\gamma \frac{e^{3z}}{1 + e^z} \, dz \right| \le \frac{2\pi e^{3R}}{e^R - 1}.$$

(c) If Γ is the arc of the circle $|z| = 1$ that lies in the first quadrant, then

$$\left| \int_\Gamma \text{Log } z \, dz \right| \le \frac{\pi^2}{4}.$$

(d) If γ is the line segment from $z = 0$ to $z = i$, then

$$\left| \int_\gamma e^{\sin z} \, dz \right| \le 1.$$

14. Let $f(t)$ be a continuous complex-valued function on the real interval $[a, b]$. Prove that

$$\left| \int_a^b f(t) \, dt \right| \le \int_a^b |f(t)| \, dt.$$

[HINT: Consider the Riemann sums of f over $[a, b]$.]

15. Let γ be a directed smooth curve with initial point α and terminal point β. Use formula (6) and Theorem 3 to show that

$$\int_\gamma z \, dz = \frac{\beta^2 - \alpha^2}{2}.$$

16. Using the result of Prob. 15, prove that for any closed contour Γ

$$\int_\Gamma z \, dz = 0.$$

17. Let Γ_1 be a closed contour parametrized by $z = z_1(t), a \le t \le b$. We can shift the initial-terminal point of Γ_1 by choosing a number c in the interval (a, b) and letting Γ_2 be the contour parametrized by

$$z_2(t) = \begin{cases} z_1(t) & \text{if } c \le t \le b, \\ z_1(t - b + a) & \text{if } b \le t \le b - a + c. \end{cases}$$

Prove that

$$\int_{\Gamma_1} f(z)\, dz = \int_{\Gamma_2} f(z)\, dz$$

for any function f continuous on the underlying curve of Γ_1.

4.3 INDEPENDENCE OF PATH

One of the important results in the theory of complex analysis is the extension of the Fundamental Theorem of Calculus to *contour* integrals. It implies that in certain situations the integral of a function is independent of the particular path joining the initial and terminal points; in fact, it completely characterizes the conditions under which this property holds. In this section we shall explore this phenomenon in detail. We begin with the Fundamental Theorem, which enables us to evaluate integrals when an antiderivative of the integrand is known.

THEOREM 6 *Suppose that the function $f(z)$ is continuous in a domain D and has an antiderivative $F(z)$ throughout D; i.e., $dF(z)/dz = f(z)$ at each z in D. Then for any contour Γ lying in D, with initial point z_1 and terminal point z_T, we have*

(16)
$$\int_{\Gamma} f(z)\, dz = F(z_T) - F(z_1).$$

[Notice that the conditions of the theorem imply that $F(z)$ is analytic and hence continuous in D. The function Log z, for example, is *not* an antiderivative for $1/z$ in any domain containing points of the negative real axis.]

Before proceeding with the proof we shall show how Theorem 6 can greatly facilitate the computation of certain contour integrals.

EXAMPLE 9 Compute the integral $\int_{\Gamma} \cos z\, dz$, where Γ is the contour shown in Fig. 4.19.

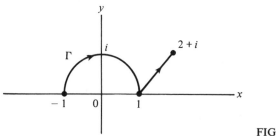

FIGURE 4.19

SOLUTION   There is no need to parametrize Γ since the integrand has the antiderivative $F(z) = \sin z$ for all z. Hence by Theorem 6, the value of the integral can be computed using only the endpoints of Γ:

$$\int_\Gamma \cos z \, dz = \sin z \Big|_{-1}^{2+i} = \sin(2+i) - \sin(-1). \quad \blacksquare$$

Proof of Theorem 6   The demonstration is quite straightforward; once we write the integral in terms of the real parameter t, the conclusion will follow as a result of the chain rule and Theorem 3 of the last section.

So suppose that Γ is a contour in D joining z_I to z_T. We select a suitable parametrization $z = z(t)$, $a \le t \le b$, for Γ and, as in the previous section, let $\{\tau_k\}_0^n$ denote the values of t corresponding to the endpoints of the smooth components $\{\gamma_j\}_1^n$ of Γ [in particular, $z(\tau_0) = z_I$ and $z(\tau_n) = z_T$]. Then we have

$$(17) \qquad \int_\Gamma f(z) \, dz = \sum_{j=1}^n \int_{\gamma_j} f(z) \, dz = \sum_{j=1}^n \int_{\tau_{j-1}}^{\tau_j} f(z(t)) z'(t) \, dt.$$

Using the fact that F is an antiderivative of f, it is possible to rewrite the integrands appearing in Eq. (17). For this purpose we recall that on each separate interval $[\tau_{j-1}, \tau_j]$ the derivative dz/dt exists and is continuous. Hence the chain rule implies that

$$\frac{d}{dt}[F(z(t))] = \frac{dF}{dz}\frac{dz}{dt} = f(z(t))z'(t) \qquad (\tau_{j-1} \le t \le \tau_j),$$

and so by Theorem 3

$$\int_{\tau_{j-1}}^{\tau_j} f(z(t))z'(t) \, dt = \int_{\tau_{j-1}}^{\tau_j} \frac{d}{dt}[F(z(t))] \, dt = F(z(\tau_j)) - F(z(\tau_{j-1})).$$

Therefore we have

$$(18) \quad \int_\Gamma f(z) \, dz = \sum_{j=1}^n [F(z(\tau_j)) - F(z(\tau_{j-1}))]$$

$$= [F(z(\tau_1)) - F(z(\tau_0))] + [F(z(\tau_2)) - F(z(\tau_1))]$$

$$+ \cdots + [F(z(\tau_n)) - F(z(\tau_{n-1}))].$$

But this sum telescopes, leaving

$$\int_\Gamma f(z) \, dz = F(z(\tau_n)) - F(z(\tau_0))$$

$$= F(z_T) - F(z_I). \quad \blacksquare$$

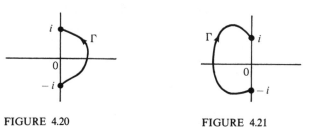

FIGURE 4.20   FIGURE 4.21

EXAMPLE 10 Compute $\int_\Gamma 1/z\ dz$, where (a) Γ is the contour shown in Fig. 4.20 and (b) Γ is the contour indicated in Fig. 4.21.

SOLUTION   (a) At each point of the contour Γ of Fig. 4.20 the function $1/z$ is the derivative of the principal branch of log z (cf. Sec. 3.2). Hence

$$\int_\Gamma \frac{dz}{z} = \text{Log } z \Big|_{-i}^{i} = \frac{\pi}{2}i - \left(-\frac{\pi}{2}i\right) = \pi i.$$

(b) For the contour Γ of Fig. 4.21 we cannot employ the function Log z since its branch cut intersects Γ. We use instead the function $\mathscr{L}_0(z) = \text{Log } |z| + i \arg z, 0 < \arg z < 2\pi$, which is a branch of the logarithm analytic in a domain containing Γ. Thus

$$\int_\Gamma \frac{dz}{z} = \mathscr{L}_0(z) \Big|_{-i}^{i} = \frac{\pi}{2}i - \frac{3\pi}{2}i = -\pi i. \quad \blacksquare$$

Since the endpoints of a *loop*, i.e., a closed contour, are equal, we have as an immediate consequence of Theorem 6

COROLLARY 2   *If $f(z)$ is continuous in a domain D and has an antiderivative throughout D, then*

$$\int_\Gamma f(z)\ dz = 0$$

for all loops Γ lying in D.

Corollary 2 provides an alternative solution to the problem of evaluating the integral $\int_{C_r} (z - z_0)^n\ dz$ of Example 5 in Sec. 4.2 when $n \neq -1$. For if we set $f(z) = (z - z_0)^n$, then $f(z)$ is the derivative of the function $F(z) = (z - z_0)^{n+1}/(n + 1)$, which is analytic in the domain D consisting of all points in the plane except $z = z_0$. (Actually the point z_0 need be excluded only in the case when n is negative. Why?) Since C_r is a closed contour which lies in D, we deduce from the corollary that $\int_{C_r} (z - z_0)^n\ dz = 0$, $n \neq -1$.

Another important conclusion that can be drawn from Eq. (16) is that when a function $f(z)$ has an antiderivative, its contour integral depends only on the endpoints z_I and z_T; i.e., the integral is independent of the path Γ joining these two points! For instance, in Fig. 4.22 all the integrals $\int_{\Gamma_1} f(z)\, dz$, $\int_{\Gamma_2} f(z)\, dz$, and $\int_{\Gamma_3} f(z)\, dz$ are equal under this condition. As a matter of fact, we shall establish that the three properties we have discussed in this section amount to equivalent statements about a continuous function $f(z)$.

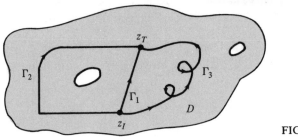

FIGURE 4.22

THEOREM 7   *Let $f(z)$ be continuous in a domain D. Then the following are equivalent:*

(i) *$f(z)$ has an antiderivative in D.*

(ii) *Every loop integral of f in D vanishes* [i.e., *if Γ is any loop in D,* $\int_\Gamma f(z)\, dz = 0$].

(iii) *The contour integrals of f are independent of path in D* [i.e., *if Γ_1 and Γ_2 are any two contours in D sharing the same initial and terminal points,* $\int_{\Gamma_1} f(z)\, dz = \int_{\Gamma_2} f(z)\, dz$].

Proof We have already seen from Theorem 6 that statement (*i*) implies (*ii*) [as well as (*iii*)]. Thus Theorem 7 will be proved if we can show that (*ii*) implies (*iii*) and that (*iii*) implies (*i*).

So assume that statement (*ii*) is true, and let Γ_1 and Γ_2 be any two contours in D sharing the same initial point z_I and terminal point z_T. Now define Γ to be the contour generated by proceeding first along Γ_1 from z_I to z_T and then backwards from z_T to z_I along $-\Gamma_2$. Then by Eq. (11) of Sec. 4.2, we have

$$\int_\Gamma f(z)\, dz = \int_{\Gamma_1} f(z)\, dz + \int_{-\Gamma_2} f(z)\, dz = \int_{\Gamma_1} f(z)\, dz - \int_{\Gamma_2} f(z)\, dz.$$

On the other hand, since Γ is closed, (*ii*) implies that

$$\int_\Gamma f(z)\, dz = 0.$$

Thus we deduce statement (*iii*).

We now show that whenever property (*iii*) holds, so does (*i*). To prove that *f* has an antiderivative, we must define some function $F(z)$ and show that its derivative is $f(z)$. The clue as to where to look for $F(z)$ is provided by the earlier considerations; If *f had* an antiderivative, Eq. (16) would hold. So we use Eq. (16) to *define* the function $F(z)$, and show that it is, indeed, an antiderivative.

Accordingly, we fix some point z_0 in *D*. Then for any point *z* in *D*, let $F(z)$ be the integral of *f* along some contour Γ in *D* joining z_0 to *z*. Since *D* is connected, we know that there will be at least one such contour (a directed polygonal line), and by condition (*iii*) it does not matter which contour we choose; all the possible paths will yield the same value for $F(z)$. Hence $F(z)$ is a well-defined single-valued function in *D*. To prove (*i*) we compute $F(z + \Delta z) - F(z)$.

We prudently elect to evaluate $F(z + \Delta z)$ by first integrating *f* along the contour Γ from z_0 to *z* and then along the straight-line segment from *z* to $z + \Delta z$. This segment will lie in *D* if Δz is small enough, because *D* is an open set. But now the difference $F(z + \Delta z) - F(z)$ is simply the integral of *f* along this segment. Parametrizing the latter by $z(t) = z + t \Delta z$, $0 \le t \le 1$, we have

$$F(z + \Delta z) - F(z) = \int_0^1 f(z + t\,\Delta z)\,\Delta z\,dt,$$

and thus

$$\frac{F(z + \Delta z) - F(z)}{\Delta z} = \int_0^1 f(z + t\,\Delta z)\,dt.$$

Since *f* is continuous, it is easy to see (Prob. 8) that as $\Delta z \to 0$ the last integral approaches $\int_0^1 f(z)\,dt = f(z)$. Thus $F'(z)$ exists and equals $f(z)$. This concludes the proof of the equivalence. ∎

Theorem 7 probably appears useless to the reader at present; he may wonder how in the world one tests whether the integral of a function around *every* closed curve is zero. In the next section our efforts will be vindicated thanks to a surprising result known as Cauchy's Theorem, which gives a simple condition for this property to hold. For now, we shall simply summarize by saying that a given continuous function has an antiderivative in *D if and only if* its integral around every loop in *D* is zero.

Exercises 4.3

1. If $P(z)$ is a polynomial and Γ is any closed contour, explain why $\int_\Gamma P(z)\,dz = 0$.

2. Calculate each of the following integrals along the indicated contours. (Observe that a standard table of integrals can be used. Explain why.)

(a) $\int_\Gamma (3z^2 - 5z + i)\, dz$ along the line segment from $z = i$ to $z = 1$

(b) $\int_\Gamma e^z\, dz$ along the upper half of the circle $|z| = 1$ from $z = 1$ to $z = -1$

(c) $\int_\Gamma 1/z\, dz$ for any contour in the right half-plane from $z = -3i$ to $z = 3i$

(d) $\int_\Gamma \csc^2 z\, dz$ for any closed contour which avoids the points 0, $\pm\pi, \pm 2\pi, \ldots$.

(e) $\int_\Gamma \sin^2 z \cos z\, dz$ along the contour in Fig. 4.23

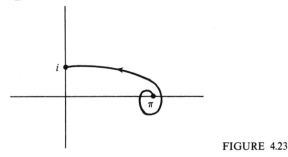

FIGURE 4.23

(f) $\int_\Gamma e^z \cos z\, dz$ along the contour in Fig. 4.23

(g) $\int_\Gamma z^{1/2}\, dz$ for the principal branch of $z^{1/2}$ along the contour in Fig. 4.23

(h) $\int_\Gamma (\text{Log } z)^2\, dz$ along the line segment from $z = 1$ to $z = i$

(i) $\int_\Gamma 1/(1 + z^2)\, dz$ along the line segment from $z = 1$ to $z = 1 + i$

3. Explain why Example 10 shows that the function $f(z) = 1/z$ has no antiderivative in the punctured plane $|z| > 0$.

4. Although Corollary 2 does not apply to the function $1/(z - z_0)$ in the plane punctured at z_0, Theorem 6 can be used as follows to show that

$$\int_C \frac{dz}{z - z_0} = 2\pi i$$

for any circle C traversed once in the positive direction surrounding the point z_0. Introduce a horizontal branch cut from z_0 to ∞ as in Fig. 4.24. In the resulting "slit plane" the function $1/(z - z_0)$ has the antiderivative $\text{Log}(z - z_0)$. Apply the theorem to integrate along the portion of C from α to β as indicated in Fig. 4.24. Now let α and β approach the cut to evaluate the given integral over all of C.

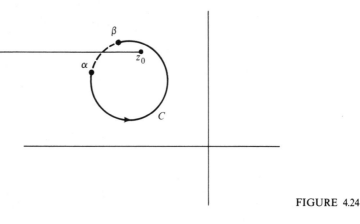

FIGURE 4.24

5. Show that if C is a positively oriented circle and z_0 lies outside C, then

$$\int_C \frac{dz}{z - z_0} = 0.$$

6. Show directly that property (*iii*) implies (*ii*) in Theorem 7.

7. From Prob. 7 in Exercises 2.4, we immediately deduce that the antiderivative of any function is unique up to a constant. How does this fact come into play in the proof that (*iii*) implies (*i*) in Theorem 7?

8. Verify the statement made in the text that if f is continuous at the point z, then

$$\int_0^1 f(z + t\,\Delta z)\,dt \to f(z) \qquad \text{as } \Delta z \to 0.$$

[HINT: Estimate the difference

$$\int_0^1 f(z + t\,\Delta z)\,dt - f(z) = \int_0^1 [f(z + t\,\Delta z) - f(z)]\,dt.]$$

9. Prove the *integration-by-parts formula*: If f and g have continuous first derivatives in a domain containing the contour Γ, then

$$\int_\Gamma f'(z)g(z)\,dz = f(z)g(z)\Big|_{z_I}^{z_T} - \int_\Gamma f(z)g'(z)\,dz,$$

where z_I and z_T are the initial and terminal points of Γ. [HINT: Use Theorem 6 on the function $d(fg)/dz$.]

10. Let $f(z)$ be an analytic function whose continuous derivative satisfies $|f'(z)| \le M$ for all z in the disk D: $|z| < 1$. Show that

$$|f(z_2) - f(z_1)| \le M|z_2 - z_1| \qquad (z_1, z_2 \text{ in } D).$$

[HINT: Observe that $f(z_2) - f(z_1) = \int f'(z)\,dz$, where the integration can be taken along the line segment from z_1 to z_2.]

4.4 CAUCHY'S INTEGRAL THEOREM

The essential content of this section is the Cauchy Integral Theorem. We feel that a clear and intuitive approach to this subject is provided by the concept of continuous deformations of one contour into another. On the other hand, some teachers may feel that the theorem is better handled by appealing to vector analysis, in particular Green's Theorem. Accordingly we have provided the reader with two alternative sections, 4.4a and 4.4b, and either one may be studied without affecting the subsequent development.

In order that each section may be self-contained, some duplication of text appears; for instance Theorem 12 in Sec. 4.4b restates Theorem 9 in Sec. 4.4a, and many of the same examples occur in both sections (though the methods of solution are different).

Exercises 4.4 is divided into three parts: problems appropriate to Sec. 4.4a, problems appropriate to Sec. 4.4b, and problems for all readers.

4.4a Deformation of contours approach

In the last section we saw that if a continuous function $f(z)$ possesses an (analytic) antiderivative in a domain D, its integral around any loop in D is zero, and vice versa. Now we are going to show how this property ties in with the analyticity of $f(z)$ itself. Our first task will be to develop the necessary geometry.

The critical notion in this regard is the *continuous deformation of one loop into another, in a given domain* of the plane. Deformations are quite easily visualized but somewhat harder to express in precise mathematical language. It is the visualization, however, that will suffice for most of our purposes. With this in mind, we first give an intuitive definition of deformations.

We say that the loop Γ_0 can be continuously deformed into the loop Γ_1 in the domain D if Γ_0 (considered as an elastic string with indicated direction) can be continuously moved about the plane *without leaving D* in such a manner that it ultimately coincides with Γ_1 (in position as well as direction).

The following examples serve to illustrate this notion:

(a) Let D be the annulus and let Γ_0 and Γ_1 be the circular contours indicated in Fig. 4.25(a). Since both circles are positively oriented, the "elastic" circle Γ_0 can be continuously deformed to Γ_1 in D by expanding the radius of Γ_0 from 1 to 2; i.e., we visualize a continuum of concentric circles varying in radii from 1 to 2. The dashed lines in Fig. 4.25(b) depict some of the intermediate loops; notice that all of them lie in D.

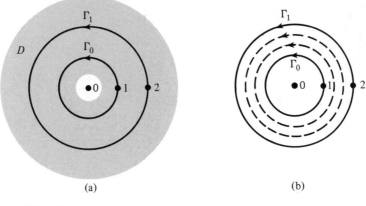

(a)   (b)

FIGURE 4.25

(b) Let D be the annulus, Γ_0 the triangular contour, and Γ_1 the circular contour of Fig. 4.26. Then Γ_0 can be deformed to Γ_1 in D by expanding Γ_0 and simultaneously making its sides more circular. Some intermediate loops are sketched in Fig. 4.26. Again they all remain in D.

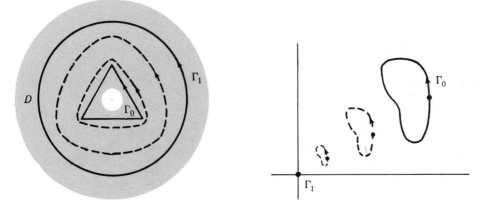

FIGURE 4.26   FIGURE 4.27

(c) Let D be the whole plane, Γ_0 the loop indicated in Fig. 4.27, and Γ_1 the point contour $z = 0$. Then Γ_0 can be continuously deformed to Γ_1 in D by simply shrinking and shifting, as indicated in Fig. 4.27.

(d) Let D be the first quadrant and let Γ_0 and Γ_1 be the circular contours in Fig. 4.28(a). Notice that merely moving Γ_0 to the right will not yield the desired deformation, for while Γ_0 will eventually coincide in position with Γ_1, the orientations will be different. To circumvent this difficulty we first shrink Γ_0 to a point (which has no direction) and then

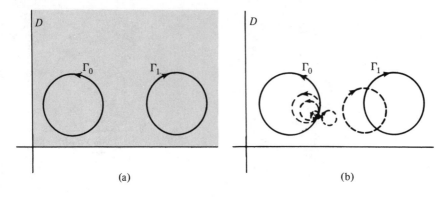

(a)   (b)

FIGURE 4.28

expand the point to Γ_1, always remaining in D, as indicated in Fig. 4.28(b).

(e) Let D be the plane minus the points $\pm i$, and let Γ_0 be the circular contour and Γ_1 be the "barbell" contour of Fig. 4.29. Then Γ_0 can be continuously deformed to Γ_1 in D, as illustrated by the dashed-line intermediate loops in Fig. 4.29.

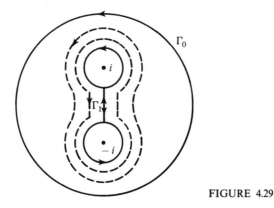

FIGURE 4.29

Now let us be more precise. The above examples show that Γ_0 can be continuously deformed to Γ_1 in D if Γ_0 and Γ_1 belong to a continuum of loops $\{\Gamma_s\}$, $0 \le s \le 1$, each lying in D, such that any pair $\Gamma_{s'}$ and $\Gamma_{s''}$ can be made "arbitrarily close" by taking s' sufficiently near s''. Thus there must be parametrizations $\{z_s(t)\}$ for the contours $\{\Gamma_s\}$ which are continuous in the variable s. Using the standard $[0, 1]$ parametric interval for the loops and rewriting $z_s(t)$ as $z(s, t)$, we formalize these ideas in

DEFINITION 5   The loop Γ_0 is said to be **continuously deformable**† *to the loop Γ_1* **in the domain** D *if there exists a function $z(s, t)$ continuous on*

122   †The word *homotopic* is sometimes used.

the unit square $0 \le s \le 1$, $0 \le t \le 1$, which satisfies the following conditions:

(i) *For each fixed s in* [0, 1], *the function z(s, t) parametrizes a loop lying in D.*

(ii) *The function z(0, t) parametrizes the loop* Γ_0.

(iii) *The function z(1, t) parametrizes the loop* Γ_1. (*See Fig. 4.30.*)

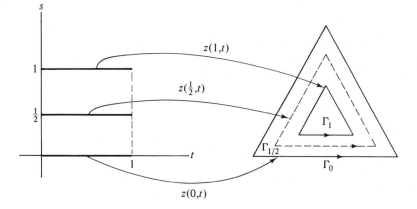

FIGURE 4.30

EXAMPLE 11 By exhibiting a deformation function $z(s, t)$, *prove* that the loop Γ_0: $z = e^{2\pi i t}$, $0 \le t \le 1$, can be continuously deformed to the loop Γ_1: $z = 2e^{2\pi i t}$, $0 \le t \le 1$, in the domain D consisting of the annulus $\frac{1}{2} < |z| < 3$.

SOLUTION This is precisely the problem illustrated in Fig. 4.25; the intermediate loops Γ_s, $0 \le s \le 1$, are concentric circles with radii varying from 1 to 2. The function

$$z(s, t) = (1 + s)e^{2\pi i t} \qquad (0 \le s \le 1, 0 \le t \le 1),$$

therefore effects the deformation. ∎

EXAMPLE 12 Exhibit a deformation function that shows that in the domain consisting of the whole plane any loop can be shrunk to the point contour $z = 0$.

SOLUTION This is the situation of Fig. 4.27. If Γ_0 is parametrized by $z = z_0(t)$, $0 \le t \le 1$, then the shrinking can be accomplished by multiplying $z_0(t)$ by a scaling factor which varies from 1 to 0. The deformation function is therefore given by

$$z(s, t) = (1 - s)z_0(t) \qquad (0 \le s \le 1, 0 \le t \le 1). \quad ∎$$

A few elementary observations about continuous deformations are now in order. First, notice that if $z(s, t)$ generates a deformation of loop Γ_0 into loop Γ_1, then $z(1 - s, t)$ deforms Γ_1 into Γ_0. Furthermore, if in a given domain Γ_0 can be deformed into a single point and Γ_1 can also be deformed into a point, then Γ_0 can be deformed into Γ_1 in the domain (see Prob. 1).

As we have observed in Example 12, in the domain D consisting of the entire complex plane *any* loop can be deformed into the single point $z = 0$. Consequently, any two loops can be deformed one into the other in this domain. There are many other domains with this property, e.g., interiors of circles, interiors of regular polygons, half-planes, etc. We categorize such domains in

DEFINITION 6   Any domain D possessing the property that every loop in D can be continuously deformed in D to a point is called a **simply connected domain.**

Roughly speaking, we say that a simply connected domain cannot have any "holes," for if there were a hole in D, then a loop surrounding it could not be shrunk to a point without leaving D. It is shown in topology that if γ is any simple closed contour, then its interior is a simply connected domain. Indeed, this fact is often regarded as part of the Jordan Curve Theorem. These considerations provide us with a quick method of identifying some simply connected domains (see Fig. 4.31).

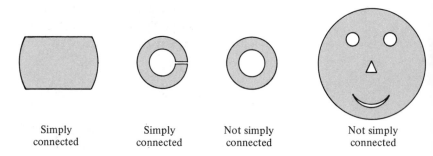

Simply connected	Simply connected	Not simply connected	Not simply connected

FIGURE 4.31

Interesting situations arise when the domain is not simply connected (such a domain is called *multiply connected*). For example, let D be the complex plane with the origin deleted, and let Γ be the unit circle $|z| = 1$, traversed once in the counterclockwise direction starting from the point $z = 1$. We list some loops which are not deformable to Γ in D:

(a) The circle parametrized by $z(t) = 4 + e^{2\pi it}$, $0 \le t \le 1$, cannot be deformed into Γ, because some intermediate loop would have to pass through $z = 0$ (see Fig. 4.32).

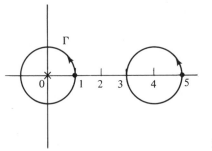

FIGURE 4.32

(b) Γ cannot be shrunk to a point in D. [However, the circle in (a) can be so deformed.]

(c) Γ cannot be continuously deformed into the opposite contour, $-\Gamma$. (The reader should mentally try to devise a family of loops linking these two, to see why the quarantining of the origin inhibits the deformation.)

(d) Γ cannot be deformed into the same unit circle circumscribed *twice* in the positive direction.

At this point we would like to insert a word of comfort to the reader, who is probably developing some anxiety concerning his ability to construct the deformation function $z(s, t)$. Theorem 8 below will show that, in practice, only the *existence* of the continuous deformation is important (at least for the theory of analytic functions). Consequently, we shall be content to merely visualize the deformation in most cases, without officially verifying its existence.

We are now ready to state the main theorem of this section.

THEOREM 8 (Deformation Invariance Theorem) *Let $f(z)$ be a function analytic in a domain D containing the loops Γ_0 and Γ_1. If these loops can be continuously deformed into one another in D, then*

$$\int_{\Gamma_0} f(z)\, dz = \int_{\Gamma_1} f(z)\, dz.$$

A rigorous proof of Theorem 8 involves procedures which take us far afield from the basic techniques of complex analysis. Here we shall only prove a weaker version of this theorem for the special case when

Γ_0 and Γ_1 are linked by a deformation function $z(s, t)$ whose second-order partial derivatives are continuous. We shall also assume that $f'(z)$ is continuous (recall that analyticity merely requires that f' exist).

The fact that one need *not* assume continuity for f' was first demonstrated by the mathematician Edouard Goursat; see Ref. [2]. The subsequent extension of the restricted theorem to the general statement of Theorem 8 can be effected by techniques of approximation theory (Ref. [4]).

Proof of Weak Version of Theorem 8 As we mentioned above, we shall assume that the deformation function $z(s, t)$ has continuous partial derivatives up to order 2 for $0 \le s \le 1$, $0 \le t \le 1$, and that $f'(z)$ is continuous. Now, for each fixed s the equation $z = z(s, t)$, $0 \le t \le 1$, defines a loop Γ_s in D. Let $I(s)$ be the integral of $f(z)$ along this loop, so that

$$(19) \qquad I(s) = \int_{\Gamma_s} f(z)\, dz = \int_0^1 f(z(s, t))\frac{\partial z(s, t)}{\partial t}\, dt.$$

We wish to take the derivative of $I(s)$ with respect to s. The assumptions guarantee that the integrand in Eq. (19) is continuously differentiable in s, so *Leibniz's rule for integrals* (Ref. [5]) sanctions differentiation under the integral sign. Using the chain rule we obtain

$$(20) \qquad \frac{dI(s)}{ds} = \int_0^1 \left[f'(z(s, t))\frac{\partial z}{\partial s}\cdot\frac{\partial z}{\partial t} + f(z(s, t))\frac{\partial^2 z}{\partial s\, \partial t} \right] dt.$$

On the other hand, observe that

$$\frac{\partial}{\partial t}\left[f(z(s, t))\frac{\partial z}{\partial s} \right] = f'(z(s, t))\frac{\partial z}{\partial t}\cdot\frac{\partial z}{\partial s} + f(z(s, t))\frac{\partial^2 z}{\partial t\, \partial s}.$$

Because of the continuity conditions the mixed partials of $z(s, t)$ are equal, so the last expression is the same as the integrand in Eq. (20). Thus

$$\frac{dI(s)}{ds} = \int_0^1 \frac{\partial}{\partial t}\left[f(z(s, t))\frac{\partial z}{\partial s} \right] dt$$

$$= f(z(s, 1))\frac{\partial z}{\partial s}(s, 1) - f(z(s, 0))\frac{\partial z}{\partial s}(s, 0).$$

But since each Γ_s is closed, we have $z(s, 1) = z(s, 0)$ for all s, so dI/ds is zero and, consequently, $I(s)$ is constant. In particular $I(0) = I(1)$, which is merely a disguised form of the conclusion

$$\int_{\Gamma_0} f(z)\, dz = \int_{\Gamma_1} f(z)\, dz. \qquad \blacksquare$$

An easy consequence of Theorem 8 is the following, familiarly known as *Cauchy's Integral Theorem*:

THEOREM 9 *If $f(z)$ is analytic in a simply connected domain D and Γ is any loop (closed contour) in D, then*

$$(21) \qquad\qquad \int_{\Gamma} f(z)\, dz = 0.$$

Proof The proof is immediate; in a simply connected domain any loop can be shrunk to a point. The integral of $f(z)$ over a loop consisting of a single point is, of course, zero. ∎

It can be shown by topological methods that if Γ is a simple closed contour and f is analytic at each point on and inside Γ, then f must be analytic in some simply connected domain containing Γ. Thus, by Theorem 9, the integral along Γ must vanish whenever the integrand is analytic "inside and on Γ."

Cauchy's theorem links the considerations of the last section with the property of analyticity. We can conclude

THEOREM 10 *In a simply connected domain, an analytic function has an antiderivative, its contour integrals are independent of path, and its loop integrals vanish.*

The possibilities delineated by the theory in this and the preceding sections are neatly exemplified by the functions z^n for integer values of n. If n is positive or zero, z^n is analytic in the whole plane, which is simply connected; thus Theorem 10 applies, z^n does have an antiderivative [the function $z^{n+1}/(n+1)$], and its loop integrals are zero.

If n is negative, z^n is analytic in the plane with the origin deleted. This domain is not simply connected, so Theorem 10 does *not* apply. In fact, for $n = -1$, the function z^n does not have an antiderivative in the punctured plane (recall that any branch of log z will be discontinuous on the branch cut), and its integral around the unit circle is not zero (Example 5, Sec. 4.2). On the other hand, for $n = -2, -3, \ldots$, z^n *does* have an antiderivative [again $z^{n+1}/(n+1)$], and its loop integrals are zero. Thus either case can happen in a multiply connected domain.

The main value of Theorem 8 is that it allows us to replace complicated contours with more familiar ones, for the purpose of integration. We shall illustrate this point with several examples.

EXAMPLE 13 Evaluate $\int_{\Gamma} 1/z\, dz$, where Γ is the ellipse defined by $x^2 + 4y^2 = 1$, traversed once in the positive sense, as indicated in Fig. 4.33.

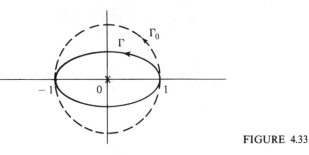

FIGURE 4.33

SOLUTION The integrand $1/z$ is analytic in the plane with the origin deleted. Furthermore, it is obvious that Γ can be continuously deformed without passing through the origin into the unit circle Γ_0, oriented positively. Thus

$$\int_\Gamma \frac{1}{z}\,dz = \int_{\Gamma_0} \frac{1}{z}\,dz = 2\pi i$$

by Example 5. ∎

EXAMPLE 14 Evaluate

$$\oint_{|z|=2} \frac{e^z}{z^2-9}\,dz.$$

SOLUTION The notation employed signifies that the contour of integration is the circle $|z|=2$ traversed once counterclockwise. The integrand $e^z/(z^2-9)$ is analytic everywhere except at $z=\pm3$, where the denominator vanishes. From Fig. 4.34 we immediately see that the contour can be shrunk to a point in the domain of analyticity, and thus

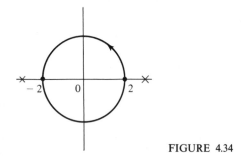

FIGURE 4.34

the integral is zero. (Alternatively, Cauchy's Theorem can be applied to this example.) ∎

EXAMPLE 15 Characterize $\int_\Gamma 1/(z-a)\,dz$, where Γ is any circle not passing through $z=a$, traversed once in the counterclockwise direction.

SOLUTION   The integrand is analytic in the domain D consisting of the plane with the point $z = a$ deleted. If this point lies exterior to Γ, then Γ can be continuously deformed to a point in D, and so the integral vanishes. If a lies in the interior of Γ, the contour can be continuously deformed in D to a positively oriented circle centered at $z = a$, and thus the integral is $2\pi i$ (by Example 5). Summarizing we have

(22)
$$\int_\Gamma \frac{dz}{z - a} = \begin{cases} 0 & \text{if } a \text{ lies outside } \Gamma, \\ 2\pi i & \text{if } a \text{ lies inside } \Gamma. \end{cases}$$ ∎

EXAMPLE 16   Find $\int_\Gamma (3z - 2)/(z^2 - z)\, dz$, where Γ is the simple closed contour indicated in Fig. 4.35.

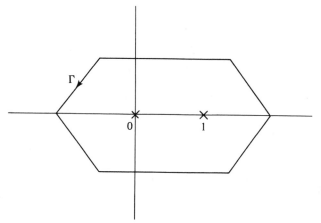

FIGURE 4.35

SOLUTION   We don't need an exact description of Γ; since the integrand $f(z) = (3z - 2)/(z^2 - z)$ is analytic except at $z = 0$ and $z = 1$, the contour can be deformed to the "barbell"-shaped contour of Fig. 4.36 without affecting the value of the integral. This can be further

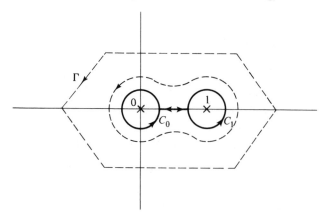

FIGURE 4.36

simplified by observing that the integration along the line segment proceeds forwards and backwards, the results canceling each other. Thus

$$\int_\Gamma f(z)\, dz = \int_{C_0} f(z)\, dz + \int_{C_1} f(z)\, dz,$$

where the circles C_0 and C_1 are as indicated in the Fig. 4.36.

Now we use partial fractions to write the integrand as

(23)
$$\frac{3z - 2}{z^2 - z} = \frac{A}{z} + \frac{B}{z - 1}.$$

In Eq. (23) the constants A and B are determined by recombining the right-hand side:

$$\frac{A}{z} + \frac{B}{z - 1} = \frac{Az - A + Bz}{z(z - 1)} = \frac{3z - 2}{z(z - 1)}.$$

Thus $A = 2$, $B = 1$, and

$$\int_\Gamma \frac{3z - 2}{z(z - 1)}\, dz = \int_{C_0} \left(\frac{2}{z} + \frac{1}{z - 1}\right) dz + \int_{C_1} \left(\frac{2}{z} + \frac{1}{z - 1}\right) dz.$$

The right-hand side of the last equation can be viewed as the sum of four integrals, each of the form of Example 15. So by Eq. (22), the integral is

$$2(2\pi i) + 0 + 2 \cdot 0 + 2\pi i = 6\pi i. \quad \blacksquare$$

EXAMPLE 17   Evaluate $\int_\Gamma 1/(z^2 - 1)\, dz$, where Γ is depicted in Fig. 4.37.

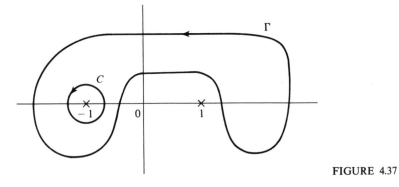

FIGURE 4.37

SOLUTION   Observing that $1/(z^2 - 1)$ fails to be analytic only at $z = \pm 1$, we see that without passing through these points Γ can be continuously deformed into a small positively oriented circle C around $z = -1$. Again we use partial fractions to find

$$\frac{1}{z^2 - 1} = \frac{1}{2(z - 1)} - \frac{1}{2(z + 1)}.$$

Hence

$$\int_\Gamma \frac{1}{z^2 - 1}\, dz = \int_C \left[\frac{1}{2(z - 1)} - \frac{1}{2(z + 1)} \right] dz,$$

which, by Example 15, equals

$$\frac{1}{2} \cdot 0 - \frac{1}{2}(2\pi i) = -\pi i. \quad \blacksquare$$

Note: Exercises appear at end of Sec. 4.4b.

4.4b   Vector analysis approach

In Sec. 4.3 we deduced that a continuous function $f(z)$ possesses an (analytic) antiderivative in a domain D if, and only if, its integral around every loop in D is zero. Now we are going to show how this property ties in with the analyticity of $f(z)$ itself. To do this we shall employ some concepts and theorems from vector analysis (Ref. [7]). First we demonstrate that our definition of the integral of $f(z)$ over a contour Γ can be related to the vector concept of a *line integral.*

Suppose that we have a two-dimensional vector $\mathbf{V} = (V_1, V_2)$ defined at every point (x, y) in some domain D in the plane; i.e., $\mathbf{V}$ is a *vector field*

$$\mathbf{V} = \mathbf{V}(x, y) = (V_1(x, y), V_2(x, y)).$$

For our purposes we require $V_1(x, y)$ and $V_2(x, y)$ to be continuous functions. Suppose furthermore that the (oriented) contour Γ, lying in D, has the parametrization

(24) $$x = x(t), \qquad y = y(t) \qquad (a \le t \le b).$$

Then the *line integral of* $\mathbf{V}$ *along* Γ, denoted by

$$\int_\Gamma (V_1\, dx + V_2\, dy),$$

is given by

$$\int_\Gamma (V_1\, dx + V_2\, dy) = \int_a^b \left[V_1(x(t), y(t))\frac{dx}{dt} + V_2(x(t), y(t))\frac{dy}{dt} \right] dt.$$

Students of physics can interpret this as the work done by a force $\mathbf{V}(x, y)$ exerted on a particle as it traverses the contour Γ.

To see how line integrals relate to complex integration we shall write out $\int_\Gamma f(z)\, dz$ in terms of its real and imaginary parts, utilizing the parametrization (24). With the usual notation $f(z) = u(x, y) + iv(x, y)$, we have

$$\int_\Gamma f(z)\, dz = \int_a^b f(z(t)) \frac{dz(t)}{dt}\, dt$$

$$= \int_a^b [u(x(t), y(t)) + iv(x(t), y(t))] \left(\frac{dx}{dt} + i\frac{dy}{dt}\right)\, dt$$

$$= \int_a^b \left[u(x(t), y(t)) \frac{dx}{dt} - v(x(t), y(t)) \frac{dy}{dt}\right]\, dt$$

$$+ i \int_a^b \left[v(x(t), y(t)) \frac{dx}{dt} + u(x(t), y(t)) \frac{dy}{dt}\right]\, dt;$$

i.e.,

(25) $$\int_\Gamma f(z)\, dz = \int_\Gamma (u\, dx - v\, dy) + i \int_\Gamma (v\, dx + u\, dy).$$

From this equation we can see that *the real part of $\int_\Gamma f(z)\, dz$ equals the line integral over Γ of the vector field $(u, -v)$, and that its imaginary part equals the line integral of the vector field (v, u).*†

Now that we can express complex integrals in terms of line integrals, we can translate many of the theorems of vector analysis into theorems about complex analysis. In Probs. 5, 6, and 7 the reader is guided in rediscovering some of the earlier theorems by this route. Our immediate goal is to uncover the consequences of *Green's Theorem* in the context of the complex integral.

To apply Green's Theorem it is convenient to introduce one new geometric concept: the simply connected domain. Roughly speaking, a domain is said to be simply connected if it has no holes, e.g., the inside of a simple closed contour (recall the Jordan Curve Theorem). One way of characterizing such domains is

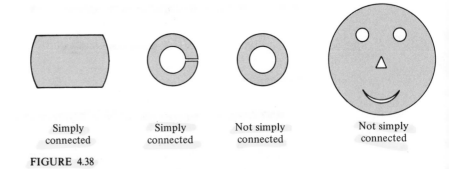

| Simply connected | Simply connected | Not simply connected | Not simply connected |

FIGURE 4.38

†Observe that the vector $(u, -v)$ corresponds to the complex number $u - iv = \bar{f}$ and that (v, u) corresponds to $i\bar{f}$.

*DEFINITION 7† A **simply connected domain** D is a domain having the following property: If Γ is any simple closed contour lying in D, then the domain interior to Γ lies wholly in D. (See Fig. 4.38.)*

In this context, one statement of *Green's Theorem* is as follows (Ref. [6]):

THEOREM 11 Let $\mathbf{V} = (V_1, V_2)$ be a continuously differentiable‡ vector field defined on a simply connected domain D, and let Γ be a positively oriented simple closed contour in D. Then the line integral of $\mathbf{V}$ around Γ equals the integral of $(\partial V_2/\partial x - \partial V_1/\partial y)$, integrated with respect to area over the domain D′ interior to Γ; i.e.,

$$\int_{\Gamma} (V_1 \ dx + V_2 \ dy) = \iint_{D'} \left(\frac{\partial V_2}{\partial x} - \frac{\partial V_1}{\partial y} \right) dx \ dy.$$

Let's apply this to the line integrals which occur in $\int_{\Gamma} f(z) \ dz$. Using Eq. (25), we have

$$(26) \quad \int_{\Gamma} f(z) \ dz = \int_{\Gamma} (u \ dx - v \ dy) + i \int_{\Gamma} (v \ dx + u \ dy)$$

$$= \iint_{D'} \left(-\frac{\partial v}{\partial x} - \frac{\partial u}{\partial y} \right) dx \ dy + i \iint_{D'} \left(\frac{\partial u}{\partial x} - \frac{\partial v}{\partial y} \right) dx \ dy.$$

Observe that we have assumed that u and v are continuously differentiable.

Now we take the big step. If $f(z)$ is *analytic* in D, the double integrals in Eq. (26) are zero because of the Cauchy-Riemann equations! In other words, we have shown that if a function is analytic in a simply connected domain and *if its derivative $f'(z)$ is continuous* (recall that analyticity only stipulates that f' exist), then its integral around any simple closed contour in the domain is zero. This result, in a somewhat more general form, is known as *Cauchy's Integral Theorem:*

THEOREM 12§ If $f(z)$ is analytic in a simply connected domain D and Γ is any loop (closed contour) in D, then

$$\int_{\Gamma} f(z) \ dz = 0.$$

---

†Definition 7 is equivalent to Definition 6.
‡Recall that this means the partials $\partial V_1/\partial x$, $\partial V_1/\partial y$, $\partial V_2/\partial x$, $\partial V_2/\partial y$ exist and are continuous.
§Theorem 12 is the same as Theorem 9.

Observe that we have generalized in two directions. First, we require only that Γ be a loop, not necessarily a simple closed curve. This is justified by the geometrically obvious fact that integration over a loop can always be decomposed into integrations over simple closed curves—see Fig. 4.39 for an illustration.

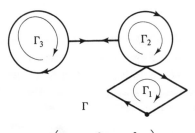

$$\int_\Gamma f(z)\,dz = \left(\int_{\Gamma_1} + \int_{\Gamma_2} + \int_{\Gamma_3}\right) f(z)\,dz$$

FIGURE 4.39

The second generalization is that we have dropped the assumption that $f'(z)$ is continuous. The fact that this is possible was first demonstrated by the mathematician Edouard Goursat; see Ref. [2].

We remark that it can be shown by topological methods that if Γ is a simple closed contour and f is analytic at each point on and inside Γ, then f must be analytic in some simply connected domain containing Γ. Thus, by Theorem 12, the integral along Γ must vanish whenever the integrand is analytic "inside and on Γ."

Cauchy's theorem links the considerations of Sec. 4.3 with the property of analyticity. Combining Theorems 7 and 12 we conclude

THEOREM 13† *In a simply connected domain, an analytic function has an antiderivative, its contour integrals are independent of path, and its loop integrals vanish.*

The possibilities delineated by the theory in this and the preceding sections are neatly exemplified by the functions z^n for integer values of n. If n is positive or zero, z^n is analytic in the whole plane, which is simply connected; thus Theorem 13 applies, z^n does have an antiderivative [the function $z^{n+1}/(n + 1)$], and its loop integrals are zero.

If n is negative, z^n is analytic in the plane with the origin deleted. This domain is not simply connected, so Theorem 13 does *not* apply. In fact, for $n = -1$, the function z^n does not have an antiderivative in the punctured plane (recall that any branch of log z will be discontinuous on the branch cut), and its integral around the unit circle is not zero (Example 5). On the other hand, for $n = -2, -3, \ldots,$ z^n *does* have an

†Theorem 13 is the same as Theorem 10.

antiderivative [again $z^{n+1}/(n + 1)$] and its loop integrals are zero. Thus either case can happen in a domain which is not simply connected (such a domain is called *multiply connected*).

EXAMPLE 18   Evaluate

$$\oint_{|z| = 2} e^z/(z^2 - 9)\,dz.$$

SOLUTION   The notation employed signifies that the contour of integration is the circle $|z| = 2$ traversed once counterclockwise. The integrand is analytic everywhere except at $z = \pm 3$, where the denominator vanishes. Since these points lie exterior to the contour, the integral is zero, by Cauchy's Integral Theorem.   ∎

Theorems 12 and 13 can often be used to change the contour of integration, as the following examples demonstrate:

EXAMPLE 19   Evaluate $\int_\Gamma 1/z\,dz$, where Γ is the ellipse defined by $x^2 + 4y^2 = 1$, traversed once in the positive sense, as indicated in Fig. 4.40(a).

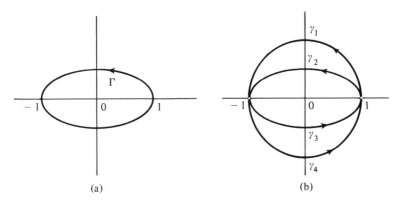

(a)   (b)

FIGURE 4.40

SOLUTION   We shall show that one can change the contour from Γ to the positively oriented unit circle without changing the integral. With reference to Fig. 4.40(b), observe that the complex plane, slit down the negative y-axis from the origin, constitutes a simply connected domain in which the function $1/z$ is analytic. Hence, by Theorem 13,

$$\int_{\gamma_2} \frac{1}{z}\,dz = \int_{\gamma_1} \frac{1}{z}\,dz.$$

Similarly, by considering the plane slit along the positive y-axis we have

$$\int_{\gamma_3} \frac{1}{z} dz = \int_{\gamma_4} \frac{1}{z} dz.$$

Hence

$$\int_\Gamma \frac{1}{z} dz = \int_{\gamma_2 + \gamma_3} \frac{1}{z} dz = \int_{\gamma_1 + \gamma_4} \frac{1}{z} dz = \oint_{|z|=1} \frac{1}{z} dz,$$

and by Example 5 of Sec. 4.2 the answer is $2\pi i$.   ∎

This technique is easily generalized in

EXAMPLE 20   Characterize $\int_\Gamma 1/(z-a) \, dz$, where Γ is any positively oriented simple closed contour not passing through $z = a$.

SOLUTION   Observe that the integrand is analytic everywhere except at the point $z = a$. Thus if a lies exterior to Γ, Cauchy's Theorem yields the answer zero for the integral. If a lies inside Γ, we choose a small circle C_r centered at a and lying within Γ, and we draw two segments from Γ to C_r; see Fig. 4.41 for the construction. Then the endpoints P_1 and P_2

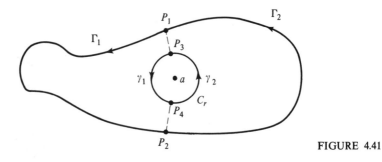

FIGURE 4.41

of the segments divide Γ into two contours Γ_1 and Γ_2, and the endpoints P_3 and P_4 divide C_r into γ_1 and γ_2. Now observe that both the contour Γ_1 and the composite contour consisting of the directed segment $P_1 P_3$, γ_1, and the segment $P_4 P_2$ can be enclosed in a simply connected domain which excludes the point a. Thus we deduce from Theorem 13 that the integral is the same along these contours:

$$\int_{\Gamma_1} \frac{dz}{z-a} = \left(\int_{P_1 P_3} + \int_{\gamma_1} + \int_{P_4 P_2} \right) \frac{dz}{z-a}.$$

Similarly,

$$\int_{\Gamma_2} \frac{dz}{z-a} = \left(\int_{P_2 P_4} + \int_{\gamma_2} + \int_{P_3 P_1} \right) \frac{dz}{z-a}.$$

Adding these and taking account of the cancellations along the line segments we find that

$$\int_{\Gamma} \frac{dz}{z-a} = \left(\int_{\Gamma_1} + \int_{\Gamma_2}\right) \frac{dz}{z-a} = \left(\int_{\gamma_1} + \int_{\gamma_2}\right) \frac{dz}{z-a} = \oint_{C_r} \frac{dz}{z-a} = 2\pi i$$

(recall Example 5 again).

Summarizing, we have

(27)
$$\int_{\Gamma} \frac{dz}{z-a} = \begin{cases} 0 & \text{if } a \text{ lies outside } \Gamma, \\ 2\pi i & \text{if } a \text{ lies inside } \Gamma. \end{cases} \qquad \blacksquare$$

EXAMPLE 21 Find $\int_{\Gamma} (3z-2)/(z^2-z) \, dz$, where Γ is the simple closed contour indicated in Fig. 4.42(a).

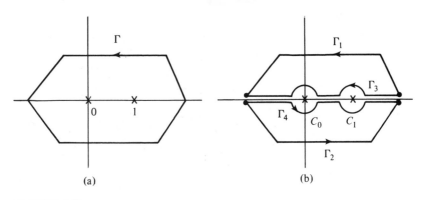

(a)   (b)

FIGURE 4.42

SOLUTION The integrand $f(z) = (3z-2)/(z^2-z)$ is, of course, analytic everywhere except for the zeros of the denominator, $z = 0$ and $z = 1$. Referring to Fig. 4.42(b), we begin by enclosing these points in small circles C_0 and C_1, respectively, and observe by Theorem 13 that the integral over Γ_1, the upper portion of Γ, equals the integral over the contour indicated as Γ_3. Similarly, integration over Γ_2 can be replaced by integration over Γ_4. Combining these and taking into account the cancellations along the segments of the real axis we find

$$\int_{\Gamma} f(z) \, dz = \left(\int_{\Gamma_1} + \int_{\Gamma_2}\right) f(z) \, dz = \oint_{C_0} f(z) \, dz + \oint_{C_1} f(z) \, dz.$$

Now we use partial fractions to write the integrand as

(28)
$$\frac{3z-2}{z^2-z} = \frac{A}{z} + \frac{B}{z-1}.$$

In Eq. (28) the coefficients A and B are determined by recombining the right-hand side:

$$\frac{A}{z} + \frac{B}{z-1} = \frac{Az - A + Bz}{z(z-1)} = \frac{3z - 2}{z(z-1)}.$$

Thus $A = 2$, $B = 1$, and

$$\int_\Gamma \frac{3z-2}{z(z-1)}\,dz = \oint_{C_0}\left(\frac{2}{z} + \frac{1}{z-1}\right)\,dz + \oint_{C_1}\left(\frac{2}{z} + \frac{1}{z-1}\right)\,dz.$$

The right-hand side of the last equation can be viewed as the sum of four integrals, each of the form of Example 20. So by Eq. (27), the integral is

$$2(2\pi i) + 0 + 2 \cdot 0 + 2\pi i = 6\pi i. \qquad \blacksquare$$

EXAMPLE 22 Evaluate $\int_\Gamma 1/(z^2 - 1)\,dz$, where Γ is depicted in Fig. 4.43(a).

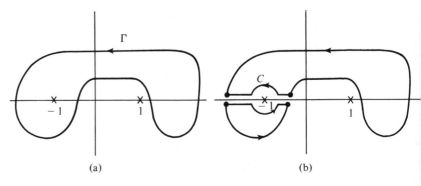

(a) (b)

FIGURE 4.43

  SOLUTION Observing that $1/(z^2 - 1)$ fails to be analytic only at $z = \pm 1$, the reader should be able by now to use the construction indicated in Fig. 4.43(b) to argue that the integral over Γ is the same as the integral over the small circle C enclosing -1. Using partial fractions again we find

$$\frac{1}{z^2 - 1} = \frac{1}{2(z-1)} - \frac{1}{2(z+1)}.$$

Hence

$$\int_\Gamma \frac{dz}{z^2 - 1} = \int_C \frac{dz}{z^2 - 1} = \int_C \left[\frac{1}{2(z-1)} - \frac{1}{2(z+1)}\right]\,dz,$$

which, by Example 20, equals

$$\frac{1}{2} \cdot 0 - \frac{1}{2}(2\pi i) = -\pi i. \qquad \blacksquare$$

Problems 1–4 refer to Sec. 4.4a.

1. Prove the statement made in the text: If the contours Γ_0 and Γ_1 can each be shrunk to points in the domain D, then Γ_0 can be continuously deformed into Γ_1 in D. (Do not assume that Γ_0 and Γ_1 are deformable to the *same* point.)

2. Write down a function $z(s, t)$ deforming Γ_0 into Γ_1 in the domain D, where Γ_0 is the ellipse $x^2/4 + y^2/9 = 1$ traversed once counterclockwise starting from $(2, 0)$, Γ_1 is the circle $|z| = 1$ traversed once counterclockwise starting from $(1, 0)$, and D is the annulus $\frac{1}{2} < |z| < 4$.

3. Let Γ_0 be the unit circle $|z| = 1$ traversed once counterclockwise and then once clockwise, starting from $z = 1$. Construct a function $z(s, t)$ which deforms Γ_0 to the single point $z = 1$ in *any* domain D containing the unit circle. Verify directly that the conclusion of Theorem 8 is true for these two contours.

4. Let D be the annulus $1 < |z| < 5$, and let Γ be the circle $|z - 3| = 1$ traversed once in the positive direction starting from the point $z = 4$. Decide which of the following contours are continuously deformable to Γ in D.

(a) the circle $|z - 3| = 1$ traversed once in the positive direction starting from the point $z = 2$

(b) the point $z = 3i$

(c) the circle $|z| = 2$ traversed once in the positive direction starting from the point $z = 2$

(d) the circle $|z + 3| = 1$ traversed once in the positive direction starting from the point $z = -2$

(e) the circle $|z - 3| = 1$ traversed twice in the negative direction starting from the point $z = 4$.

Problems 5–7 refer to Sec. 4.4b

5. It is well known from potential theory that if the line integrals of a vector field $\mathbf{V}(x, y)$ are independent of path (i.e., if $\mathbf{V}$ is a "conservative" field), then there is a scalar function of position $\phi(x, y)$ such that $V_1 = \partial\phi/\partial x$ and $V_2 = \partial\phi/\partial y$ (under such conditions we say that ϕ is a *potential* for $\mathbf{V}$). Apply this result to the vector fields $\overline{f(z)}$ and $i\overline{f(z)}$ to prove that property (*iii*) implies (*i*) in Theorem 7. What is the relationship between the (analytic) antiderivative $F(z)$ and the potentials?

6. In vector analysis a vector field $\mathbf{V} = (V_1, V_2)$ is said to be *irrotational* if its components satisfy

$$\frac{\partial V_1}{\partial y} = \frac{\partial V_2}{\partial x};$$

it is called *solenoidal* if

$$\frac{\partial V_1}{\partial x} = -\frac{\partial V_2}{\partial y}.$$

(a) Show that, if $f(z)$ is an analytic function, the vector field corresponding to $\overline{f(z)}$ is both irrotational and solenoidal.

(b) Prove the converse to part (a), if the vector field is continuously differentiable.

7. An important result from potential theory says that if a vector field $\mathbf{V}$ is irrotational (see Prob. 6) in a simply connected domain D, then there is a potential function for $\mathbf{V}(x, y)$ in D. Applying this fact to $\overline{f(z)}$ and $\overline{if(z)}$, prove the first assertion in Theorem 13.

Problems 8–17 are for both Secs. 4.4a and 4.4b.

8. Which of the following domains are simply connected?

(a) the horizontal strip $|\operatorname{Im} z| < 1$

(b) the annulus $1 < |z| < 2$

(c) the set of all points in the plane except those on the nonpositive *x*-axis

(d) the interior of the ellipse $4x^2 + y^2 = 1$

(e) the exterior of the ellipse $4x^2 + y^2 = 1$

(f) the domain D in Fig. 4.44

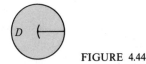

FIGURE 4.44

9. Determine the domain of analyticity for each of the given functions f and explain why

$$\oint_{|z|=2} f(z)\, dz = 0.$$

(a) $f(z) = \dfrac{z}{z^2 + 25}$ (b) $f(z) = e^{-z}(2z + 1)$

(c) $f(z) = \dfrac{\cos z}{z^2 - 6z + 10}$ (d) $f(z) = \operatorname{Log}(z + 3)$

(e) $f(z) = \sec\left(\dfrac{z}{2}\right)$

10. Explain why the function e^{z^2} has an antiderivative in the whole plane.

11. Given that D is a domain containing the closed contour Γ, that z_0 is a point not in D, and that $\int_\Gamma (z - z_0)^{-1} \, dz \neq 0$, explain why D is not simply connected.

12. Consider the shaded domain D in Fig. 4.45 which is bounded by the simple closed positively oriented contours $C, C_1, C_2,$ and C_3. If $f(z)$ is analytic on D and on its boundary, explain why

$$\int_C f(z) \, dz = \int_{C_1} f(z) \, dz + \int_{C_2} f(z) \, dz + \int_{C_3} f(z) \, dz.$$

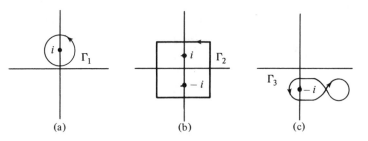

FIGURE 4.45

test

rep

13. Evaluate $\int 1/(z^2 + 1) \, dz$ along the three closed contours $\Gamma_1, \Gamma_2, \Gamma_3$ in Fig. 4.46.

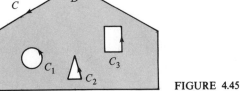

(a)   (b)   (c)

FIGURE 4.46

test   rep

⟶ **14.** Evaluate

$$\int_\Gamma \frac{z}{(z + 2)(z - 1)} \, dz,$$

rep.

where Γ is the circle $|z| = 4$ traversed twice in the clockwise direction.

test

15. Evaluate

$$\int_\Gamma \frac{2z^2 - z + 1}{(z - 1)^2(z + 1)} \, dz$$

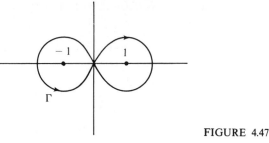

FIGURE 4.47

over the contour in Fig. 4.47. [HINT: Use the partial fraction expansion $A/(z-1)^2 + B/(z-1) + C/(z+1)$.]

16. Show that if f is of the form

$$f(z) = \frac{A_k}{z^k} + \frac{A_{k-1}}{z^{k-1}} + \cdots + \frac{A_1}{z} + g(z) \qquad (k \geq 1),$$

where g is analytic inside and on the circle $|z| = 1$, then

$$\oint_{|z|=1} f(z)\, dz = 2\pi i A_1.$$

17. Let

$$I = \oint_{|z|=2} \frac{dz}{z^2(z-1)^3}.$$

Below is an outline of a proof that $I = 0$. Justify each step.

(a) For every $R > 2$, $I = I(R)$, where

$$I(R) = \oint_{|z|=R} 1/[z^2(z-1)^3]\, dz.$$

(b) $|I(R)| \leq 2\pi/R(R-1)^3$ for $R > 2$.

(c) $\lim_{R \to +\infty} I(R) = 0$.

(d) $I = 0$.

4.5 CAUCHY'S INTEGRAL FORMULA AND ITS CONSEQUENCES

Given $f(z)$ analytic inside and on the simple closed contour Γ, we know from Cauchy's Theorem that $\int_\Gamma f(z)\, dz = 0$. However, if we consider the integral $\int_\Gamma f(z)/(z - z_0)\, dz$, where z_0 is a point in the interior of Γ, then there is no reason to expect that this integral is zero, because the integrand has a singularity inside the contour Γ. In fact, as the primary result of this section, we shall show that for all z_0 inside Γ the value of this integral is proportional to $f(z_0)$.

THEOREM 14 (Cauchy's Integral Formula) *Let Γ be a simple closed positively oriented contour. If $f(z)$ is analytic in some simply connected domain D containing Γ and z_0 is any point inside Γ, then*

(29)
$$f(z_0) = \frac{1}{2\pi i} \int_\Gamma \frac{f(z)}{z - z_0} \, dz.$$

Proof The function $f(z)/(z - z_0)$ is analytic everywhere in D except for the point z_0. Hence by the methods of Sec. 4.4 the integral over Γ can be equated to the integral over some small positively oriented circle C_r: $|z - z_0| = r$; Fig. 4.48(a) illustrates the continuous deformation method of Sec. 4.4a, while Fig. 4.48(b) shows the construction appropriate to Sec. 4.4b. So we can write

$$\int_\Gamma \frac{f(z)}{z - z_0} \, dz = \int_{C_r} \frac{f(z)}{z - z_0} \, dz.$$

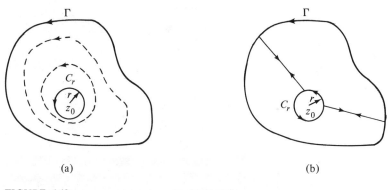

(a)　　　　　　　　　　　　(b)

FIGURE 4.48

It is now convenient to express the right-hand side as the sum of two integrals:

$$\int_{C_r} \frac{f(z)}{z - z_0} \, dz = \int_{C_r} \frac{f(z_0)}{z - z_0} \, dz + \int_{C_r} \frac{f(z) - f(z_0)}{z - z_0} \, dz.$$

From Example 5 we know that

$$\int_{C_r} \frac{f(z_0)}{z - z_0} \, dz = f(z_0) \int_{C_r} \frac{dz}{z - z_0} = f(z_0) 2\pi i;$$

consequently

(30)
$$\int_\Gamma \frac{f(z)}{z - z_0} \, dz = f(z_0) 2\pi i + \int_{C_r} \frac{f(z) - f(z_0)}{z - z_0} \, dz.$$

Now observe that the first two terms in Eq. (30) are constants independent of r, and so the value of the last term does not change if we allow r to decrease to zero; i.e.,

(31) $$\int_\Gamma \frac{f(z)}{z - z_0}\, dz = f(z_0)2\pi i + \lim_{r \to 0+} \int_{C_r} \frac{f(z) - f(z_0)}{z - z_0}\, dz.$$

Therefore Cauchy's formula will follow if we can prove that the last limit is zero.

For this purpose set $M_r = \max[\,|f(z) - f(z_0)|\,;\, z \text{ on } C_r]$. Then for z on C_r we have

$$\left| \frac{f(z) - f(z_0)}{z - z_0} \right| = \frac{|f(z) - f(z_0)|}{r} \le \frac{M_r}{r},$$

and hence by Theorem 5

$$\left| \int_{C_r} \frac{f(z) - f(z_0)}{z - z_0}\, dz \right| \le \frac{M_r}{r} l(C_r) = \frac{M_r}{r} 2\pi r = 2\pi M_r .$$

But since f is continuous at the point z_0, $\lim_{r \to 0+} M_r = 0$. Thus

$$\lim_{r \to 0+} \int_{C_r} \frac{f(z) - f(z_0)}{z - z_0}\, dz = 0,$$

and so Eq. (31) reduces to formula (29). ∎

One remarkable consequence of Cauchy's formula is the fact that by merely knowing the values of the analytic function f on Γ we can compute the integral in Eq. (29) and hence all the values of f inside Γ. In other words, the behavior of a function analytic in a region is completely determined by its behavior on the boundary.

EXAMPLE 23   Compute the integral

$$\int_\Gamma \frac{e^z + \sin z}{z}\, dz,$$

where Γ is the circle $|z - 2| = 3$ traversed once in the counterclockwise direction.

SOLUTION   Observe that the function $f(z) = e^z + \sin z$ is analytic inside and on Γ, and that the point $z_0 = 0$ lies inside this circle. Hence by formula (29) the desired value is

$$2\pi i f(0) = 2\pi i [e^0 + \sin 0] = 2\pi i. ∎$$

EXAMPLE 24   Evaluate the integral

$$\int_\Gamma \frac{\cos z}{z^2 - 4}\, dz$$

along the contour sketched in Fig. 4.49.

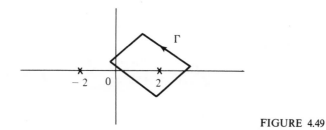

FIGURE 4.49

SOLUTION   We first notice that the integrand fails to be analytic at the points $z = \pm 2$. However, only one of these, $z = 2$, occurs inside Γ. Thus if we write

$$\frac{\cos z}{z^2 - 4} = \frac{(\cos z)/(z + 2)}{z - 2}$$

the numerator of the last fraction is analytic inside and on Γ. Now applying Cauchy's formula we obtain

$$\int_\Gamma \frac{\cos z}{z^2 - 4}\, dz = 2\pi i \left. \frac{\cos z}{(z + 2)}\right|_{z = 2} = \frac{2\pi i \cos 2}{4}. \qquad \blacksquare$$

EXAMPLE 25   Compute

$$\int_C \frac{z^2 e^z}{2z + i}\, dz,$$

where C is the unit circle $|z| = 1$ traversed once in the clockwise direction.

SOLUTION   Two minor difficulties inhibit an immediate application of Cauchy's formula. First, the denominator is not of the form $z - z_0$, and second, the contour is *negatively* oriented. The former difficulty is easily resolved by writing

$$\frac{z^2 e^z}{2z + i} = \frac{\frac{1}{2}z^2 e^z}{z + \dfrac{i}{2}}$$

(notice that the singular point $z = -i/2$ lies *inside* C). And to compensate for the negative orientation of C we have to introduce a minus sign in formula (29). Thus

$$\int_C \frac{z^2 e^z}{2z + i} \, dz = \int_C \frac{\frac{1}{2}z^2 e^z}{z + \frac{i}{2}} \, dz = -2\pi i \cdot \frac{1}{2} z^2 e^z \Big|_{z = -i/2} = \frac{\pi i}{4} e^{-i/2}. \quad \blacksquare$$

If in Cauchy's formula (29) we replace z by ξ and z_0 by z, then we obtain

$$(32) \qquad f(z) = \frac{1}{2\pi i} \int_\Gamma \frac{f(\xi)}{\xi - z} \, d\xi \qquad (z \text{ inside } \Gamma).$$

The advantage of this representation is that it suggests a formula for the derivative $f'(z)$, obtained by formally differentiating with respect to z under the integral sign. Thus we are led to suspect that

$$(33) \qquad f'(z) = \frac{1}{2\pi i} \int_\Gamma \frac{f(\xi)}{(\xi - z)^2} \, d\xi \qquad (z \text{ inside } \Gamma).$$

In verifying this equation we shall actually establish the more general

THEOREM 15 *Let g be continuous on the contour Γ, and for each z not on Γ set*

$$(34) \qquad G(z) \equiv \int_\Gamma \frac{g(\xi)}{\xi - z} \, d\xi.$$

Then the function G is analytic, and its derivative is given by

$$(35) \qquad G'(z) = \int_\Gamma \frac{g(\xi)}{(\xi - z)^2} \, d\xi$$

for all z not on Γ.

(Observe that we have generalized in two directions; we have not assumed that Γ is closed, nor that g is analytic. In fact, even if Γ *is* a simple closed contour the limiting values of G need not agree with values of g on Γ in this general case; see Prob. 12.)

Proof Let z be any fixed point not on Γ. To prove the existence of $G'(z)$ and the formula (35) we need to show that

$$\lim_{\Delta z \to 0} \frac{G(z + \Delta z) - G(z)}{\Delta z} = \int_\Gamma \frac{g(\xi)}{(\xi - z)^2} \, d\xi,$$

or, equivalently, that the difference

$$(36) \qquad J \equiv \frac{G(z + \Delta z) - G(z)}{\Delta z} - \int_\Gamma \frac{g(\xi)}{(\xi - z)^2} \, d\xi$$

approaches zero as $\Delta z \to 0$. This is accomplished by first writing J in a convenient form obtained as follows:

Using Eq. (34) we have

$$\frac{G(z + \Delta z) - G(z)}{\Delta z} = \frac{1}{\Delta z} \int_\Gamma \left[\frac{1}{\xi - (z + \Delta z)} - \frac{1}{\xi - z} \right] g(\xi) \, d\xi$$

$$= \int_\Gamma \frac{g(\xi) \, d\xi}{(\xi - z - \Delta z)(\xi - z)},$$

where Δz is chosen sufficiently small so that $z + \Delta z$ also lies off of Γ. Then from Eq. (36) and some elementary algebra we find

$$(37) \qquad J = \int_\Gamma \frac{g(\xi) \, d\xi}{(\xi - z - \Delta z)(\xi - z)} - \int_\Gamma \frac{g(\xi)}{(\xi - z)^2} \, d\xi$$

$$= \Delta z \int_\Gamma \frac{g(\xi) \, d\xi}{(\xi - z - \Delta z)(\xi - z)^2}.$$

To verify that $J \to 0$ as $\Delta z \to 0$, we estimate the last term in Eq. (37). In this regard, let M equal the maximum value of $|g(\xi)|$ on Γ, and set d equal to the shortest distance from z to Γ, so that $|\xi - z| \geq d > 0$ for all ξ on Γ. Since we are letting Δz approach zero, we may assume that $|\Delta z| < d/2$. Then, by the triangle inequality,

$$|\xi - z - \Delta z| \geq |\xi - z| - |\Delta z| \geq d - \frac{d}{2} = \frac{d}{2} \qquad (\xi \text{ on } \Gamma)$$

(see Fig. 4.50) and so

$$\left| \frac{g(\xi)}{(\xi - z - \Delta z)(\xi - z)^2} \right| \leq \frac{M}{\frac{d}{2} \cdot d^2} = \frac{2M}{d^3}$$

for all ξ on Γ. Hence from Theorem 5 we see that

$$|J| = \left| \Delta z \int_\Gamma \frac{g(\xi) \, d\xi}{(\xi - z - \Delta z)(\xi - z)^2} \right| \leq \frac{|\Delta z| \, 2Ml(\Gamma)}{d^3},$$

where $l(\Gamma)$ denotes the length of Γ. Thus J must approach zero as $\Delta z \to 0$. This implies that formula (35) is valid and completes the proof. ∎

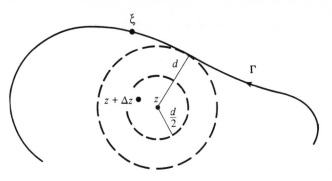

FIGURE 4.50

The above argument can be carried a step further. Namely, starting with the function

(38) $$H(z) = \int_\Gamma \frac{g(\xi)}{(\xi - z)^2} \, d\xi \qquad (z \text{ not on } \Gamma),$$

it can be shown that H is analytic off of Γ and that H' is given by the formula

(39) $$H'(z) = 2 \int_\Gamma \frac{g(\xi)}{(\xi - z)^3} \, d\xi \qquad (z \text{ not on } \Gamma),$$

obtained formally from (38) by differentiation under the integral sign. The proof of this fact parallels the proof of Theorem 15 and is left as an exercise.

One important consequence of these results is that the derivative of an analytic function is again analytic. For suppose that f is analytic at the point z_0. We wish to argue that f' is also analytic at z_0; i.e., f' itself has a derivative in some neighborhood of z_0. To this end, we choose a positively oriented circle $C: |\xi - z_0| = r$ so small that f is analytic inside and on C. Since f has the Cauchy integral representation

$$f(z) = \frac{1}{2\pi i} \int_C \frac{f(\xi)}{\xi - z} \, d\xi \qquad (z \text{ inside } C),$$

it follows from Theorem 15 that

$$f'(z) = \frac{1}{2\pi i} \int_C \frac{f(\xi)}{(\xi - z)^2} \, d\xi \qquad (z \text{ inside } C).$$

But the right-hand side is a function of the form (38) and hence has a derivative at each point inside C. Thus f' must be analytic at z_0.

The same reasoning can be applied to the function f' to deduce that its derivative, f'', is also analytic at z_0. More generally, the analyticity of $f^{(n)}$ implies that of $f^{(n+1)}$, and so by induction we obtain

THEOREM 16 *If f is analytic in a domain D, then all its derivatives f', f'', $\ldots, f^{(n)}, \ldots$ exist and are analytic in D.*

This theorem is particularly surprising in light of the fact that its analogue in calculus fails to be true; for example, the function $f(x) = x^{5/3}$, $-\infty < x < \infty$, is differentiable for all real x, but $f'(x) = 5x^{2/3}/3$ has no derivative at $x = 0$.

Recall that if the analytic function f is written in the form $f(z) = u(x, y) + iv(x, y)$, then as explained in Sec. 2.4 we have the alternative expressions

(40) $$f'(z) = \frac{\partial u}{\partial x} + i\frac{\partial v}{\partial x} \quad \text{and} \quad f'(z) = \frac{\partial v}{\partial y} - i\frac{\partial u}{\partial y}.$$

We now know that f' is analytic and hence continuous. Therefore, from (40), all the first-order partial derivatives of u and v must be continuous. Similarly, since f'' exists, the formulas (40) together with the Cauchy-Riemann equations for f' lead to the expressions

$$f''(z) = \frac{\partial^2 u}{\partial x^2} + i\frac{\partial^2 v}{\partial x^2} = \frac{\partial^2 v}{\partial y\, \partial x} - i\frac{\partial^2 u}{\partial y\, \partial x},$$

$$f''(z) = \frac{\partial^2 v}{\partial x\, \partial y} - i\frac{\partial^2 u}{\partial x\, \partial y} = -\frac{\partial^2 u}{\partial y^2} - i\frac{\partial^2 v}{\partial y^2},$$

and so the continuity of f'' implies that all second-order partial derivatives of u and v are continuous at the points where f is analytic. Continuing with this process we obtain

THEOREM 17 *If $f = u + iv$ is analytic in a domain D, then all partial derivatives of u and v exist and are continuous in D.*

(Recall that we presupposed this result in Sec. 2.5, where we argued that the real and imaginary parts of analytic functions are harmonic.)

Another way to phrase the results on the analyticity of derivatives is to say that whenever a given function f has an antiderivative in a domain D, then f must itself be analytic in D. Now by Theorem 7 of Sec. 4.3 the existence of an antiderivative for a continuous function is equivalent to the property that all loop integrals vanish. Hence we deduce the following result, known as *Morera's theorem*:

THEOREM 18 *If $f(z)$ is continuous in a domain D and if*

$$\int_\Gamma f(z)\, dz = 0$$

for every closed contour Γ in D, then $f(z)$ is analytic in D.

Observe that in proving Eqs. (35) and (39) we actually verify that for certain types of integrands the process of differentiation with respect to z can be interchanged with the process of integration with respect to ζ. In fact, starting with Cauchy's integral formula, it can be shown inductively that repeated differentiation with respect to z under the integral sign yields valid formulas for the successive derivatives of f. Keeping track of the exponents, we have

> **THEOREM 19** *If f is analytic inside and on the simple closed positively oriented contour Γ and if z is any point inside Γ, then*
>
> (41) $$f^{(n)}(z) = \frac{n!}{2\pi i} \int_\Gamma \frac{f(\xi)}{(\xi - z)^{n+1}} \, d\xi \qquad (n = 1, 2, 3, \ldots).\dagger$$

For purposes of application it is convenient to write Eq. (41) in the equivalent form

(42) $$\int_\Gamma \frac{f(z)}{(z - z_0)^m} \, dz = \frac{2\pi i f^{(m-1)}(z_0)}{(m-1)!} \qquad (z_0 \text{ inside } \Gamma).$$

EXAMPLE 26   Compute $\int_\Gamma e^{5z}/z^3 \, dz$, where Γ is the circle $|z| = 1$ traversed once counterclockwise.

SOLUTION   Observe that $f(z) = e^{5z}$ is analytic inside and on Γ. Therefore, from Eq. (42) with $z_0 = 0$ and $m = 3$ we have

$$\int_\Gamma \frac{e^{5z}}{z^3} \, dz = \frac{2\pi i f''(0)}{2!} = 25\pi i. \quad \blacksquare$$

EXAMPLE 27   Compute

$$\int_C \frac{2z + 1}{z(z - 1)^2} \, dz$$

along the contour C sketched in Fig. 4.51.

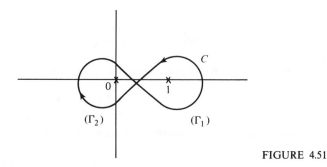

(Γ_2) (Γ_1)

FIGURE 4.51

---

†When $n = 0$, this reduces to Cauchy's formula. In fact, (41) is sometimes called the *Generalized Cauchy Integral Formula*.

SOLUTION   Notice that integration along C is equivalent to integrating once around the positively oriented right lobe Γ_1 and then integrating once around the negatively oriented left lobe Γ_2; i.e.,

$$\int_C \frac{2z + 1}{z(z - 1)^2} \, dz = \int_{\Gamma_1} \frac{(2z + 1)/z}{(z - 1)^2} \, dz + \int_{\Gamma_2} \frac{(2z + 1)/(z - 1)^2}{z} \, dz,$$

where we have written the integrand in each term so as to display the relevant singularity. These integrals along Γ_1 and Γ_2 can be evaluated by using Eqs. (42) and (29); the desired value is

$$\frac{2\pi i}{1!} \frac{d}{dz} \left(\frac{2z + 1}{z} \right)\bigg|_{z=1} - 2\pi i \frac{2z + 1}{(z - 1)^2} \bigg|_{z=0} = -2\pi i - 2\pi i = -4\pi i. \quad \blacksquare$$

Exercises 4.5

1. Let f be analytic inside and on the simple closed contour Γ. What is the value of

$$\frac{1}{2\pi i} \int_\Gamma \frac{f(z)}{z - z_0} \, dz$$

when z_0 lies outside Γ?

2. Let f and g be analytic inside and on the simple loop Γ. Prove that if $f(z) = g(z)$ for all z on Γ, then $f(z) = g(z)$ for all z inside Γ.

3. Let C be the circle $|z| = 2$ traversed once in the positive sense. Compute each of the following integrals.

(a) $\displaystyle\int_C \frac{\sin 3z}{z - \dfrac{\pi}{2}} \, dz$ (b) $\displaystyle\int_C \frac{ze^z}{2z - 3} \, dz$

(c) $\displaystyle\int_C \frac{\cos z}{z^3 + 9z} \, dz$ (d) $\displaystyle\int_C \frac{5z^2 + 2z + 1}{(z - i)^3} \, dz$

(e) $\displaystyle\int_C \frac{e^{-z}}{(z + 1)^2} \, dz$ (f) $\displaystyle\int_C \frac{\sin z}{z^2(z - 4)} \, dz$

4. Compute

$$\int_C \frac{z + i}{z^3 + 2z^2} \, dz,$$

where C is

(a) the circle $|z| = 1$ traversed once counterclockwise,
(b) the circle $|z + 2 - i| = 2$ traversed once counterclockwise,
(c) the circle $|z - 2i| = 1$ traversed once counterclockwise.

test

test .

5. Let C be the ellipse $x^2/4 + y^2/9 = 1$ traversed once in the positive direction, and define

$$G(z) = \int_C \frac{\xi^2 - \xi + 2}{\xi - z} \, d\xi \qquad (z \text{ inside } C).$$

Find $G(1)$, $G'(i)$, and $G''(-i)$.

6. Evaluate

$$\int_\Gamma \frac{e^{iz}}{(z^2 + 1)^2} \, dz,$$

where Γ is the circle $|z| = 3$ traversed once counterclockwise. [HINT: Show that the integral can be written as the sum of two integrals around small circles centered at the singularities.]

7. Compute

$$\int_\Gamma \frac{\cos z}{z^2(z - 3)} \, dz$$

along the contour indicated in Fig. 4.52.

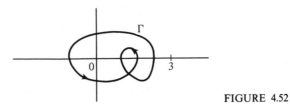

FIGURE 4.52

8. Use Cauchy's formula to prove that if f is analytic inside and on the circle $|z - z_0| = r$, then

$$f(z_0) = \frac{1}{2\pi} \int_0^{2\pi} f(z_0 + re^{i\theta}) \, d\theta.$$

Prove more generally that

$$f^{(n)}(z_0) = \frac{n!}{2\pi r^n} \int_0^{2\pi} f(z_0 + re^{i\theta}) e^{-in\theta} \, d\theta.$$

9. Let f be analytic inside and on the simple closed contour Γ. Verify from the theorems in this section that

$$\int_\Gamma \frac{f'(z)}{z - z_0} \, dz = \int_\Gamma \frac{f(z)}{(z - z_0)^2} \, dz$$

for all z_0 not on Γ.

10. Let $f = u + iv$ be analytic in a domain D. Explain why $\partial^2 u/\partial x^2$ is harmonic in D.

11. Prove that the function H defined in Eq. (38) is analytic inside Γ and that its derivative is given by formula (39).

12. According to Theorem 15, when Γ is a simple closed contour the function G defined by

$$G(z) = \frac{1}{2\pi i} \int_{\Gamma} \frac{f(\xi)}{\xi - z} \, d\xi$$

is analytic in the domain enclosed by Γ (assuming f is continuous on Γ). Show that the limiting values of $G(z)$ as z approaches Γ need not coincide with the values of f, by considering the situation where Γ is the positively oriented circle $|z| = 1$ and $f(z) = 1/z$. Does this violate Cauchy's Formula? [HINT: Use partial fractions to evaluate G.]

13. Suppose that $f(z)$ is analytic at each point of the closed disk $|z| \leq 1$ and that $f(0) = 0$. Prove that the function

$$F(z) = \begin{cases} \dfrac{f(z)}{z} & z \neq 0, \\[2mm] f'(0) & z = 0, \end{cases}$$

is analytic on $|z| \leq 1$. [HINT: To show that F is analytic at $z = 0$ note that by Theorem 15 the function

$$G(z) = \frac{1}{2\pi i} \oint_{|\xi| = 1} \frac{f(\xi)/\xi}{\xi - z} \, d\xi$$

is analytic at this point. Using partial fractions deduce that $G(z) = F(z)$ for $|z| < 1$.]

14. Below is an outline of a proof of the fact that for any analytic function $f(z)$ that is never zero in a simply connected domain D there exists a single-valued branch of $\log f(z)$ analytic in D. Justify each step in the proof.

(a) $f'(z)/f(z)$ is analytic in D.

(b) $f'(z)/f(z)$ has an (analytic) antiderivative in D, say $H(z)$.

(c) The function $f(z)e^{-H(z)}$ is constant in D, so that $f(z) = ce^{H(z)}$.

(d) Letting α be a value of $\log c$, the function $H(z) + \alpha$ is a branch of $\log f(z)$ analytic in D.

Many interesting facts about analytic functions are uncovered when one considers upper bounds on their moduli. We already have one result in this direction, namely, the integral estimate of Sec. 4.2. When this is judiciously applied to the Cauchy integral formulas we obtain the *Cauchy estimates* for the derivatives of an analytic function.

$C_r: |z - z_0| = R.$

THEOREM 20   *Let $f(z)$ be analytic inside and on a circle C_R of radius R centered about z_0. If $|f(z)| \le M$ for all z on C_R, then the derivatives of f at z_0 satisfy*

(43)
$$|f^{(n)}(z_0)| \le \frac{n!\,M}{R^n} \qquad (n = 1, 2, 3, \ldots).$$

Proof   Giving C_R a positive orientation, we have by Theorem 19

$$f^{(n)}(z_0) = \frac{n!}{2\pi i} \int_{C_R} \frac{f(\xi)}{(\xi - z_0)^{n+1}}\, d\xi.$$

For ξ on C_R, the integrand is bounded by M/R^{n+1}; the length of C_R is $2\pi R$. Thus from Theorem 5, there follows

$$|f^{(n)}(z_0)| \le \frac{n!}{2\pi} \frac{M}{R^{n+1}} 2\pi R,$$

which reduces to (43).   ∎

This innocuous looking theorem actually places rather severe restrictions on the behavior of analytic functions. Suppose, for instance, that $f(z)$ is analytic and bounded by some number M over the whole plane. Then the conditions of the theorem hold for any z_0 and for any R. Taking $n = 1$ in (43) and letting $R \to \infty$, we conclude that f' vanishes everywhere; i.e., f must be constant. This result is known as *Liouville's theorem*:

THEOREM 21   *The only bounded entire functions are the constant functions.*

Of course, nonconstant polynomials are not bounded over the whole plane. Loosely speaking, we expect a polynomial of degree n to behave like z^n for large $|z|$, because the leading term will dominate the lower powers. To be precise, we establish the following technical lemma:

LEMMA 1   *Let $P(z) = z^n + a_{n-1} z^{n-1} + \cdots + a_1 z + a_0$, where $n \ge 1$, and set $A \equiv \max\{1, |a_{n-1}|, \ldots, |a_1|, |a_0|\}$. Then for $|z| \ge 2nA$, we have $|P(z)| \ge |z|^n/2$.*

Proof Write $P(z) = z^n(1 + a_{n-1}/z + \cdots + a_1/z^{n-1} + a_0/z^n)$. By the triangle inequality, we have

$$(44) \quad \left| 1 + \frac{a_{n-1}}{z} + \cdots + \frac{a_1}{z^{n-1}} + \frac{a_0}{z^n} \right| \geq 1 - \left| \frac{a_{n-1}}{z} + \cdots + \frac{a_1}{z^{n-1}} + \frac{a_0}{z^n} \right|.$$

But for $|z| \geq 2nA \ (\geq 1)$,

$$\left| \frac{a_{n-k}}{z^k} \right| \leq \frac{A}{|z|^k} \leq \frac{A}{|z|} \leq \frac{1}{2n},$$

so

$$\left| \frac{a_{n-1}}{z} + \cdots + \frac{a_1}{z^{n-1}} + \frac{a_0}{z^n} \right| \leq \frac{1}{2n} + \frac{1}{2n} + \cdots + \frac{1}{2n}.$$

Since there are n terms, this sum is exactly $\frac{1}{2}$. Using this in (44) we obtain

$$|P(z)| = |z|^n \left| 1 + \frac{a_{n-1}}{z} + \cdots + \frac{a_1}{z^{n-1}} + \frac{a_0}{z^n} \right|$$

$$\geq |z|^n \left(1 - \frac{1}{2} \right)$$

$$= \frac{|z|^n}{2}. \quad \blacksquare$$

Probably the reader has heard about the *Fundamental Theorem of Algebra*, which states that every equation of the form $P(z) = 0$, where P is a nonconstant polynomial, has a root. We can now prove this result by applying Liouville's Theorem to $1/P(z)$, which would be entire if $P(z)$ had no zeros.

THEOREM 22   *Every nonconstant polynomial with complex coefficients has at least one zero.*

Proof Assume, to the contrary, that $P(z) = a_n z^n + a_{n-1} z^{n-1} + \cdots + a_1 z + a_0$ is a nonconstant polynomial $(n \geq 1)$ having no zeros. Clearly we may take $a_n = 1$. As we just remarked, $1/P(z)$ must be entire. We shall show that it is bounded over the whole plane:

(i) For $|z| \geq 2nA$ as in the lemma, we have

$$\left| \frac{1}{P(z)} \right| \leq \frac{1}{|z|^n/2} \leq \frac{2}{(2nA)^n}.$$

(ii) For $|z| \leq 2nA$, the situation becomes that of a continuous function, $|1/P(z)|$, on a closed disk. Under such circumstances it is known from calculus that the function is bounded there (indeed, it achieves its maximum!).

But then $1/P(z)$ is bounded and entire so it must be constant. Thus $P(z)$, itself, is constant, violating our assumptions. ∎

Actually any polynomial of degree n must possess *exactly n zeros*, counting multiplicities. These refinements are discussed in the exercises.

Let us now return to the Cauchy formula for the function $f(z)$, analytic inside and on the circle C_R of radius R around z_0. We have

$$(45) \qquad f(z_0) = \frac{1}{2\pi i} \oint_{C_R} \frac{f(z)}{z - z_0} \, dz.$$

Parametrizing C_R by $z = z_0 + Re^{it}$, $0 \le t \le 2\pi$, we write Eq. (45) as

$$f(z_0) = \frac{1}{2\pi i} \int_0^{2\pi} \frac{f(z_0 + Re^{it})}{Re^{it}} \, iRe^{it} \, dt,$$

or

$$(46) \qquad f(z_0) = \frac{1}{2\pi} \int_0^{2\pi} f(z_0 + Re^{it}) \, dt.$$

Equation (46), which is known as the *Mean-Value Property*, displays $f(z_0)$ as the average of its values around the circle C_R. Clearly, if $|f(z)| \le M$ on C_R, then $|f(z_0)|$ is bounded by M also (this verifies the case $n = 0$ of Theorem 20, with $0! = 1$). But more importantly, we can utilize Eq. (46) to establish

LEMMA 2 Suppose that $f(z)$ is analytic in a disk centered at z_0 and that the maximum value of $|f(z)|$ over this disk is $|f(z_0)|$. Then $|f(z)|$ is constant in the disk.

Proof Assume to the contrary that $|f(z)|$ is not constant. Then there must exist a point z_1 in the disk such that $|f(z_0)| > |f(z_1)|$. Let C_R denote the circle centered at z_0 which passes through z_1. Then by hypothesis $|f(z_0)| \ge |f(z)|$ for all z on C_R. Moreover, by the continuity of f, the strict inequality $|f(z_0)| > |f(z)|$ must hold for z on a portion of C_R containing z_1. This leads to a contradiction of Eq. (46), because the portion containing z_1 would contribute " less than its share " to the average in Eq. (46), and the "deficit" cannot be made up anywhere else on C_R. (We invite the reader to provide a rigorous version of this argument in Prob. 6.) ∎

Observe that the lemma says that the modulus of an analytic function cannot achieve its maximum at the center of the disk unless $|f|$ is constant. We shall use the lemma to extend this idea in the following version of the *Maximum Modulus Principle*:

THEOREM 23 *Let f be analytic in a domain D, with its modulus,* $|f|$, *bounded by M there. If, for some* z_0 *in D,* $|f(z_0)| = M$, *then f is constant in D.*

Proof We shall prove that $|f|$ is constant in D; by Prob. 12 in Exercises 2.4, we can then conclude that f, itself, is constant.

So for the moment let us suppose that $|f(z)|$ is not constant. Then there must be a point z_1 in D such that $|f(z_1)| < |f(z_0)|$. Let γ be a path in D running from z_0 to z_1. Now we consider the values of $|f(z)|$ for z on γ, starting at z_0. Intuitively, we expect to encounter a point w where $|f(z)|$ first starts to decrease on γ. That is, there should be a point w on γ with the following properties:

(*i*) $|f(z)| = |f(z_0)|$ for all z preceding w on γ.
(*ii*) There are points z on γ, arbitrarily close to w, where $|f(z)| < |f(z_0)|$.

(It is possible that w may coincide with z_0.) The existence of w is a fact which can be rigorously demonstrated from the axioms of the real number system, but we shall omit the details here. Naturally, from property (*i*) and the continuity of f, we have $|f(w)| = |f(z_0)| = M$.

Now since every point of a domain is an interior point, there must be a disk centered at w which lies in D. But Lemma 2 applies and says that $|f|$ is constant in this disk, contradicting property (*ii*) above. We are forced to conclude, therefore, that our initial supposition about the existence of z_1 is erroneous. Consequently $|f|$, and hence f itself, is constant in D. ∎

Next, consider a function $f(z)$ analytic in a *bounded* domain D and continuous on D and its boundary. From calculus we know that the continuous function $|f(z)|$ must achieve its maximum on this closed bounded region; on the other hand, Theorem 23 says that the maximum cannot occur at an interior point unless f is constant. In any case, we deduce the following modification of the Maximum Modulus Principle:

THEOREM 24 *A function analytic in a bounded domain and continuous up to and including its boundary attains its maximum modulus on the boundary.*

EXAMPLE 28 Find the maximum value of $|z^2 + 3z - 1|$ in the disk $|z| \leq 1$.

SOLUTION   Using the triangle inequality we have immediately

(47) $|z^2 + 3z - 1| \leq |z|^2 + 3|z| + 1 \leq 5$ (for $|z| \leq 1$).

However, the maximum is actually smaller than this, as the following analysis shows. Since the polynomial is analytic, its maximum modulus will be attained on the unit circle $|z| = 1$. We have

$$|z^2 + 3z - 1|^2 = (z^2 + 3z - 1)(\bar{z}^2 + 3\bar{z} - 1)$$

$$= z^2\bar{z}^2 + 3z\bar{z}(z + \bar{z}) + 9z\bar{z} - (z^2 + \bar{z}^2) - 3(z + \bar{z}) + 1,$$

and for $|z| = 1$, this becomes $11 - (z^2 + \bar{z}^2) = 11 - 2\text{Re}(z^2)$. This expression attains its maximum, 13, at $z = \pm i$. Thus

$$\max_{|z| \leq 1} |z^2 + 3z - 1| = \sqrt{13},$$

a finer estimate than (47).   ∎

Exercises 4.6

1. Let $f(z) = 1/(1 - z)^2$,   and   let $0 < R < 1$.   Verify   that $\max_{|z| = R} |f(z)| = 1/(1 - R)^2$, and also show $f^{(n)}(0) = (n + 1)!$, so that by the Cauchy estimates

    $$(n + 1)! \leq \frac{n!}{R^n(1 - R)^2}.$$

2. Suppose that f is analytic in $|z| < 1$ and that $|f(z)| < 1/(1 - |z|)$. Prove that

    $$|f^{(n)}(0)| \leq \frac{n!}{R^n(1 - R)}   \quad (0 < R < 1),$$

    and show that the upper bound is smallest when $R = n/(n + 1)$.

3. Let f be analytic and bounded by M in $|z| \leq r$. Prove that

    $$|f^{(n)}(z)| \leq \frac{n! M}{(r - |z|)^n}   \quad (|z| < r).$$

4. Let f be entire and suppose that $\text{Re } f(z) \leq M$ for all z. Prove that f must be a constant function. [HINT: Apply Liouville's Theorem to the function e^f.]

5. Suppose that f is entire and that $|f(z)| \leq |z|^2$ for all sufficiently large values of $|z|$, say $|z| > r_0$. Prove that f must be a polynomial of degree at most 2. [HINT: Show by the Cauchy estimates that for R sufficiently large $|f^{(3)}(z_0)| \leq 3!(R + |z_0|)^2/R^3$, and thereby conclude that $f^{(3)}$ vanishes everywhere.] Generalize this result.

6. Using formula (46) prove that if the analytic function f satisfies $|f(z_0)| \geq |f(z)|$ for all z on $C_R: z_0 + Re^{it}, 0 \leq t \leq 2\pi$, then there is no point z_1 on C_R for which $|f(z_0)| > |f(z_1)|$. [HINT: If z_1 exists, then by the continuity of f there is an $\varepsilon > 0$ such that $|f(z_0 + Re^{it})| \leq |f(z_0)| - \varepsilon$ over some interval of t. Using this interval divide up the integration in Eq. (46) to reach the contradiction $|f(z_0)| < |f(z_0)|$.]

7. Find all functions f analytic in $D: |z| < R$ which satisfy $f(0) = i$ and $|f(z)| \leq 1$ for all z in D.

8. Suppose that f is analytic inside and on the simple closed curve C and that $|f(z) - 1| < 1$ for all z on C. Prove that f has no zeros inside C.

9. It is proved in Chapter 7 that every nonconstant analytic function maps open sets onto open sets. Using this fact give another proof of the Maximum Principle (Theorem 23).

10. Let f and g be analytic in the bounded domain D and continuous up to and including its boundary B. Suppose that g never vanishes. Prove that if the inequality $|f(z)| \leq |g(z)|$ holds for all z on B, then it must hold for all z in D.

11. Prove the *Minimum Modulus Principle*: Let f be analytic in a bounded domain D and continuous up to and including its boundary. Then *if f is nonzero in D*, the modulus $|f(z)|$ attains its minimum value on the boundary of D. [HINT: Consider the function $1/f(z)$.] Give an example to show why the italicized condition is essential.

12. Let the nonconstant function f be analytic in the bounded domain D and continuous up to and including its boundary B. Prove that if $|f(z)|$ is constant on B, then f must have at least one zero in D.

Polynomial Problems

13. Modify the proof of Lemma 1 to show that, with the same hypotheses, $|P(z)| \leq 3|z|^n/2$ for $|z| \geq 2nA$.

14. Show that $\max_{|z| \leq 1} |az^n + b| = |a| + |b|$.

15. Find $\max_{|z| \leq 1} |(z - 1)(z + \frac{1}{2})|$.

16. Prove that if $P(z)$ is a polynomial of degree n and $P(\alpha) = 0$, then $P(z) = (z - \alpha)Q(z)$, where $Q(z)$ is a polynomial of degree $n - 1$.

[HINT: Observe that $P(z) = P(z) - P(\alpha)$ and that for each integer $k \geq 1$,

$$z^k - \alpha^k = (z - \alpha)(z^{k-1} + z^{k-2}\alpha + \cdots + z\alpha^{k-2} + \alpha^{k-1}).]$$

17. Prove that every polynomial of degree n (≥ 1) has exactly n zeros (counting multiplicity). [HINT: Proceed by induction using the result of Prob. 16.]

18. Prove that if $P(z)$ is a polynomial of degree $\leq n$ which vanishes on the n distinct points $z_1, z_2, \ldots, z_n$, then

$$P(z) = c(z - z_1)(z - z_2) \cdots (z - z_n),$$

where c is a constant.

19. Prove that if $p(z)$ and $q(z)$ are polynomials of degree $\leq n$ which agree on $n + 1$ distinct points, then $q(z) = p(z)$ for all z.

20. Let P be a polynomial which has no zeros on the simple closed positively oriented contour Γ. Prove that the number of zeros of P (counting multiplicity) which lie *inside* Γ is given by the integral

$$\frac{1}{2\pi i} \int_\Gamma \frac{P'(z)}{P(z)} \, dz.$$

[HINT: First show that

$$\frac{P'(z)}{P(z)} = \sum_{k=1}^{n} \frac{1}{z - z_k},$$

where $z_1, z_2, \ldots, z_n$ are all the zeros of P listed according to multiplicity.]

21. Prove that for any polynomial P of the form $P(z) = z^n + a_{n-1}z^{n-1} + \cdots + a_1 z + a_0$, we have $\max_{|z|=1} |P(z)| \geq 1$. [HINT: Consider the polynomial $Q(z) = z^n P(1/z)$, and note that $Q(0) = 1$ and that

$$\max_{|z|=1} |Q(z)| = \max_{|z|=1} |P(z)|.]$$

*4.7 APPLICATIONS TO HARMONIC FUNCTIONS

In the last part of Chapter 2 we discussed the harmonic functions, which are twice-continuously differentiable solutions of Laplace's equation

$$\frac{\partial^2 \phi}{\partial x^2} + \frac{\partial^2 \phi}{\partial y^2} = 0.$$

In particular, we showed that the real and imaginary parts of analytic functions are harmonic, subject to an assumption about their differentiability. This assumption has now been vindicated by Theorem 17. Conversely, we showed how a given harmonic function can be regarded as the real (or imaginary) part of an analytic function by giving a method which constructs the harmonic conjugate, at least in certain elementary domains such as disks. We shall now exploit this interpretation to derive some more facts about harmonic functions.

Our first step will be to extend the harmonic-analytic dualism to simply connected domains, via

THEOREM 25 *Let ϕ be a function harmonic on a simply connected domain D. Then there is an analytic function F such that $\phi = \operatorname{Re} F$ on D.*

Proof For motivation, suppose that we *had* such an analytic function, say, $F = \phi + i\psi$. Then one expression for $F'(z)$ would be given by $F' = \partial\phi/\partial x - i\,\partial\phi/\partial y$ (using the Cauchy-Riemann equations), and F would be an antiderivative of this analytic function.

Accordingly, we begin the proof by defining $f(z) = \partial\phi/\partial x - i\,\partial\phi/\partial y$. We now claim that f satisfies the Cauchy-Riemann equations in D:

$$\frac{\partial}{\partial x}\frac{\partial\phi}{\partial x} = \frac{\partial}{\partial y}\left(-\frac{\partial\phi}{\partial y}\right)$$

because ϕ is harmonic, and

$$\frac{\partial}{\partial y}\frac{\partial\phi}{\partial x} = -\frac{\partial}{\partial x}\left(-\frac{\partial\phi}{\partial y}\right)$$

because of the equality of mixed second partial derivatives. Of course, these partials are continuous since ϕ is harmonic. Hence $f(z)$ is analytic, and by Theorem 10 (or 13) it has an analytic antiderivative $G = u + iv$ in the simply connected domain D. Since $G' = f$, we can write

$$\frac{\partial u}{\partial x} - i\frac{\partial u}{\partial y} = \frac{\partial\phi}{\partial x} - i\frac{\partial\phi}{\partial y},$$

showing that u and ϕ have identical first partial derivatives. Thus by arguing as in the proof of Theorem 5 in Sec. 2.4, we conclude that $\phi - u$ is constant in D; i.e., $\phi = u + c$. It follows that $F(z) \equiv G(z) + c$ is an analytic function of the kind predicted by the theorem. ∎

With Theorem 25 in hand we are fully equipped to study harmonic functions in simply connected domains, using the theory of analytic functions. Varying the notation of the above theorem slightly, we let $\phi(x, y)$ be a harmonic function and $f = \phi + i\psi$ be an "analytic completion" for ϕ

in the simply connected domain D. Now observe the following about the function $e^{f(z)}$:

(48) $$|e^f| = |e^{\phi + i\psi}| = |e^\phi| |e^{i\psi}| = e^\phi.$$

Because the exponential is a monotonically increasing function of a real variable, Eq. (48) implies that the maximum points of ϕ coincide with the maximum points of the modulus of the analytic function e^f. Thus we immediately have a maximum principle for harmonic functions! Furthermore, since the minimum points of ϕ are the same as the maximum points of the harmonic function $-\phi$, we can state the following versions of the *Maximum-Minimum Principle for Harmonic Functions*:

THEOREM 26 *Let ϕ be harmonic in a simply connected domain D and bounded from above (or below) by the number M there. If, for some z_0 in D, $\phi(z_0) = M$, then ϕ is constant in D.*

THEOREM 27 *A function harmonic in a bounded simply connected domain and continuous up to and including the boundary attains its maximum and minimum on the boundary.*

Actually, these principles can easily be extended to multiply connected domains by appropriately modifying the proof of Theorem 23; see Prob. 3. We shall utilize this extended form hereafter.

An important problem that arises in electromagnetism, fluid mechanics, and heat transfer is the

DIRICHLET PROBLEM *Find a function $\phi(x, y)$ continuous on a domain D and its boundary, harmonic in D, and taking specified values on the boundary of D.*

The function ϕ can be interpreted as electric potential, velocity potential, or steady-state temperature. In studying the Dirichlet problem we are concerned with two main questions: Does a solution exist, and, if so, is it uniquely determined by the given boundary values? For the case of bounded domains the question of uniqueness is answered by

THEOREM 28 *Let $\phi_1(x, y)$ and $\phi_2(x, y)$ each be harmonic in a bounded domain D, and continuous on D and its boundary. Furthermore, suppose that $\phi_1 = \phi_2$ on the boundary of D. Then $\phi_1 = \phi_2$ throughout D.*

Proof Consider the harmonic function $\phi = \phi_1 - \phi_2$. It must attain its maximum and minimum on the boundary, but it vanishes there! Hence $\phi = 0$ throughout D. ∎

A solution to the Dirichlet problem could be expressed by a formula giving ϕ inside D in terms of its (specified) values on the boundary. Surely this suggests experimenting with the Cauchy Integral Formula. We are, in fact, able to solve the Dirichlet problems for the disk and the half-plane by this technique. Leaving the latter as an exercise for the reader, we proceed with the former.

For simplicity we consider the disk which is bounded by the positively oriented circle C_R: $|z| = R$. The Cauchy Integral Formula gives the values of an analytic function f inside the disk in terms of its values on the circle:

$$(49) \qquad f(z) = \frac{1}{2\pi i} \int_{C_R} \frac{f(\xi)}{\xi - z} d\xi \qquad (|z| < R)$$

(assuming the domain of analyticity includes the circle C_R as well as its interior). We wish to transform Eq. (49) into a formula which involves only the real part of f. To this end, we observe that for fixed z, with $|z| < R$, the function

$$\frac{f(\xi)\bar{z}}{R^2 - \xi\bar{z}}$$

is an *analytic function of* ξ inside and on C_R (think about this; the denominator does not vanish). Hence by the Cauchy theorem

$$\frac{1}{2\pi i} \int_{C_R} \frac{f(\xi)\bar{z}}{R^2 - \xi\bar{z}} d\xi = 0.$$

The utility of this relationship will become apparent when we add it to Eq. (49):

$$(50) \qquad f(z) = \frac{1}{2\pi i} \int_{C_R} \left(\frac{1}{\xi - z} + \frac{\bar{z}}{R^2 - \xi\bar{z}} \right) f(\xi) \, d\xi$$

$$= \frac{1}{2\pi i} \int_{C_R} \frac{R^2 - |z|^2}{(\xi - z)(R^2 - \xi\bar{z})} f(\xi) \, d\xi.$$

If we parametrize C_R by $\xi = Re^{it}$, $0 \le t \le 2\pi$, Eq. (50) becomes

$$f(z) = \frac{1}{2\pi i} \int_0^{2\pi} \frac{R^2 - |z|^2}{(Re^{it} - z)(R^2 - Re^{it}\bar{z})} f(Re^{it}) Rie^{it} \, dt$$

$$= \frac{R^2 - |z|^2}{2\pi} \int_0^{2\pi} \frac{f(Re^{it})}{(Re^{it} - z)(Re^{-it} - \bar{z})} \, dt$$

$$= \frac{R^2 - |z|^2}{2\pi} \int_0^{2\pi} \frac{f(Re^{it})}{|Re^{it} - z|^2} \, dt.$$

Finally, by taking the real part of this equation, identifying Re f as the harmonic function ϕ, and writing z in the polar form $z = re^{i\theta}$, we arrive at *Poisson's integral formula*, which we state as

THEOREM 29 *Let ϕ be harmonic in a domain containing the disk $|z| \leq R$. Then for $0 \leq r < R$, we have*

$$(51) \qquad \phi(re^{i\theta}) = \frac{R^2 - r^2}{2\pi} \int_0^{2\pi} \frac{\phi(Re^{it})}{R^2 + r^2 - 2rR\cos(t - \theta)} dt.$$

Actually, Poisson's formula is more general than is indicated in this statement. We direct the interested reader to Ref. [8] for a proof of

THEOREM 30 *Let U be a real function defined on the circle C_R: $|z| = R$ and continuous there except for a finite number of jump discontinuities. Then the function*

$$(52) \qquad u(re^{i\theta}) = \frac{R^2 - r^2}{2\pi} \int_0^{2\pi} \frac{U(Re^{it})}{R^2 + r^2 - 2rR\cos(t - \theta)} dt$$

is harmonic inside C_R, and as $re^{i\theta}$ approaches any point on C_R where U is continuous, $u(re^{i\theta})$ approaches the value of U at that point.

Naturally at the points of discontinuity the behavior is more complicated. As an example, consider the harmonic function $\text{Arg}(z + 1)$, which is the imaginary part of $\text{Log}(z + 1)$, in the domain depicted in Fig. 4.53. Clearly, $\text{Arg}(z + 1)$ approaches its boundary values in a reasonable manner except at $z = -1$, where it is erratic (the boundary value jumps from $\pi/2$ to $-\pi/2$ there).

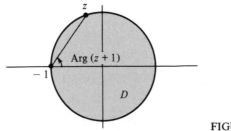

FIGURE 4.53

Thus Poisson's formula solves the Dirichlet problem for the disk under very general circumstances.

EXAMPLE 29 Find the steady-state temperature T at each point inside the unit disk if the temperature is prescribed to be $+1$ on the part of the rim in the first quadrant, and 0 elsewhere. (Fig. 4.54).

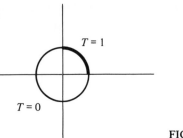

FIGURE 4.54

SOLUTION   The temperature is given by the harmonic function taking the prescribed boundary values. Here formula (52) becomes

$$T(re^{i\theta}) = \frac{1 - r^2}{2\pi} \int_0^{\pi/2} \frac{1}{1 + r^2 - 2r \cos(t - \theta)} \, dt$$

$$= \frac{1}{\pi} \arctan \left[\frac{1 + r}{1 - r} \tan\left(\frac{t - \theta}{2}\right) \right] \Big|_{t=0}^{\pi/2}.$$

For example, at the origin

$$T(0) = \frac{1}{\pi}\left(\frac{\pi}{4} - 0\right) = \frac{1}{4}. \quad \blacksquare$$

Exercises 4.7

1. Imitating the proof of Theorem 5 in Sec. 2.4, show that if the first partial derivatives of $u(x, y)$ and $\phi(x, y)$ agree in a domain D, then u and ϕ differ by a constant.

2. Show by example that a harmonic function need not have an analytic completion in a multiply connected domain. [HINT: Consider $\text{Log}|z|$.]

3. Prove the Maximum-Minimum Principle for harmonic functions in an arbitrary domain. [HINT: Theorem 26 can be used to establish a lemma analogous to Lemma 2 in Sec. 4.6. Then argue as in Theorem 23.]

4. What is the physical interpretation of the Maximum-Minimum Principle in steady-state heat flow?

5. Show by example that the solution to the Dirichlet problem need not be unique for unbounded domains. [HINT: Construct two functions which are harmonic in the upper half-plane, each vanishing on the x-axis.]

6. Prove the *Circumferential Mean-Value Theorem* for harmonic functions: If ϕ is harmonic in a domain containing the disk $|z| \leq \rho$, then

$$\phi(0) = \frac{1}{2\pi} \int_0^{2\pi} \phi(\rho e^{it}) \, dt.$$

7. Prove the following version of the *Solid Mean-Value Theorem* for harmonic functions: If ϕ is harmonic in a domain containing the closed disk $D: |z| \leq R$, then

$$\phi(0) = \frac{1}{\pi R^2} \iint_D \phi \, dx \, dy.$$

[HINT: Multiply the equation in Prob. 6 by $\rho \, d\rho$ and integrate.]

8. Explain why

$$\frac{R^2 - r^2}{2\pi} \int_0^{2\pi} \frac{1}{R^2 + r^2 - 2rR \cos(t - \theta)} \, dt = 1 \qquad \text{for } 0 \leq r < R.$$

9. Prove *Harnack's Inequality*: If $\phi(z)$ is harmonic and nonnegative in a domain containing the disk $|z| \leq R$, then for $0 \leq r < R$

$$\phi(0) \frac{R - r}{R + r} \leq \phi(re^{i\theta}) \leq \phi(0) \frac{R + r}{R - r}.$$

[HINT: Use Poisson's formula and the mean-value property of Prob. 6, observing also that

$$(R - r)^2 \leq R^2 + r^2 - 2rR \cos(t - \theta) \leq (R + r)^2.]$$

10. Prove *Liouville's Theorem for harmonic functions*: If ϕ is harmonic in the whole plane and bounded from above or below there, then ϕ is constant. [HINT: Modify ϕ so that Harnack's inequality (Prob. 9) can be applied.]

11. The temperature of the rim of the unit disk is maintained at the levels indicated in Fig. 4.55. What is the temperature at the center?

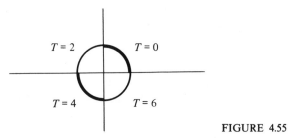

FIGURE 4.55

12. Consider the following version of Poisson's integral formula for the *half-plane*: If $f(z) = \phi + i\psi$ is analytic in a domain containing the x-axis and the upper half-plane and $|f(z)| \le K$ in this domain, then the values of the harmonic function ϕ in the upper half-plane are given in terms of its values on the x-axis by

$$\phi(x, y) = \frac{y}{\pi} \int_{-\infty}^{\infty} \frac{\phi(\xi, 0)\, d\xi}{(\xi - x)^2 + y^2} \qquad (y > 0).$$

Here is an outline of the derivation; justify the steps.

(a) For the situation depicted in Fig. 4.56,

$$f(z) = \frac{1}{2\pi i} \int_{\Gamma_R} \frac{f(\xi)}{\xi - z}\, d\xi.$$

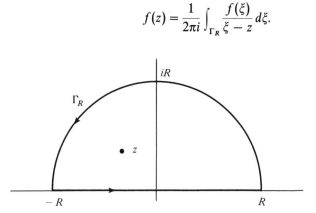

FIGURE 4.56

(b) For the same situation,

$$0 = \frac{1}{2\pi i} \int_{\Gamma_R} \frac{f(\xi)}{\xi - \bar{z}}\, d\xi.$$

(c) Subtract these two equations to conclude that

$$f(z) = \frac{1}{2\pi i} \int_{-R}^{R} f(\xi) \frac{2i\, \mathrm{Im}\, z}{|\xi - z|^2}\, d\xi$$

$$+ \frac{1}{2\pi i} \int_{C_R} f(\xi) \frac{2i\, \mathrm{Im}\, z}{(\xi - z)(\xi - \bar{z})}\, d\xi,$$

where C_R is the semicircular portion of Γ_R.

(d) Show that the integral along C_R is bounded by

$$\frac{K}{\pi} \frac{\mathrm{Im}\, z}{(R - |z|)^2}\, \pi R.$$

(e) Let $R \to \infty$ in the last equation and take the real part.

13. Use the Poisson formula of Prob. 12 to find the temperature in the upper half-plane if the temperature on the x-axis is maintained as in Fig. 4.57.

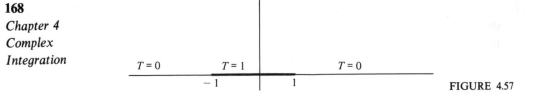

FIGURE 4.57

14. (For advanced physics students) In electrostatics, the Maximum-Minimum Principle provides some information about the stability of equilibrium electrical charge configurations. Give the interpretation.

SUMMARY

Integration in the complex plane takes place along contours, which are continuous chains of directed smooth curves. The definite integral over each smooth part is defined by imitating the Riemann sum definition used in calculus, and the contour integral $\int_\Gamma f(z)\, dz$ is the sum of the integrals over the smooth components of Γ.

A "brute-force" technique for computing an integral along a contour Γ involves finding a parametrization $z = z(t)$, $a \le t \le b$, for Γ; then the contour integral can be obtained by performing an integration with respect to the real variable t, in accordance with the formula

$$\int_\Gamma f(z)\, dz = \int_a^b f(z(t))z'(t)\, dt.$$

When the integrand $f(z)$ is analytic in a domain containing Γ, the following considerations may be useful in evaluating $\int_\Gamma f(z)\, dz$:

(i) (Fundamental Theorem of Calculus.) If f has an antiderivative F in a domain containing Γ, then

$$\int_\Gamma f(z)\, dz = F(z_T) - F(z_I),$$

where z_I and z_T are the initial and terminal points of Γ.

(ii) (Cauchy's Theorem.) If Γ is a simple closed contour (i.e., $z_I = z_T$ but no other self-intersections occur) and f is analytic inside and on Γ, then $\int_\Gamma f(z)\, dz = 0$.

(iii) (Cauchy's Integral Formula.) If Γ is a simple closed positively oriented contour and f has the form $f(z) = g(z)/(z - z_0)$, with g analytic inside and on Γ and z_0 lying inside Γ, then

$$\int_\Gamma f(z)\, dz = \int_\Gamma \frac{g(z)}{z - z_0}\, dz = 2\pi i g(z_0).$$

More generally,

$$\int_\Gamma \frac{g(z)\,dz}{(z-z_0)^m} = \frac{2\pi i g^{(m-1)}(z_0)}{(m-1)!}.$$

(*iv*) If f is of the form $f(z) = g(z)/(z-z_1)^{m_1}(z-z_2)^{m_2}$, with g as in (*iii*) and the points z_1 and z_2 lying inside the simple closed contour Γ, then $\int_\Gamma f(z)\,dz$ can be reduced to case (*iii*) by using partial fractions.

(*v*) (Deformation Invariance Theorem.) If the closed contour Γ can be continuously deformed to another closed contour Γ' without passing through any singularities of f, then

$$\int_\Gamma f(z)\,dz = \int_{\Gamma'} f(z)\,dz.$$

When the domain of analyticity of f is simply connected (i.e., has no "holes"), then f has an antiderivative and integrals of f are independent of path.

The Cauchy Integral Formula has many consequences for analytic functions. Among these are the infinite differentiability of analytic functions (see (*iii*) above), Liouville's theorem (bounded entire functions are constant), the Fundamental Theorem of Algebra (every nonconstant polynomial has a zero), and the Maximum Modulus Theorem (the maximum of $|f|$ is attained on the boundary of a bounded domain).

In simply connected domains harmonic functions can be identified as real parts of analytic functions. Hence these results have analogues for harmonic functions, such as infinite differentiability, maximum principles, and Poisson's formula (the analogue of Cauchy's Integral Formula).

SUGGESTED READING

The following references will be helpful to the reader interested in seeing a more detailed treatment of the topics of this chapter:

Theory of Curves, Arc Length

[1] APOSTOL, T. M. *Mathematical Analysis.* Addison-Wesley Publishing Company, Inc., Reading, Mass., 1957.

[2] NEVANLINNA, R., and PAATERO, V. *Introduction to Complex Analysis.* Addison-Wesley Publishing Company, Inc., Reading, Mass., 1969.

Jordan Curve Theorem

[3] PEDERSON, R. N. "The Jordan Curve Theorem for Piecewise Smooth Curves," *American Mathematical Monthly,* 76 (1969), 605–610.

Riemann Sums, Integration

Reference 2 above.

Continuous Deformation of Curves, Cauchy's Theorem

[4] FLATTO, L., and SHISHA, O. "A Proof of Cauchy's Integral Theorem," *Journal of Approximation Theory*, 7 (1973), 386–390.
Reference 2 above.

Leibniz's Rule for Integrals

[5] TAYLOR, A. E., and MANN, W. R. *Advanced Calculus*, 2nd ed. Xerox College Publishing, Lexington, Mass., 1972.

Green's Theorem

[6] BORISENKO, A. I., and TARAPOV, I. E. *Vector and Tensor Analysis with Applications* (trans. from Russian by R. A. Silverman). Prentice-Hall, Inc., Englewood Cliffs, N.J., 1968.

[7] DAVIS, H., and SNIDER, A. D. *Introduction to Vector Analysis*, 3rd ed. Allyn and Bacon, Inc., Boston, 1975.

Poisson's Formula

[8] HILLE, E. *Analytic Function Theory*, Vol. 2. Ginn and Company, Boston, 1962.

Additional Problems

[9] SPIEGEL, M. R. *Theory and Problems of Complex Variables*. Schaum's Outline Series, McGraw-Hill Book Company, New York, 1964.

5

Series Representations for Analytic Functions

5.1 SEQUENCES AND SERIES

In Chapter 2 we defined what is meant by convergence of a sequence of complex numbers; recall that the sequence $\{A_n\}_{n=1}^{\infty}$ approaches A as a limit if $|A - A_n|$ can be made arbitrarily small by taking n large enough. For computational convenience it is often advantageous to use an element A_n of the sequence as an approximation to A. Indeed, when we calculate the area of a circle we usually use an element of the sequence 3.14, 3.141, 3.1415, 3.14159, ... as an approximation to π. The use of sequences, and in particular the kind of sequences associated with *series*, is an important tool in both the theory and applications of analytic functions, and the present chapter is devoted to the development of this subject.

The possibility of summing an infinite string of numbers must have occurred to anyone who has toyed with adding

$$\frac{1}{2} + \frac{1}{4} = \frac{3}{4}, \qquad \frac{1}{2} + \frac{1}{4} + \frac{1}{8} = \frac{7}{8}, \qquad \frac{1}{2} + \frac{1}{4} + \frac{1}{8} + \frac{1}{16} = \frac{15}{16}, \qquad \text{etc.}$$

The sequence of sums thus derived obviously approaches 1 as a limit, and it seems sensible to say $\frac{1}{2} + \frac{1}{4} + \frac{1}{8} + \cdots = 1$. We are motivated to generalize this by saying that any *infinite series* of the form $c_0 + c_1 + c_2 + \cdots$ has the sum S if the sums of the first n terms approach S as a limit as n goes to infinity. The customary nomenclature is summarized in

DEFINITION 1 *A* **series** *is a formal expression of the form* $c_0 + c_1 + c_2 + \cdots$, *or equivalently* $\sum_{j=0}^{\infty} c_j$, *where the* **terms** c_j *are complex numbers. The* **nth partial sum** *of the series, usually denoted* S_n, *is the sum of the first $n + 1$ terms, i.e.,* $S_n = \sum_{j=0}^{n} c_j$. *If the sequence of partial sums* $\{S_n\}_{n=0}^{\infty}$ *has a limit S, the series is said to* **converge**, *or* **sum**, *to S, and we write $S = \sum_{j=0}^{\infty} c_j$. A series which does not converge is said to* **diverge**.

Notice that the notion of convergence for a series has been defined in terms of convergence for a sequence. As an illustration, observe that π is the sum of the series $3.14 + 0.001 + 0.0005 + 0.00009 + \cdots$.

Clearly one way to demonstrate that a series converges to S is to show that the *remainder* after summing the first $n + 1$ terms, $S - \sum_{j=0}^{n} c_j$, goes to zero as $n \to \infty$. We use this technique in proving

LEMMA 1 *The series* $\sum_{j=0}^{\infty} c^j$ *converges to* $1/(1 - c)$ *if* $|c| < 1$. (A series of this form is called a **geometric series**. In Prob. 7 we shall see that such a series diverges if $|c| \geq 1$.)

Proof   Observe that

$$(1 - c)(1 + c + c^2 + \cdots + c^{n-1} + c^n)$$

$$= 1 + c + c^2 + \cdots + c^{n-1} + c^n - c - c^2 - \cdots - c^{n-1} - c^n - c^{n+1}$$

$$= 1 - c^{n+1}.$$

Rearranging this yields

$$(1) \qquad \frac{1}{1 - c} - (1 + c + c^2 + \cdots + c^{n-1} + c^n) = \frac{c^{n+1}}{1 - c}.$$

Since $|c| < 1$, the lemma follows immediately; Eq. (1) displays the remainder as $c^{n+1}/(1 - c)$, which certainly goes to zero as $n \to \infty$ (recall Prob. 7 in Exercises 2.2).   ∎

Another important way to establish the convergence of a series involves comparing it with another series whose convergence is known. The following theorem, which generalizes a result from calculus, seems so transparent that we shall spare our trusting readers the proof and refer the skeptics to the (optional) Sec. 5.4.

THEOREM 1 (Comparison Test) *If $\sum_{j=0}^{\infty} M_j$ is a convergent series with real nonnegative terms and if, for all j larger than some number J, $|c_j| \le M_j$, then the series $\sum_{j=0}^{\infty} c_j$ converges also.*

EXAMPLE 1 Show that the series $\sum_{j=0}^{\infty} (3 + 2i)/(j + 1)^j$ converges.

SOLUTION We compare the series

(2) $$\sum_{j=0}^{\infty} \frac{3 + 2i}{(j + 1)^j} = (3 + 2i) + \frac{(3 + 2i)}{2} + \frac{(3 + 2i)}{9} + \frac{(3 + 2i)}{64} + \cdots$$

with the *convergent* geometric series

(3) $$\sum_{j=0}^{\infty} \frac{1}{2^j} = 1 + \frac{1}{2} + \frac{1}{4} + \frac{1}{8} + \cdots.$$

Since $|3 + 2i| = \sqrt{13} < 4$, the reader can easily verify that for $j \ge 3$

$\sqrt{9^{5}+4}$

$$\left| \frac{3 + 2i}{(j + 1)^j} \right| < \frac{4}{(j + 1)^j} \le \frac{1}{2^j};$$

i.e., the terms of (3) dominate those of (2); hence (2) converges. ∎

Sometimes the *Ratio Test* can be applied to a series to establish convergence:

THEOREM 2 (Ratio Test) *Suppose that the terms of the series $\sum_{j=0}^{\infty} c_j$ have the property that the ratios $|c_{j+1}/c_j|$ approach a limit L as $j \to \infty$. Then the series converges if $L < 1$ and diverges if $L > 1$.*

The proof involves comparing the given series with a series obtained by judiciously modifying the geometric series $\sum_{j=0}^{\infty} L^j$. See Prob. 16.

EXAMPLE 2 Show that the series $\sum_{j=0}^{\infty} 4^j/j!$ converges.

SOLUTION We have

$$\left| \frac{c_{j+1}}{c_j} \right| = \frac{4^{j+1}}{(j + 1)!} \frac{j!}{4^j} = \frac{4}{j + 1}.$$

This ratio approaches zero as $j \to \infty$; thus the series converges. ∎

For completeness, we add the remark that a series $\sum_{j=0}^{\infty} c_j$ is said to be *absolutely convergent* if the series $\sum_{j=0}^{\infty} |c_j|$ converges. Any absolutely convergent series is convergent, by a trivial application of the Comparison Test.

The kinds of sequences and series which arise in complex analysis are those where the terms are functions of a complex variable z. Thus if we have a sequence of functions $F_1(z)$, $F_2(z)$, $F_3(z)$, ..., we must consider the possibility that for some values of z the sequence converges, while for others it diverges. As an example, the sequence $(z/2i)^n$, $n = 1, 2, 3, ...$, approaches zero for $|z| < |2i| = 2$, approaches 1 for $z = 2i$ (obviously!), and has no limit otherwise (see Prob. 5). Similarly, a *series* of complex functions $\sum_{j=0}^{\infty} f_j(z)$ may converge for some values of z and diverge for others.

EXAMPLE 3   If z_0 ($\neq 0$) is fixed, show that the series $\sum_{j=0}^{\infty} (z/z_0)^j$ [which is quite distinct from the *sequence* $(z/z_0)^j$] converges for $|z| < |z_0|$.

SOLUTION   This is merely a thinly disguised resurrection of Lemma 1. In fact, setting $c = z/z_0$ in Eq. (1) yields

$$(4) \qquad \frac{1}{1 - \dfrac{z}{z_0}} - \left[1 + \frac{z}{z_0} + \left(\frac{z}{z_0}\right)^2 + \cdots + \left(\frac{z}{z_0}\right)^n\right] = \frac{\left(\dfrac{z}{z_0}\right)^{n+1}}{1 - \dfrac{z}{z_0}}.$$

We conclude, as before, that for $|z| < |z_0|$ the series sums to the function $1/(1 - z/z_0)$. ∎

In applying this theory to analytic functions we need a somewhat stronger notion of convergence. To see this, let us carefully restate the definition of convergence for a sequence of functions: We said that the sequence $\{F_n(z)\}_{n=1}^{\infty}$ converges to $F(z)$ at the point $z = z_1$ if, for any prespecified $\varepsilon > 0$, we can find an integer N such that $|F(z_1) - F_n(z_1)|$ will be no greater than ε for all n larger than N. So for definiteness, let us say that the sequence $F_n(z)$ approaches $F(z)$ for each z in a set $\mathcal{S}$ (i.e., $F_n(z) \to F(z)$ pointwise on $\mathcal{S}$), and we wish to use the members of the sequence to approximate F on $\mathcal{S}$. We encounter the following difficulty: Although at any particular point z_1 we can choose N as above to achieve a given accuracy ε, we may have to take a higher N to get ε-accuracy at another point z_2—and higher still at each succeeding point. In fact, if $\mathcal{S}$ contains an infinite number of points, we may be unable to find a satisfactory N which will assure ε-accuracy over *all* of $\mathcal{S}$. The kind of convergence we really need is given in

DEFINITION 2   *The sequence* $\{F_n(z)\}_{n=1}^{\infty}$ *is said to* **converge uniformly to** $F(z)$ **on the set** $\mathcal{S}$ *if for any $\varepsilon > 0$ there exists an integer N such that when* $n > N$, $|F(z) - F_n(z)| < \varepsilon$ *for* **all** z *in* $\mathcal{S}$. *[Accordingly, the series* $\sum_{j=0}^{\infty} f_j(z)$ *converges uniformly to* $f(z)$ *on* $\mathcal{S}$ *if the sequence of partial sums converges uniformly to* $f(z)$ *there.]*

EXAMPLE 4 Show that the series of Example 3, $\sum_{j=0}^{\infty} (z/z_0)^j$, is uniformly convergent in every closed disk $|z| \le r < |z_0|$.

SOLUTION Given $\varepsilon > 0$, we have to show that the remainder after $n + 1$ terms will be less than ε for all z in the disk, when n is large enough. This is easy; from Eq. (4) we can find an upper bound, independent of z, for the remainder:

$$\left| \frac{\left(\dfrac{z}{z_0}\right)^{n+1}}{1 - \dfrac{z}{z_0}} \right| \le \frac{\left(\dfrac{r}{|z_0|}\right)^{n+1}}{1 - \dfrac{r}{|z_0|}}, \qquad \text{for } |z| \le r.$$

This can be made arbitrarily small since $r < |z_0|$. ∎

Exercises 5.1

1. Find the sum of the following convergent series.

(a) $\displaystyle\sum_{j=0}^{\infty} \left(\frac{i}{3}\right)^j$

(b) $\displaystyle\sum_{k=0}^{\infty} \frac{3}{(1+i)^k}$

(c) $\displaystyle\sum_{j=0}^{\infty} (-1)^j \left(\frac{2}{3}\right)^j$

(d) $\displaystyle\sum_{k=14}^{\infty} \left(\frac{1}{2i}\right)^k$

(e) $\displaystyle\sum_{j=0}^{\infty} \left(\frac{1}{3}\right)^{2j}$

(f) $\displaystyle\sum_{j=0}^{\infty} \left[\frac{1}{j+2} - \frac{1}{j+1}\right]$

2. How many terms of the series $\sum_{j=0}^{\infty} (z/z_0)^j$ would have to be used to compute $z_0/(z_0 - z)$ to four decimal places if $z_0 = 3$ and $|z| \le \frac{3}{2}$?

3. Using the Ratio Test, show that the following series converge.

(a) $\displaystyle\sum_{j=1}^{\infty} \frac{1}{j!}$

(b) $\displaystyle\sum_{k=1}^{\infty} \frac{(3+i)^k}{k!}$

(c) $\displaystyle\sum_{j=0}^{\infty} \frac{j^2}{4^j}$

(d) $\displaystyle\sum_{k=1}^{\infty} \frac{k!}{k^k}$

4. Prove that if the sequence $\{z_n\}_{n=1}^{\infty}$ converges, then $(z_n - z_{n-1}) \to 0$ as $n \to \infty$.

5. Let $z_0 \ne 0$. Prove that the sequence $(z/z_0)^n$, $n = 1, 2, \ldots$, diverges if $|z| \ge |z_0|$, $z \ne z_0$. [HINT: For $|z| = |z_0|$, observe that

$$\left| \left(\frac{z}{z_0}\right)^n - \left(\frac{z}{z_0}\right)^{n-1} \right| = \left| \frac{z}{z_0} - 1 \right| > 0.]$$

6. Prove that if the series $\sum_{j=0}^{\infty} c_j$ converges, then $c_j \to 0$ as $j \to \infty$. [HINT: Consider the difference $S_n - S_{n-1}$ of consecutive partial sums.]

7. Prove that the series $\sum_{j=0}^{\infty} c^j$ diverges if $|c| \geq 1$. [HINT: See Prob. 6.]

8. For each of the following determine if the given series converges or diverges.

(a) $\sum_{k=0}^{\infty} \left(\frac{1+2i}{1-i}\right)^k$ (b) $\sum_{j=1}^{\infty} \frac{1}{j^2 3^j}$

(c) $\sum_{n=1}^{\infty} \frac{ni^n}{2n+1}$ (d) $\sum_{j=1}^{\infty} \frac{j!}{5^j}$

(e) $\sum_{k=1}^{\infty} \frac{(-1)^k k^3}{(1+i)^k}$

9. Prove the following statements.
(a) If $\sum_{j=0}^{\infty} c_j$ sums to S, then $\sum_{j=0}^{\infty} \bar{c}_j$ sums to $\bar{S}$.
(b) If $\sum_{j=0}^{\infty} c_j$ sums to S and λ is any complex number, then $\sum_{j=0}^{\infty} \lambda c_j$ sums to λS.
(c) If $\sum_{j=0}^{\infty} c_j$ sums to S and $\sum_{j=0}^{\infty} d_j$ sums to T, then $\sum_{j=0}^{\infty} (c_j + d_j)$ sums to $S + T$.

10. Prove that the series $\sum_{j=0}^{\infty} z_j$ converges if and only if both of the series $\sum_{j=0}^{\infty} \text{Re}(z_j)$ and $\sum_{j=0}^{\infty} \text{Im}(z_j)$ converge.

11. Show that the sequence of functions $F_n(z) = z^n/(z^n - 3^n)$, $n = 1$, $2, \ldots$, converges to zero for $|z| < 3$ and to 1 for $|z| > 3$.

12. Using the Ratio Test, find a domain in which convergence holds for each of the following series of functions.

(a) $\sum_{j=1}^{\infty} jz^j$ (b) $\sum_{k=0}^{\infty} \frac{(z-i)^k}{2^k}$

(c) $\sum_{j=0}^{\infty} \frac{z^j}{j!}$ (d) $\sum_{k=0}^{\infty} (z+5i)^{2k}(k+1)^2$

13. Let $F_n(z) = [nz/(n+1)] + (3/n)$, $n = 1, 2, \ldots$. Prove that the sequence $\{F_n(z)\}_1^{\infty}$ converges uniformly to $F(z) = z$ on every closed disk $|z| \leq R$.

14. Prove that $\sum_{j=1}^{\infty} 1/j^p$ converges if $p > 1$. [HINT: Interpret the integral $\int_1^N (1/x^p)\, dx$ as an area; then interpret $\sum_{j=2}^{N} 1/j^p$ as an area, and compare.]

15. Using the Comparison Test and the result of Prob. 14, show that the following series converge.

(a) $\displaystyle\sum_{j=1}^{\infty} \frac{1}{j(j+i)}$ (b) $\displaystyle\sum_{k=1}^{\infty} \frac{\sin(k^2)}{k^{3/2}}$

(c) $\displaystyle\sum_{k=1}^{\infty} \frac{k^2 i^k}{k^4+1}$ (d) $\displaystyle\sum_{k=2}^{\infty} (-1)^k \left(\frac{5k+8}{k^3-1}\right)$

16. Prove the Ratio Test (Theorem 2). [HINT: If $L < 1$, choose $\varepsilon > 0$ and J so that $|c_{j+1}/c_j| < L + \varepsilon < 1$ for $j \geq J$. Then show that $|c_k| \leq |c_J|(L+\varepsilon)^{k-J}$ for $k > J$ and use the Comparison Test.]

17. Prove that the sequence $\{z_n\}_1^{\infty}$ converges if and only if the series $\sum_{k=1}^{\infty} (z_{k+1} - z_k)$ converges.

18. Assume that the sequence of functions $\{F_n(z)\}_1^{\infty}$ converges uniformly to $F(z)$ on a set $\mathcal{S}$. Prove that if $|F(z)| \geq \rho > 0$ for all z on $\mathcal{S}$, then there exists an integer N such that for $n > N$ the inequality $|F_n(z)| > \rho/2$ holds for all z on $\mathcal{S}$. [HINT: Take $\varepsilon = \rho/2$ in Definition 2 and apply the triangle inequality.]

19. It will be shown in the next section that the series $\sum_{k=0}^{\infty} z^k/k!$ converges uniformly to e^z on every disk $|z| \leq R$. Accepting this fact, prove that, for n sufficiently large, none of the polynomials $S_n(z) = \sum_{k=0}^{n} z^k/k!$ has zeros on $|z| \leq 5$. [HINT: Use Prob. 18.]

5.2 TAYLOR SERIES

Suppose that we wish to find a polynomial $p_n(z)$ of degree at most n which approximates an analytic function $f(z)$ in a neighborhood of a point z_0. Naturally there are differing criteria as to how well the polynomial approximates the function. We shall construct a polynomial which "looks like" $f(z)$ at the point z_0 in the sense that its derivatives match those of f at z_0, insofar as possible:

$$p_n(z_0) = f(z_0)$$
$$p_n'(z_0) = f'(z_0)$$
$$\vdots$$
$$p_n^{(n)}(z_0) = f^{(n)}(z_0).$$

[Of course, the $(n+1)$st derivative of any polynomial of degree n must equal *zero*.]

The constant polynomial $p_0(z)$ which matches f at z_0 is simply $f(z_0)$:

$$p_0(z) \equiv f(z_0).$$

The polynomial of degree 1 which matches f and f' at z_0 is

$$p_1(z) = f(z_0) + f'(z_0)(z - z_0).$$

The second-degree polynomial matching f, f', and f'' at z_0 is

$$p_2(z) = f(z_0) + f'(z_0)(z - z_0) + \frac{f''(z_0)}{2!}(z - z_0)^2.$$

From this we can see a pattern emerging; the third-degree polynomial is

$$p_3(z) = f(z_0) + f'(z_0)(z - z_0) + \frac{f''(z_0)}{2!}(z - z_0)^2 + \frac{f'''(z_0)}{3!}(z - z_0)^3,$$

and *the nth degree polynomial which matches $f, f', f'', \ldots, f^{(n)}$ at z_0 is*

$$(5) \qquad p_n(z) = f(z_0) + f'(z_0)(z - z_0) + \frac{f''(z_0)}{2!}(z - z_0)^2 + \cdots$$

$$+ \frac{f^{(n)}(z_0)}{n!}(z - z_0)^n.\dagger$$

Naturally we conjecture that as n tends to infinity, $p_n(z)$ becomes a better and better approximation to $f(z)$ near z_0. In fact, the astute reader may have noticed that $p_n(z)$ looks like a partial sum of a series, and he might anticipate that this series converges to $f(z)$. The precise state of affairs is given in the following definition and theorem:

DEFINITION 3 *If $f(z)$ is analytic at z_0, then the series*

$$(6) \quad f(z_0) + f'(z_0)(z - z_0) + \frac{f''(z_0)}{2!}(z - z_0)^2 + \cdots = \sum_{j=0}^{\infty} \frac{f^{(j)}(z_0)}{j!}(z - z_0)^j$$

*is called the **Taylor series** for f around z_0. When $z_0 = 0$, (6) is also known as the **Maclaurin series** for f.*

THEOREM 3 *If $f(z)$ is analytic in the disk $|z - z_0| < R$, then the Taylor series (6) converges to $f(z)$ for all z in this disk. Furthermore, the convergence of the series is uniform in any closed subdisk $|z - z_0| \le R' < R$.*

Proof Notice that if we prove uniform convergence in *every* closed subdisk $|z - z_0| \le R' < R$, we will have convergence for each z in the open disk $|z - z_0| < R$ (why?); thus we deal only with the closed subdisk statement. Let C be the circle $|z - z_0| = (R + R')/2$, positively oriented (see Fig. 5.1). Then by Cauchy's Integral Formula we have, for any z within the closed subdisk,

$$f(z) = \frac{1}{2\pi i}\int_C \frac{f(\xi)}{\xi - z}\, d\xi.$$

---

†Actually $p_n(z)$ will have degree less than n if $f^{(n)}(z_0) = 0$.

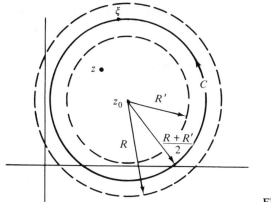

FIGURE 5.1

We modify the integrand by writing

(7)
$$\frac{1}{\xi - z} = \frac{1}{(\xi - z_0) - (z - z_0)} = \frac{1}{\xi - z_0} \cdot \frac{1}{1 - \dfrac{z - z_0}{\xi - z_0}}$$

$$= \frac{1}{\xi - z_0}\left[1 + \frac{z - z_0}{\xi - z_0} + \frac{(z - z_0)^2}{(\xi - z_0)^2} + \cdots + \frac{(z - z_0)^n}{(\xi - z_0)^n} + \frac{\dfrac{(z - z_0)^{n+1}}{(\xi - z_0)^{n+1}}}{1 - \dfrac{z - z_0}{\xi - z_0}} \right],$$

using identity (1) with $c = (z - z_0)/(\xi - z_0)$. Putting this into the integral formula we find

(8)
$$f(z) = \frac{1}{2\pi i} \int_C \frac{f(\xi)}{\xi - z_0}\, d\xi + \frac{z - z_0}{2\pi i} \int_C \frac{f(\xi)}{(\xi - z_0)^2}\, d\xi$$

$$+ \frac{(z - z_0)^2}{2\pi i} \int_C \frac{f(\xi)}{(\xi - z_0)^3}\, d\xi + \cdots + \frac{(z - z_0)^n}{2\pi i} \int_C \frac{f(\xi)}{(\xi - z_0)^{n+1}}\, d\xi$$

$$+ T_n(z),$$

where $T_n(z)$ can be expressed (with a little algebra) as

(9)
$$T_n(z) = \frac{1}{2\pi i} \int_C \frac{f(\xi)}{(\xi - z)} \frac{(z - z_0)^{n+1}}{(\xi - z_0)^{n+1}}\, d\xi.$$

Now by Cauchy's integral formula for derivatives

$$\frac{1}{2\pi i} \int_C \frac{f(\xi)}{\xi - z_0} = f(z_0), \qquad \frac{1}{2\pi i} \int_C \frac{f(\xi)}{(\xi - z_0)^2}\, d\xi = f'(z_0),$$

$$\frac{1}{2\pi i} \int_C \frac{f(\xi)}{(\xi - z_0)^3}\, d\xi = \frac{f''(z_0)}{2!}, \text{ etc.,}$$

and so the first $n + 1$ terms on the right-hand side of Eq. (8) yield the nth partial sum of the series in (6). Thus $T_n(z)$ is the remainder, and we must show that it can be made "uniformly small" for all z in the subdisk, by taking n sufficiently large.

This is simply a matter of applying the integral inequality of Chapter 4; for the terms in the integrand of Eq. (9) we have

$$|z - z_0| \leq R', \qquad |\xi - z_0| = \frac{R + R'}{2},$$

and, from Fig. 5.1,

$$|\xi - z| \geq \frac{R + R'}{2} - R' = \frac{R - R'}{2}.$$

The length of C is $2\pi(R + R')/2$. Thus for z in the subdisk

$$|T_n(z)| \leq \frac{1}{2\pi} \cdot \max_{\xi \text{ on } C} |f(\xi)| \frac{2}{R - R'} \left(\frac{2R'}{R + R'} \right)^{n+1} 2\pi \left(\frac{R + R'}{2} \right).$$

The right-hand side is independent of z, and since $2R' < R + R'$, it can be made less than any positive ε by taking n large enough. ■

Notice that the theorem implies that *the Taylor series will converge to $f(z)$ everywhere inside the largest open disk, centered at z_0, over which f is analytic.*

EXAMPLE 5 Compute and state the convergence properties of the Taylor series for (a) Log z around $z_0 = 1$ and (b) $1/(1 - z)$ around $z_0 = 0$.

SOLUTION (a) The consecutive derivatives of Log z are Log z, z^{-1}, $-z^{-2}$, $2z^{-3}$, $-3 \cdot 2z^{-4}$, etc.; in general,

$$\frac{d^j \text{ Log } z}{dz^j} = (-1)^{j+1}(j - 1)! z^{-j} \qquad (j = 1, 2, \ldots).$$

Evaluating these at $z = 1$ we find

$$(10) \qquad \text{Log } z = 0 + (z - 1) - \frac{(z - 1)^2}{2!} + 2! \frac{(z - 1)^3}{3!} - 3! \frac{(z - 1)^4}{4!} + \cdots$$

$$= \sum_{j=1}^{\infty} \frac{(-1)^{j+1}(z - 1)^j}{j}.$$

This is valid for $|z - 1| < 1$, the largest open disk centered at $+1$ over which Log z is analytic.

(b) The consecutive derivatives of $(1 - z)^{-1}$ are $(1 - z)^{-1}$, $(1 - z)^{-2}$, $2(1 - z)^{-3}$, $3 \cdot 2(1 - z)^{-4}$, and, in general,

$$\frac{d^j}{dz^j} (1 - z)^{-1} = j! (1 - z)^{-j-1}.$$

Evaluating these at $z = 0$ gives the Taylor series

(11)
$$\frac{1}{1-z} = 1 + z + \frac{2!z^2}{2!} + \frac{3!z^3}{3!} + \cdots = \sum_{j=0}^{\infty} z^j,$$

valid for $|z| < 1$ (why?). In fact, this is just the geometric series considered in Lemma 1. ∎

Let's experiment with Eq. (10), the expansion for Log z. If we differentiate the *series* term by term, we get

(12)
$$1 - (z-1) + (z-1)^2 - (z-1)^3 + \cdots = \sum_{j=0}^{\infty} (1-z)^j,$$

properly incorporating the negative signs. But the series (12) has the same form as the series in Eq. (11), with $(1-z)$ in place of z. Thus (12) actually converges to the function $1/[1 - (1-z)] = 1/z$; in other words, by differentiating the *series* for Log z we obtained the *series* for $1/z$, which is, in fact, the derivative of Log z. We are led to conjecture

THEOREM 4 *If f is analytic at z_0, the Taylor series for f' around z_0 can be obtained by termwise differentiation of the Taylor series for f around z_0, and converges in the same disk as the series for f.*

Proof The jth derivative of f' is, of course, the $(j+1)$st derivative of f. Thus the Taylor series for f' is given by

(13)
$$f'(z_0) + f''(z_0)(z-z_0) + \frac{f'''(z_0)}{2!}(z-z_0)^2 + \cdots.$$

On the other hand, termwise differentiation of the Taylor series (6) yields

$$f'(z_0) + f''(z_0)(z-z_0) + \frac{f'''(z_0)}{2!}(z-z_0)^2 + \cdots,$$

which is the same as (13). Furthermore, application of Theorem 3 to the function $f'(z)$ establishes that (13) converges in the largest open disk around z_0 over which f' is analytic. But according to the theory of Chapter 4, f' is analytic wherever f is analytic. This completes the proof. ∎

EXAMPLE 6 Find the Maclaurin series for sin z and cos z.

SOLUTION We expand sin z as usual. The sequence of derivatives is sin z, cos z, $-\sin z$, $-\cos z$, sin z, ... Evaluating at the origin yields

(14)
$$\sin z = z - \frac{z^3}{3!} + \frac{z^5}{5!} - \frac{z^7}{7!} + \cdots,$$

which holds for all z since $\sin z$ is entire. To get $\cos z$, we differentiate Eq. (14):

$$\text{(15)} \qquad \cos z = 1 - \frac{z^2}{2!} + \frac{z^4}{4!} - \frac{z^6}{6!} + \cdots.$$

The reader should by now be able to predict what would result if we differentiate Eq. (15). ∎

The next two theorems may sometimes simplify the computation of a Taylor series.

THEOREM 5 *Let f and g be analytic functions with Taylor series $f(z) = \sum_{j=0}^{\infty} a_j(z - z_0)^j$ and $g(z) = \sum_{j=0}^{\infty} b_j(z - z_0)^j$ around the point z_0 [i.e., $a_j = f^{(j)}(z_0)/j!$ and $b_j = g^{(j)}(z_0)/j!$]. Then*

(i) *The Taylor series for $cf(z)$, c a constant, is $\sum_{j=0}^{\infty} ca_j(z - z_0)^j$, and*

(ii) *The Taylor series for $f(z) \pm g(z)$ is $\sum_{j=0}^{\infty} (a_j \pm b_j)(z - z_0)^j$.*

The proof is left as an easy exercise. The disk of convergence for $f \pm g$ is, of course, at least as big as the smaller of the convergence disks for f and g.

EXAMPLE 7 Find the Maclaurin series for $\sin z + i \cos z$.

SOLUTION Using the expansions (14) and (15) we find

$$\sin z + i \cos z = i + z - \frac{iz^2}{2!} - \frac{z^3}{3!} + \frac{iz^4}{4!} + \frac{z^5}{5!} - \cdots$$

for all z. ∎

Theorem 5 naturally leads us to cogitate the corresponding statement for products. First we must find a sensible way of multiplying two Taylor series. The *Cauchy product* of two Taylor series around a point z_0 is defined in the manner suggested by applying the distributive law and then grouping the terms in powers of $(z - z_0)$. Thus, if $z_0 = 0$, we find for the Cauchy product

$$\text{(16)} \qquad [a_0 + a_1 z + a_2 z^2 + a_3 z^3 + \cdots] \cdot [b_0 + b_1 z + b_2 z^2 + b_3 z^3 + \cdots]$$

$$= a_0 b_0 + (a_1 b_0 + a_0 b_1)z + (a_2 b_0 + a_1 b_1 + a_0 b_2)z^2$$

$$+ (a_3 b_0 + a_2 b_1 + a_1 b_2 + a_0 b_3)z^3 + \cdots.$$

The coefficient, c_j, of z^j is therefore given by

$$\text{(17)} \qquad c_j = a_j b_0 + a_{j-1} b_1 + a_{j-2} b_2 + \cdots + a_1 b_{j-1} + a_0 b_j = \sum_{l=0}^{j} a_{j-l} b_l.$$

DEFINITION 4   The **Cauchy product** of two Taylor series $\sum_{j=0}^{\infty} a_j(z - z_0)^j$ and $\sum_{j=0}^{\infty} b_j(z - z_0)^j$ is defined to be the (formal) series $\sum_{j=0}^{\infty} c_j(z - z_0)^j$, where c_j is given by formula (17).

THEOREM 6   Let f and g be analytic functions with Taylor series $f(z) = \sum_{j=0}^{\infty} a_j(z - z_0)^j$ and $g(z) = \sum_{j=0}^{\infty} b_j(z - z_0)^j$ around the point z_0. Then the Taylor series for fg around z_0 is given by the Cauchy product of these two series.

Actually, we anticipated this result in electing to write the Cauchy product in (16) as if it were an ordinary product. As in Theorem 5, the Taylor series for fg converges at least in the smaller of the convergence disks for f and g.

Proof of Theorem 6   We compute the consecutive derivatives of the product fg:

$$(fg)' = f'g + fg',$$

$$(fg)'' = f''g + 2f'g' + fg'',$$

$$(fg)''' = f'''g + 3f''g' + 3f'g'' + fg''',$$

and, in general, we have *Leibniz's formula* for the jth derivative of fg:

(18)
$$(fg)^{(j)} = \sum_{l=0}^{j} j! \frac{f^{(j-l)}}{(j-l)!} \cdot \frac{g^{(l)}}{l!},$$

which we invite the reader to prove in Prob. 10.

On the other hand, if we identify the constants a_k and b_k in Eq. (17) with their expressions in terms of derivatives of f and g [e.g., $a_k = f^{(k)}(z_0)/k!$], we see from Eq. (18) that $(fg)^{(j)}/j!$ evaluated at z_0 is precisely c_j. This completes the proof.   ∎

EXAMPLE 8   Use the Cauchy product to find the Maclaurin series for $\sin z \cdot \cos z$.

SOLUTION   We have

$$\left(z - \frac{z^3}{3!} + \frac{z^5}{5!} - \frac{z^7}{7!} + \cdots\right) \cdot \left(1 - \frac{z^2}{2!} + \frac{z^4}{4!} - \frac{z^6}{6!} + \cdots\right)$$

$$= z - \left(\frac{1}{3!} + \frac{1}{2!}\right)z^3 + \left(\frac{1}{5!} + \frac{1}{3!\,2!} + \frac{1}{4!}\right)z^5$$

$$- \left(\frac{1}{7!} + \frac{1}{5!\,2!} + \frac{1}{3!\,4!} + \frac{1}{6!}\right)z^7 + \cdots.$$

It is amusing to try to simplify the coefficients; the reader can verify that

$$\sin z \cdot \cos z = z - \frac{4}{3!} z^3 + \frac{16}{5!} z^5 - \frac{64}{7!} z^7 + \cdots,$$

which, when rewritten as

$$\frac{1}{2} \left[(2z) - \frac{(2z)^3}{3!} + \frac{(2z)^5}{5!} - \frac{(2z)^7}{7!} + \cdots \right],$$

will be recognized as the Taylor series for $\frac{1}{2} \sin w$, with $w = 2z$. We have reproduced a well-known trigonometric identity! ∎

EXAMPLE 9   Find the first few terms of the Maclaurin series for $\tan z$.

   SOLUTION   The expressions for the higher derivatives of $\tan z$ are cumbersome, so let's try to use the Cauchy product. First observe that $\cos z \cdot \tan z = \sin z$. Now set $\tan z = \sum_{j=0}^{\infty} a_j z^j$ for $|z| < \pi/2$ (why?). The product $\cos z \cdot \tan z$ then becomes

$$\left(1 - \frac{z^2}{2!} + \frac{z^4}{4!} - \cdots \right) \cdot (a_0 + a_1 z + a_2 z^2 + a_3 z^3 + a_4 z^4 + a_5 z^5 + \cdots)$$

$$= a_0 + a_1 z + \left(a_2 - \frac{a_0}{2!} \right) z^2 + \left(a_3 - \frac{a_1}{2!} \right) z^3 + \left(a_4 - \frac{a_2}{2!} + \frac{a_0}{4!} \right) z^4$$

$$+ \left(a_5 - \frac{a_3}{2!} + \frac{a_1}{4!} \right) z^5 + \cdots.$$

Identifying this with $\sin z = z - z^3/3! + z^5/5! - \cdots$, we solve recursively and find

$$a_0 = 0, \quad a_1 = 1, \quad a_2 = 0, \quad a_3 = \frac{1}{3}, \quad a_4 = 0, \quad a_5 = \frac{2}{15}, \quad \text{etc.}$$

Thus

$$\tan z = z + \frac{z^3}{3} + \frac{2z^5}{15} + \cdots.$$

The shrewd reader will observe that we have actually discovered an indirect method of dividing Taylor series! ∎

   In closing this section we would like to point out that the proof of the validity of the Taylor expansion verifies the claim, made in Sec. 2.3, that any *analytic* function can be displayed with a formula involving z alone, and not $\bar{z}$, x, or y.

1. Verify each of the following Taylor expansions by finding a general formula for $f^{(j)}(z_0)$.

   (a) $e^z = \sum_{j=0}^{\infty} \dfrac{z^j}{j!} = 1 + z + \dfrac{z^2}{2!} + \dfrac{z^3}{3!} + \cdots, \quad z_0 = 0$

   (b) $\cosh z = \sum_{j=0}^{\infty} \dfrac{z^{2j}}{(2j)!} = 1 + \dfrac{z^2}{2!} + \dfrac{z^4}{4!} + \cdots, \quad z_0 = 0$

   (c) $\sinh z = \sum_{j=0}^{\infty} \dfrac{z^{2j+1}}{(2j+1)!} = z + \dfrac{z^3}{3!} + \dfrac{z^5}{5!} + \cdots, \quad z_0 = 0$

   (d) $\dfrac{1}{1-z} = \sum_{j=0}^{\infty} \dfrac{(z-i)^j}{(1-i)^{j+1}}, \quad z_0 = i$

   (e) $\text{Log}(1 - z) = \sum_{j=1}^{\infty} \dfrac{-z^j}{j}, \quad z_0 = 0$

   (f) $z^3 = 1 + 3(z-1) + 3(z-1)^2 + (z-1)^3, \quad z_0 = 1$

2. Determine the disks over which the Taylor expansions in Prob. 1 are valid.

3. Let $f(z) = \sum_{j=0}^{\infty} a_j z^j$ be the Maclaurin expansion of a function $f(z)$ analytic at the origin. Prove each of the following statements.

   (a) $\sum_{j=0}^{\infty} a_j z^{2j}$ is the Maclaurin expansion of $g(z) \equiv f(z^2)$.

   (b) $\sum_{j=0}^{\infty} a_j c^j z^j$ is the Maclaurin expansion of $h(z) \equiv f(cz)$.

   (c) $\sum_{j=0}^{\infty} a_j z^{m+j}$ is the Maclaurin expansion of $H(z) \equiv z^m f(z)$.

   (d) $\sum_{j=0}^{\infty} a_j (z - z_0)^j$ is the Taylor expansion of $G(z) \equiv f(z - z_0)$ around z_0.

4. Let α be a complex number. Show that if $(1 + z)^\alpha$ is taken as $e^{\alpha \, \text{Log}(1+z)}$, then for $|z| < 1$

   $$(1 + z)^\alpha = 1 + \dfrac{\alpha}{1} z + \dfrac{\alpha(\alpha - 1)}{1 \cdot 2} z^2 + \dfrac{\alpha(\alpha - 1)(\alpha - 2)}{1 \cdot 2 \cdot 3} z^3 + \cdots.$$

   (This generalizes the Binomial Theorem.)

5. Find and state the convergence properties of the Taylor series for the following.

   (a) $\dfrac{1}{1+z}$ around $z_0 = 0$

   (b) e^{-z^2} around $z_0 = 0$

   (c) $z^3 \sin 3z$ around $z_0 = 0$

(d) $2 \cos z - ie^z$ around $z_0 = 0$

(e) $\dfrac{1 + z}{1 - z}$ around $z_0 = i$

(f) $\cos z$ around $z_0 = \dfrac{\pi}{4}$

(g) $\dfrac{z}{(1 - z)^2}$ around $z_0 = 0$

6. Prove that the Taylor expansion of $1/(\xi - z)$ around z_0 ($\neq \xi$) is given by

$$\frac{1}{\xi - z} = \sum_{j=0}^{\infty} \frac{(z - z_0)^j}{(\xi - z_0)^{j} + 1} \qquad \text{for } |z - z_0| < |\xi - z_0|.$$

[REMARK: The expansion lies at the heart of the proof of Theorem 3.]

7. Verify that the identity

$$\text{Log}\left(\frac{1 + z}{1 - z}\right) = \text{Log}(1 + z) - \text{Log}(1 - z)$$

holds when $|z| < 1$. Then using the Maclaurin expansions of $\text{Log}(1 + z)$ and $\text{Log}(1 - z)$, find the Maclaurin expansion of $\text{Log}[(1 + z)/(1 - z)]$.

8. Use Taylor series to verify the following identities.

(a) $\sin(-z) = -\sin z$ (b) $\dfrac{de^z}{dz} = e^z$

(c) $e^{iz} = \cos z + i \sin z$ (d) $e^{2z} = e^z \cdot e^z$

9. Prove Theorem 5.

10. Prove Leibniz's formula, Eq. (18). [HINT: Use mathematical induction.]

test →

11. Using Theorem 6 for computing the product of Taylor series, find the first three nonzero terms in the Maclaurin expansion of the following.

(a) $e^z \cos z$ (b) $\dfrac{e^z}{z - 1}$

(c) $\sec z = \dfrac{1}{\cos z}$ (d) $\tanh z = \dfrac{\sinh z}{\cosh z}$

12. Find an explicit formula for the analytic function $f(z)$ that has the Maclaurin expansion $\sum_{k=0}^{\infty} k^2 z^k$. [HINT: Starting with the expression $(1-z)^{-1} = \sum_{k=0}^{\infty} z^k$, differentiate, multiply by z, differentiate again, and finally multiply by z.]

13. Prove that the polynomial $p_n(z)$ of Eq. (5) is the *only* polynomial of degree at most n which matches $f, f', f'', \ldots, f^{(n)}$ at z_0.

14. Let $f(z)$ be analytic in the disk $D: |z - z_0| < R$. Prove that if $f^{(k)}(z_0) = 0$ for every $k = 0, 1, 2, \ldots$, then $f(z)$ is identically zero in D.

15. Let $f(z)$ be analytic in the unit disk $|z| < 1$. Prove that if $f'(0) = f^{(3)}(0) = f^{(5)}(0) = \cdots = 0$, then $f(-z) = f(z)$ for all z in this disk.

16. Rewrite the polynomial $p(z) = a_0 + a_1 z + \cdots + a_n z^n$ in powers of $(z - 1)$; i.e., find the coefficients c_i of the expansion $p_n(z) = c_0 + c_1(z - 1) + \cdots + c_n(z - 1)^n$ in terms of the a_j. [HINT: Do not rearrange; use Taylor series.]

17. Prove that the Taylor series $\sum_{j=0}^{\infty} z^j$ does not converge *uniformly* to $(1 - z)^{-1}$ on the open disk $D: |z| < 1$. Does this contradict Theorem 3? [HINT: By considering points close to 1, show that for each fixed n the error $\left| \sum_{j=0}^{n} z^j - (1 - z)^{-1} \right|$ is not even bounded in D.]

18. The Taylor series provides a workable method of numerically tabulating the functions of mathematical physics when the remainder term can be estimated. Establish each of the following error estimates.

(a) $\left| e^z - \sum_{k=0}^{n} \frac{z^k}{k!} \right| \leq \frac{1}{(n + 1)!} \cdot \left(1 + \frac{1}{n + 1} \right)$ for $|z| \leq 1$

(b) $\left| \sin z - \sum_{k=0}^{n} \frac{(-1)^k z^{2k+1}}{(2k + 1)!} \right| \leq \frac{1}{(2n + 3)!} \left(\frac{4n^2 + 18n + 20}{4n^2 + 18n + 19} \right)$

$$\text{for } |z| \leq 1$$

[HINT: Write the error as an infinite series, factor out the first term, and then compare with a geometric series.]

19. According to the estimate in Prob. 18(a), how many terms of the expansion $\sum_{k=0}^{\infty} z^k/k!$ are needed to compute e^z to within $\pm 10^{-5}$ for $|z| \leq 1$?

5.3 POWER SERIES

A Taylor series for an analytic function appears to be a special instance of a certain general type of series of the form $\sum_{j=0}^{\infty} a_j(z - z_0)^j$. Such series have a name:

DEFINITION 5 *A series of the form $\sum_{j=0}^{\infty} a_j(z - z_0)^j$ is called a* **power series**. *The constants a_j are the* **coefficients** *of the power series.*

Suppose that we are presented with an arbitrary power series, such as

(19)
$$\sum_{j=0}^{\infty} \frac{z^j}{(j + 1)^2} = 1 + \frac{z}{4} + \frac{z^2}{9} + \frac{z^3}{16} + \cdots.$$

Certain questions then arise naturally. For what values of z does the series converge? Is the limit an analytic function? Is the power series representation of a function unique? In short, is every power series a Taylor series? This section is devoted to answering these questions.

The issue of convergence is settled by the following result, which smacks of the Taylor expansion:

THEOREM 7 *For any power series $\sum_{j=0}^{\infty} a_j(z - z_0)^j$ there is a real number R between 0 and ∞, inclusive, which depends only on the coefficients $\{a_j\}$, such that*

   (i) *The series converges for $|z - z_0| < R$,*

   (ii) *The series converges uniformly in any closed subdisk $|z - z_0| \le R' < R$, and*

   (iii) *The series diverges for $|z - z_0| > R$.*

In particular, when $R = 0$ the power series converges only at $z = z_0$, and when $R = \infty$ the series converges for all z. For $0 < R < \infty$, the circle $|z - z_0| = R$ is called the *circle of convergence*, but no general convergence statement can be made for z lying on this circle (see Prob. 1). The situation is depicted in Fig. 5.2.

Although a rigorous proof of Theorem 7 is deferred to (optional) Sec. 5.4, we can give an informal argument here which shows why the region of convergence has to be a disk. The essential ingredient is the following lemma, which we state for the special case $z_0 = 0$:

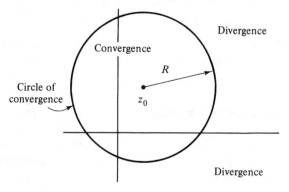

FIGURE 5.2

LEMMA 2 *If the power series $\sum_{j=0}^{\infty} a_j z^j$ converges at a point having modulus r, then it converges at every point in the disk $|z| < r$.*

Proof By hypothesis, there exists a point z_1, with $|z_1| = r$, such that the series $\sum_{j=0}^{\infty} a_j z_1^j$ converges. This implies that the sequence of terms $a_j z_1^j$ is bounded; i.e., there exists a constant M such that

$$|a_j z_1^j| = |a_j| r^j \leq M \qquad \text{(for all } j\text{)}.$$

Now for $|z| < r$ we can write

$$|a_j z^j| = |a_j| r^j \cdot \left(\frac{|z|}{r}\right)^j \leq M\left(\frac{|z|}{r}\right)^j.$$

Thus, as $|z|/r < 1$, the terms of the series $\sum_{j=0}^{\infty} a_j z^j$ are dominated by the terms of a convergent geometric series. By the Comparison Test, therefore, we conclude that the series converges. ∎

To see the existence of the number R in Theorem 7 for the power series $\sum_{j=0}^{\infty} a_j z^j$ we reason informally as follows: Consider the set of all real numbers r such that the series converges at some point having modulus r. Let R be the "largest"† of these numbers r. Then, by Lemma 2, the series converges for $|z| < R$, and from the definition of R the series diverges for all z with $|z| > R$.

If z is replaced by $(z - z_0)$ in the above argument, we deduce that the region of convergence of the general power series $\sum_{j=0}^{\infty} a_j (z - z_0)^j$ must be a disk with center z_0.

A formula for the radius of convergence R can be given, but we shall postpone it also until Sec. 5.4. However, in Prob. 2, the Ratio Test is used to show that in the special case where $|a_{j+1}/a_j|$ has a limit as j goes to infinity, R is the reciprocal of this limit. For example, the coefficients of the power series (19) satisfy

$$\lim_{j \to \infty} \left|\frac{a_{j+1}}{a_j}\right| = \lim_{j \to \infty} \frac{(j+1)^2}{(j+2)^2} = 1,$$

so its circle of convergence is $|z| = R = \frac{1}{1} = 1$.

Uniform convergence [cf. (ii) of Theorem 7] is a powerful feature of a sequence, as the next three results show. The first says that the uniform limit of continuous functions is itself continuous.

LEMMA 3 *Let $f_n(z)$ be a sequence of functions continuous on a set $\mathcal{S}$ and converging uniformly to $f(z)$ on $\mathcal{S}$. Then $f(z)$ is also continuous on $\mathcal{S}$.*

---

†More advanced students will recognize that R is precisely defined as the *least upper bound* of the numbers r.

Proof To prove that f is continuous at a point z_0 of $\mathscr{S}$, we must show that for any $\varepsilon > 0$ there is a $\delta > 0$ such that if z belongs to $\mathscr{S}$ and $|z_0 - z| < \delta$, then $|f(z_0) - f(z)| < \varepsilon$. We proceed by first choosing an integer N so large that $|f(z) - f_N(z)| < \varepsilon/3$ for all z in $\mathscr{S}$; this is possible thanks to uniform convergence. Now since f_N is continuous, there is a number $\delta' > 0$ such that $|f_N(z_0) - f_N(z)| < \varepsilon/3$ for any z in $\mathscr{S}$ satisfying $|z_0 - z| < \delta'$. But then $|f(z_0) - f(z)| < \varepsilon$ for such z, since

$$|f(z_0) - f(z)| = |f(z_0) - f_N(z_0) + f_N(z_0) - f_N(z) + f_N(z) - f(z)|$$

$$\leq |f(z_0) - f_N(z_0)| + |f_N(z_0) - f_N(z)| + |f_N(z) - f(z)|$$

$$< \frac{\varepsilon}{3} + \frac{\varepsilon}{3} + \frac{\varepsilon}{3} = \varepsilon.$$

Thus $f(z)$ is continuous at every point of $\mathscr{S}$. ∎

Knowing that the uniform limit of a sequence of continuous functions is continuous, we can integrate this limit. In fact we have

THEOREM 8   *Let $f_n(z)$ be a sequence of functions continuous on a set $\mathscr{S}$ containing the contour Γ, and suppose that $f_n(z)$ converges uniformly to $f(z)$ on $\mathscr{S}$. Then the sequence $\int_\Gamma f_n(z)\, dz$ converges to $\int_\Gamma f(z)\, dz$.*

Proof   This is easy. Let l be the length of Γ, and choose N so that $|f(z) - f_n(z)| < \varepsilon/l$ for any $n > N$ and for all z on Γ. Then

$$\left| \int_\Gamma f(z)\, dz - \int_\Gamma f_n(z)\, dz \right| = \left| \int_\Gamma [f(z) - f_n(z)]\, dz \right|$$

$$< \frac{\varepsilon}{l} \cdot l = \varepsilon. \quad ∎$$

Combining these results with Morera's theorem (Theorem 18 of Chapter 4), we have

THEOREM 9   *Let $f_n(z)$ be a sequence of functions analytic in a simply connected domain D and converging uniformly to $f(z)$ in D. Then f is analytic in D.*†

Proof   By Lemma 3, f is continuous. Let Γ be any loop contained in D. Then by Theorem 8, $\int_\Gamma f(z)\, dz$ is the limit of $\int_\Gamma f_n(z)\, dz$; but the latter is zero for all n, since each f_n is analytic inside and on Γ. Thus $\int_\Gamma f(z)\, dz = 0$, and Morera's result applies. ∎

---

†In fact this result holds in *any* domain, since a domain is a union of open disks.

Now we have everything we need to render an account of power series. Since the partial sums of a power series are analytic functions (indeed, polynomials) and since they converge uniformly in any closed subdisk interior to the circle of convergence, Theorem 9 tells us that the limit function is analytic inside every such subdisk. But *any* point within the circle of convergence lies inside such a subdisk, so we can state

THEOREM 10 *A power series sums to a function which is analytic at every point inside its circle of convergence.*

For example, the power series (19) defines an analytic function for $|z| < 1$.

Notice that Theorems 7 and 8 justify integrating a power series termwise, as long as the contour lies inside the circle of convergence. Using this fact, we can identify every power series (with $R > 0$) as a Taylor series, in accordance with

THEOREM 11 *If $\sum_{j=0}^{\infty} a_j (z - z_0)^j$ converges to $f(z)$ in some neighborhood of z_0 (i.e., the radius of its circle of convergence is nonzero), then*

$$a_j = \frac{f^{(j)}(z_0)}{j!} \qquad (j = 0, 1, 2, \ldots).$$

Consequently $\sum_{j=0}^{\infty} a_j (z - z_0)^j$ is the Taylor expansion of f around z_0.

Proof Let C be a positively oriented circle centered at z_0 and lying inside the circle of convergence. Since the limit $f(z)$ is analytic, we can write the generalized Cauchy integral formula

(20) $\qquad f^{(n)}(z_0) = \frac{n!}{2\pi i} \int_C \frac{f(\xi)}{(\xi - z_0)^{n+1}} d\xi \qquad (n = 0, 1, 2, \ldots).$

Now plug in the series $\sum_{j=0}^{\infty} a_j (\xi - z_0)^j$ for $f(\xi)$ and integrate termwise; since

$$\int_C \frac{(\xi - z_0)^j}{(\xi - z_0)^{n+1}} d\xi = \begin{cases} 2\pi i & \text{if } j = n, \\ 0 & \text{otherwise,} \end{cases}$$

the only term which survives the integration in (20) is $n! a_n$. ∎

At this point we have specified two disks around z_0 in which a power series such as (19) converges. One is the interior of the "circle of convergence" of Theorem 7. The other is the largest disk over whose interior the limit function f is analytic. Are these two disks the same? Observe that the "Taylor disk" cannot extend beyond the circle of convergence because the series is known to *diverge* outside the latter. On the other hand, the limit function is analytic inside the circle of convergence, so the Taylor disk must enclose the interior of this circle. Thus the disks actually do coincide.

Summarizing, we have shown that if a power series converges inside some circle, it is the Taylor series of its (analytic) limit function and can be integrated and differentiated term by term inside this circle; moreover, this limit function must fail to be analytic somewhere on the circle of convergence.

EXAMPLE 10   Find a function analytic at $z = 0$ which satisfies

$$(21) \qquad \frac{df(z)}{dz} = 3if(z)$$

and takes the value 1 at $z = 0$.

SOLUTION   Since f is analytic at the origin, it must have a Maclaurin series representation. We can find this series by using Eq. (21) to compute the derivatives.

We are given $f(0) = 1$, and from Eq. (21) we have $f'(0) = 3i \cdot 1 = 3i$. By differentiating Eq. (21), we see recursively that

$$f''(0) = 3if'(0) = (3i)^2,$$

$$f'''(0) = 3if''(0) = (3i)^3,$$

and in general

$$f^{(j)}(0) = 3if^{(j-1)}(0) = (3i)^2 f^{(j-2)}(0) = \cdots = (3i)^j.$$

Thus we can write the solution as

$$(22) \qquad f(z) = 1 + 3iz + \frac{(3i)^2 z^2}{2!} + \cdots = \sum_{j=0}^{\infty} \frac{(3iz)^j}{j!}.$$

In Prob. 1 of Exercises 5.2 the reader was directed to show that the exponential function can be represented by

$$(23) \qquad e^w = \sum_{j=0}^{\infty} \frac{w^j}{j!};$$

consequently, we can identify our solution (22) as

$$f(z) = e^{3iz}.$$

Direct computation quickly verifies that e^{3iz} solves the problem.   ∎

EXAMPLE 11   Let g be a continuous complex-valued function of a real variable on $[0, 2]$, and for each complex number z define

$$F(z) \equiv \int_0^2 e^{zt} g(t) \, dt.$$

Prove that F is entire, and find its power series around the origin.

SOLUTION   We first find a power series representation for F. Let z be fixed and define

$$h(t) \equiv e^{zt}, \qquad \text{for all complex numbers } t.$$

Then h is an entire function of t, and so its Maclaurin expansion

$$h(t) = e^{zt} = \sum_{k=0}^{\infty} \frac{z^k}{k!} \cdot t^k = 1 + zt + \frac{z^2 t^2}{2!} + \cdots$$

converges uniformly on every disk $|t| \leq r$; in particular, the convergence is uniform on the interval $[0, 2]$. Furthermore, termwise multiplication by the bounded function g preserves this convergence; i.e.,

$$e^{zt} g(t) = \sum_{k=0}^{\infty} \frac{z^k}{k!} \cdot t^k g(t)$$

uniformly for $0 \leq t \leq 2$. Now, by Theorem 8, we can integrate term by term with respect to t to obtain

$$F(z) = \int_0^2 e^{zt} g(t) \, dt = \sum_{k=0}^{\infty} \left[\frac{1}{k!} \int_0^2 t^k g(t) \, dt \right] z^k.$$

Since the bracketed quantity in the series is a constant dependent only on k, the above expansion is a power series in z; moreover, it converges to $F(z)$ for all z. Hence F is entire by Theorem 10.   ▮

Exercises 5.3

1. (a) Prove that the power series $\sum_{j=0}^{\infty} z^j$ converges at no point on its circle of convergence $|z| = 1$.

   (b) Prove that the power series $\sum_{j=1}^{\infty} z^j/j^2$ converges at every point on its circle of convergence $|z| = 1$. [HINT: See Prob. 14 in Exercises 5.1.]

2. Assume that for the power series $\sum_{j=0}^{\infty} a_j(z - z_0)^j$ we have $\lim_{j \to \infty} |a_{j+1}/a_j| = L$. Prove, by the Ratio Test, that the radius of convergence of the power series is given by $R = 1/L$.

3. Using the result of Prob. 2, find the circle of convergence of each of the following power series.

   (a) $\displaystyle\sum_{j=0}^{\infty} j^3 z^j$

   (b) $\displaystyle\sum_{k=0}^{\infty} 2^k(z - 1)^k$

   (c) $\displaystyle\sum_{j=0}^{\infty} j! \, z^j$

   (d) $\displaystyle\sum_{k=0}^{\infty} \frac{(-1)^k k}{3^k}(z - i)^k$

   (e) $\displaystyle\sum_{k=1}^{\infty} \frac{(3 - i)^k}{k^2}(z + 2)^k$

   (f) $\displaystyle\sum_{j=0}^{\infty} \frac{z^{2j}}{4^j}$

4. Does there exist a power series $\sum_{j=0}^{\infty} a_j z^j$ which converges at $z = 2 + 3i$ and diverges at $z = 3 - i$?

5. Let $f(z) = \sum_{k=0}^{\infty} (k^3/3^k) z^k$. Compute each of the following.

(a) $f^{(6)}(0)$

(b) $\oint_{|z|=1} \dfrac{f(z)}{z^4} \, dz$ $2\pi i$

(c) $\oint_{|z|=1} e^z f(z) \, dz$ 0 (d) $\oint_{|z|=1} \dfrac{f(z) \sin z}{z^2} \, dz$ 0

6. Define

$$f(z) = \begin{cases} \dfrac{\sin z}{z} & \text{for } z \neq 0, \\[2mm] 1 & \text{for } z = 0. \end{cases}$$

(a) Using the Maclaurin expansion for $\sin z$, show that for *all* z

$$f(z) = 1 - \frac{z^2}{3!} + \frac{z^4}{5!} - \frac{z^6}{7!} + \cdots.$$

(b) Explain why $f(z)$ is analytic at the origin.

(c) Find $f^{(3)}(0)$ and $f^{(4)}(0)$.

7. Assume that $f(z)$ is analytic at the origin and that $f(0) = f'(0) = 0$. Prove that $f(z)$ can be written in the form $f(z) = z^2 g(z)$, where $g(z)$ is analytic at $z = 0$.

8. Find the first three nonzero terms in the Maclaurin expansion of $f(z) \equiv \int_0^z e^{\xi^2} \, d\xi$. [HINT: First expand e^{ξ^2}.]

9. Suppose that g is continuous on the circle $C: |z| = 1$, and that there exists a sequence of polynomials which converge uniformly to g on C. Prove that

$$\oint_C g(z) \, dz = 0.$$

10. Explain why the two power series $\sum_{k=0}^{\infty} a_k z^k$ and $\sum_{k=0}^{\infty} k a_k z^{k-1}$ have the same radius of convergence.

11. Let $\sum_{k=0}^{\infty} a_k z^k$ and $\sum_{k=0}^{\infty} b_k z^k$ be two power series having a positive radius of convergence.

(a) Show that if $\sum_{k=0}^{\infty} a_k z^k = \sum_{k=0}^{\infty} b_k z^k$ in some neighborhood of the origin, then $a_k = b_k$ for all k.

(b) Show, more generally, that if $\sum_{k=0}^{\infty} a_k x^k = \sum_{k=0}^{\infty} b_k x^k$ for all real x in some open interval containing the origin, then $a_k = b_k$ for all k.

12. Prove by means of power series that the only solution of the initial-value problem

$$\begin{cases} \dfrac{df}{dz} = f \\ f(0) = 1 \end{cases}$$

which is analytic at $z = 0$ is $f(z) = e^z$.

13. Prove by means of power series that the only solution of the initial-value problem

$$\begin{cases} \dfrac{d^2f}{dz^2} + f = 0 \\ f(0) = 0, \qquad f'(0) = 1 \end{cases}$$

which is analytic at $z = 0$ is $f(z) = \sin z$.

14. Each of the following initial-value problems has a unique solution which is analytic at the origin. Find the power series expansion $\sum_{j=0}^{\infty} a_j z^j$ of the solution by determining a recurrence relation for the coefficients a_j.

(a) $\begin{cases} \dfrac{d^2f}{dz^2} - z\dfrac{df}{dz} - f = 0 \\ f(0) = 1, \qquad f'(0) = 0 \end{cases}$

(b) $\begin{cases} \dfrac{d^2f}{dz^2} + 4f = 0 \\ f(0) = 1, \qquad f'(0) = 1 \end{cases}$

(c) $\begin{cases} (1 - z^2)\dfrac{d^2f}{dz^2} - 6z\dfrac{df}{dz} - 4f = 0 \\ f(0) = 1, \qquad f'(0) = 0 \end{cases}$

[HINT: The technique demonstrated in Example 10 becomes laborious for more complicated equations such as (c). It is more efficient to substitute the power series expression for $f(z)$ into the equation and collect like powers of z.]

15. Let g be continuous on the real interval $[-1, 2]$, and define

$$F(z) = \int_{-1}^{2} g(t)\sin(zt)\, dt.$$

(a) Prove that F is entire and find its power series expansion around the origin.

(b) Prove that for all z

$$F'(z) = \int_{-1}^{2} tg(t)\cos(zt)\, dt.$$

16. Let g be continuous on the real interval $[0, 1]$ and define

$$H(z) = \int_{0}^{1} \frac{g(t)}{1 - zt^2}\, dt \qquad (|z| < 1).$$

Prove that H is analytic in the open disk $|z| < 1$.

17. Using the notation

$$(a)_j \equiv a(a + 1) \cdots (a + j - 1), \quad j \geq 1, \quad (a)_0 \equiv 1,$$

for any complex number a, the *Gaussian hypergeometric series* $_2F_1(b, c; d; z)$ is defined by

$$_2F_1(b, c; d; z) = \sum_{j=0}^{\infty} \frac{(b)_j (c)_j}{(d)_j} \cdot \frac{z^j}{j!},$$

and the *confluent hypergeometric series* $_1F_1(c; d; z)$ is given by

$$_1F_1(c; d; z) = \sum_{j=0}^{\infty} \frac{(c)_j}{(d)_j} \cdot \frac{z^j}{j!}.$$

(a) Verify that

$$\sum_{j=0}^{\infty} \frac{z^j}{j + 1} = {}_2F_1(1, 1; 2; z),$$

$$e^z - \sum_{k=0}^{n-1} \frac{z^k}{k!} = \frac{z^n}{n!} {}_1F_1(1; n + 1; z) \qquad (n = 1, 2, \ldots).$$

(b) Prove that if $d \neq 0, -1, -2, \ldots$, then the series $_2F_1(b, c; d; z)$ converges for $|z| < 1$ and satisfies the differential equation

$$z(1 - z)\frac{d^2f}{dz^2} + [d - (b + c + 1)z]\frac{df}{dz} - bc \cdot f = 0.$$

(c) Prove that if $d \neq 0, -1, -2, \ldots$, then the series $_1F_1(c; d; z)$ converges for all z and satisfies the differential equation

$$z\frac{d^2f}{dz^2} + (d - z)\frac{df}{dz} - cf = 0.$$

*5.4 MATHEMATICAL THEORY OF CONVERGENCE

In this section we shall backtrack somewhat and provide the mathematical details of the unproved theorems of this chapter. Applications-oriented students may wish to skip to Sec. 5.5.

So far all the conditions we have seen for convergence of a sequence involve the limit explicitly. However, there is a way of testing whether or not a sequence is convergent without mentioning a limit at all. It is known as the *Cauchy criterion* for convergence.

THEOREM 12 *A necessary and sufficient condition for the sequence $\{A_n\}_{n=1}^\infty$ to converge is the following: For any $\varepsilon > 0$ there exists an integer N such that $|A_n - A_m| < \varepsilon$ for every pair of integers m and n satisfying $m > N$, $n > N$.*

Proof (*necessity*) If the sequence does converge, say, to A, we choose N so that each A_l is within $\varepsilon/2$ of A for $l > N$. Then any two such A_l must lie within ε of each other.

The proof that the Cauchy criterion is sufficient for convergence requires a rigorous axiomatization for the real number system; indeed, the criterion can be used to define the concept of an irrational real number. We shall not explore this in the present text. ∎

A sequence which satisfies the Cauchy criterion is often called a *Cauchy sequence*. By Theorem 12, every convergent sequence is a Cauchy sequence and vice versa.

COROLLARY 1 *If $\{A_n\}_{n=1}^\infty$ is a Cauchy sequence and N is chosen so that $|A_n - A_m| < \varepsilon$ for every m and n greater than N, then each A_n with $n > N$ is within ε of the limit.*

Proof Let $m \to \infty$ in the inequality $|A_n - A_m| < \varepsilon$. The result is $|A_n - A| \leq \varepsilon$. ∎

The Cauchy criterion, applied to the sequence of partial sums of a series, reads as follows:

COROLLARY 2 *A necessary and sufficient condition for the series $\sum_{j=0}^\infty c_j$ to converge is the following: For any $\varepsilon > 0$ there exists an N such that $\left|\sum_{j=n+1}^m c_j\right| < \varepsilon$ for every pair of integers m and n satisfying $m > n > N$.*

The proof is immediate. Such a series is (naturally) called a *Cauchy series*. Corollary 2 justifies the following, almost obvious, result: If $\sum_{j=0}^\infty c_j$ converges, then $c_j \to 0$ as $j \to \infty$.

With the Cauchy criterion in hand we can give a proof of the Comparison Test, Theorem 1 of this chapter. However, it takes only a little more effort to prove the following, more general theorem, known as the *Weierstrass M-test*:

THEOREM 13 (M-test) *Suppose $\sum_{j=0}^{\infty} M_j$ is a convergent series with real nonnegative terms and suppose, for all z in some set $\mathscr{S}$ and for all j greater than some number J, that $|f_j(z)| \leq M_j$. Then the series $\sum_{j=0}^{\infty} f_j(z)$ converges uniformly on $\mathscr{S}$.*

Proof Since $\sum_{j=0}^{\infty} M_j$ is a Cauchy series, we can choose $N > J$ so that for any m and n satisfying $m > n > N$ we have $\sum_{j=n+1}^{m} M_j < \varepsilon$. But then for z in $\mathscr{S}$, $\sum_{j=0}^{\infty} f_j(z)$ is a Cauchy series also, because

$$(24) \qquad \left| \sum_{j=n+1}^{m} f_j(z) \right| \leq \sum_{j=n+1}^{m} |f_j(z)| \leq \sum_{j=n+1}^{m} M_j < \varepsilon.$$

Hence $\sum_{j=0}^{\infty} f_j(z)$ converges for each z in $\mathscr{S}$, say to the function $F(z)$. It is easy to see that the convergence is uniform; observe that inequality (24) can be rewritten in terms of the partial sums as

$$\left| \sum_{j=0}^{m} f_j(z) - \sum_{j=0}^{n} f_j(z) \right| < \varepsilon, \qquad \text{for all } z \text{ in } \mathscr{S}, m > n > N.$$

Therefore, by Corollary 1,

$$\left| F(z) - \sum_{j=0}^{n} f_j(z) \right| \leq \varepsilon, \qquad \text{for all } z \text{ in } \mathscr{S}, \text{ and } n > N.$$

This proves uniform convergence. ∎

The Comparison Test can be regarded as a special case of the M-test where each $f_j(z)$ is a constant function.

Now we are ready to analyze Theorem 7 of the previous section, specifying the convergence properties of power series. For convenience, we restate the theorem here.

THEOREM 7 *For any power series $\sum_{j=0}^{\infty} a_j(z - z_0)^j$ there is a real number R between 0 and ∞, inclusive, which depends only on the coefficients $\{a_j\}$ such that*

(i) *The series converges for $|z - z_0| < R$,*
(ii) *The series converges uniformly in any closed subdisk $|z - z_0| \leq R' < R$, and*
(iii) *The series diverges for $|z - z_0| > R$.*

To specify this number R, we must introduce the concept of the *limit superior* of an infinite sequence of real numbers; it generalizes the notion of limit. For motivation, let us first consider a *convergent* sequence of real numbers $\{x_n\}$, with limit x. Then for any $\varepsilon > 0$ there is an N such that all the elements x_n for $n > N$ will lie within ε of x. So, in particular, x has the following property: Given any $\varepsilon > 0$, for only a finite number of values of n does x_n exceed $x + \varepsilon$. Moreover, no number less than x has this property. We extend this notion to arbitrary sequences in

DEFINITION 6 The **limit superior** *of a sequence of real numbers* $\{x_n\}_{n=1}^{\infty}$, *abbreviated* lim sup x_n, *is defined to be the smallest real number* l *with the property that for any* $\varepsilon > 0$ *there are only a finite number of values of* n *such that* x_n *exceeds* $l + \varepsilon$; *if there are no such numbers with this property, we set* lim sup $x_n = \infty$; *if all real numbers have this property, we set* lim sup $x_n = -\infty$.

As we indicated above, if $\{x_n\}$ converges to x, then lim sup $x_n = x$. Other examples are lim sup$(-1)^n = 1$, lim sup$(n) = \infty$, and lim sup$(-n) = -\infty$. A less trivial example is

LEMMA 4 lim sup $\sqrt[n]{n!} = \infty$.

Proof Let v be any fixed positive integer. Then if n is greater than $2v$, say $n = 2v + \lambda$ with $\lambda > 0$,

$$\frac{n!}{v^n} = \frac{(2v)!}{v^{2v}} \frac{(2v+1)}{v} \frac{(2v+2)}{v} \cdots \frac{(2v+\lambda)}{v} > \frac{(2v)!}{v^{2v}} 2^{\lambda}.$$

Thus if we choose λ, and hence n, large enough, the ratio $n!/v^n$ will exceed 1, i.e., $n!$ will exceed v^n. Therefore, $\sqrt[n]{n!}$ will eventually exceed *any number* v. ∎

The number R in Theorem 7 is now specified as follows. From the set of coefficients $\{a_j\}$ we form the sequence $\sqrt[n]{|a_n|}$. Then R is equal to the reciprocal of the limit superior of this sequence,

(25)
$$R = \frac{1}{\text{lim sup } \sqrt[n]{|a_n|}},$$

with the usual conventions $1/0 = \infty$, $1/\infty = 0$. The proof of Theorem 7 proceeds as follows:

Proof of Theorem 7 First we address the convergence statements (*i*) and (*ii*). If $R \doteq 0$, there is nothing to prove. When $R > 0$, the convergence for $|z - z_0| < R$ follows from the uniform convergence in all closed subdisks $|z - z_0| \le R' < R$, so we attack the latter problem. Choose a number k in the interval

$$\frac{1}{R} < k < \frac{1}{R'}.$$

Then because of Eq. (25), all but a finite number of the a_j will satisfy $\sqrt[j]{|a_j|} < k$. Consequently, if z lies in the closed subdisk $|z - z_0| \le R'$, we have the inequality

(26)
$$|a_j(z - z_0)^j| = (\sqrt[j]{|a_j|} \, |z - z_0|)^j < (kR')^j,$$

valid for j sufficiently large. But inequality (26) tells us that the M-test (Theorem 13) is satisfied when we compare $\sum_{j=0}^{\infty} a_j(z - z_0)^j$ to the series $\sum_{j=0}^{\infty} (kR')^j$, which converges since $kR' < 1$. Accordingly, $\sum_{j=0}^{\infty} a_j(z - z_0)^j$ is uniformly convergent in the closed subdisk and (*ii*) is proved.

To prove divergence when $|z - z_0| > R$, we choose k in the interval

$$\frac{1}{|z - z_0|} < k < \frac{1}{R}.$$

Then it follows from the definition of lim sup and Eq. (25) that there must be an infinite number of a_j satisfying $\sqrt[j]{|a_j|} > k$ (remember that the lim sup is the *smallest* number such that so-and-so). For such a_j

$$|a_j(z - z_0)^j| = (\sqrt[j]{|a_j|}\, |z - z_0|)^j > (k|z - z_0|)^j > 1;$$

that is, an infinite number of the terms of $\sum_{j=0}^{\infty} a_j(z - z_0)^j$ exceed 1 in modulus. This is clearly incompatible with the Cauchy criterion, so the series must diverge. ∎

EXAMPLE 12 Describe the function

$$f(z) = \sum_{j=0}^{\infty} \frac{(-1)^j z^{2j}}{2^{2j}(j!)^2}.$$

SOLUTION   Keeping in mind that a_j is the coefficient of z^j, we have

$$a_j = \begin{cases} 0 & \text{if } j \text{ is odd,} \\[2ex] \dfrac{(-1)^{j/2}}{2^j \left[\left(\dfrac{j}{2}\right)!\right]^2} & \text{if } j \text{ is even.} \end{cases}$$

Obviously $\sqrt[j]{|a_j|} = 0$ for odd j. For even j,

$$\sqrt[j]{|a_j|} = \frac{1}{2\left[\left(\dfrac{j}{2}\right)!\right]^{2/j}},$$

and the analysis of Lemma 4 shows that this goes to zero. Hence lim sup $\sqrt[j]{|a_j|} = 0$, and $R = \infty$. Consequently $f(z)$ is an entire function. It is customarily denoted $J_0(z)$, the *Bessel function of order zero*. The reader should verify (Prob. 7) that $J_0(z)$ satisfies *Bessel's equation of order zero*:

$$\frac{d^2 f}{dz^2} + \frac{1}{z}\frac{df}{dz} + f = 0. \quad ∎$$

201

Section 5.4
Mathematical
Theory of
Convergence

Exercises 5.4

1. Find the limit superior of each of the following sequences $\{x_n\}_{n=1}^{\infty}$.

    (a) $x_n = (-1)^n \left(\dfrac{2n}{n+1} \right)$ (b) $x_n = (-1)^n n$

    (c) $x_n = \dfrac{1}{n^2}$ (d) $x_n = n \sin\left(\dfrac{n\pi}{2} \right)$

2. Prove that for any sequence $\{x_n\}$ of positive real numbers

    $$\limsup \sqrt[n]{x_n} \le \limsup \frac{x_{n+1}}{x_n}.$$

3. By considering the series $\sum_{j=1}^{\infty} z^j/j^2$, $\sum_{j=1}^{\infty} z^j/j$, and $\sum_{j=0}^{\infty} z^j$, show that a power series may converge on all, some, or none of the points on its circle of convergence.

4. Find the radius of convergence of each of the following power series.

    (a) $\displaystyle\sum_{j=1}^{\infty} \frac{2^j}{3^j + 4^j} z^j$ (b) $\displaystyle\sum_{j=0}^{\infty} 2^j z^{j^2}$

    (c) $\displaystyle\sum_{j=0}^{\infty} [2 + (-1)^j]^j z^j$ (d) $\displaystyle\sum_{j=1}^{\infty} \frac{j!}{j^j} z^j$

    (e) $\displaystyle\sum_{j=0}^{\infty} \frac{2}{3j} z^{2j}$ (f) $\displaystyle\sum_{j=0}^{\infty} z^{j!}$

5. Prove that if the radius of convergence for the series $\sum_{j=0}^{\infty} a_j z^j$ is R, then the radius of convergence for the series $\sum_{j=0}^{\infty} \mathrm{Re}(a_j) z^j$ is greater than or equal to R.

6. If the radius of convergence for the series $\sum_{j=0}^{\infty} a_j z^j$ is R, find the radius of convergence for the following.

    (a) $\displaystyle\sum_{j=0}^{\infty} j^3 a_j z^j$ (b) $\displaystyle\sum_{j=0}^{\infty} a_j^4 z^j$

    (c) $\displaystyle\sum_{j=0}^{\infty} a_j z^{2j}$ (d) $\displaystyle\sum_{j=0}^{\infty} a_j z^{j+7}$

    (e) $\displaystyle\sum_{j=1}^{\infty} j^{-j} a_j z^j$

7. Show that the Bessel function $J_0(z)$ satisfies Bessel's equation of order zero as claimed in Example 12.

8. Bessel's equation of order n is

$$\frac{d^2 f(z)}{dz^2} + \frac{1}{z}\frac{df(z)}{dz} + \left(1 - \frac{n^2}{z^2}\right)f(z) = 0.$$

Show that, for integers $n > 0$, the *Bessel function of order n*

$$J_n(z) = \sum_{j=0}^{\infty} \frac{(-1)^j}{j!\,(n+j)!} \cdot \left(\frac{z}{2}\right)^{2j+n}$$

is entire and satisfies the Bessel equation. (Bessel functions arise in the study of two-dimensional wave propagation in regions with cylindrical symmetry.)

9. The *Legendre polynomials* $P_j(\xi)$ are the coefficients of z^j in the Maclaurin series for

$$(1 - 2\xi z + z^2)^{-1/2} = \sum_{j=0}^{\infty} P_j(\xi)z^j$$

(regarding ξ as a parameter). Show that $P_j(\xi)$ is a polynomial in ξ of degree j, and compute P_0, P_1, P_2, and P_3. (These polynomials arise in three-dimensional potential theory.)

10. Use power series to solve the *functional equation*

$$f(z) = z + f(z^2)$$

for the analytic function f.

11. The *Fibonacci sequence* 1, 1, 2, 3, 5, 8, 13, ... arises with surprising frequency in natural phenomena. The defining relations for the terms are

$$a_0 = a_1 = 1,$$

$$a_n = a_{n-1} + a_{n-2} \qquad \text{(for } n \geq 2\text{)}.$$

Show that

$$f(z) = a_0 + a_1 z + a_2 z^2 + \cdots$$

defines an analytic function satisfying the equation

$$f(z) = 1 + zf(z) + z^2 f(z).$$

Solve for $f(z)$ and compute the Maclaurin series to derive the expression

$$a_j = \frac{1}{\sqrt{5}}\left[\left(\frac{1 + \sqrt{5}}{2}\right)^{j+1} - \left(\frac{1 - \sqrt{5}}{2}\right)^{j+1}\right].$$

12. The *Riemann zeta function* is defined by

$$\zeta(z) = \sum_{j=1}^{\infty} \frac{1}{j^z} \qquad (\text{Re } z > 1),$$

where $j^z = \exp(z \operatorname{Log} j)$. Prove that $\zeta(z)$ is analytic for $\text{Re } z > 1$. [HINT: Let $\text{Re } z \geq \lambda > 1$, and show that $|1/j^z| \leq j^{-\lambda}$. Then use the Weierstrass M-test.]

5.5 LAURENT SERIES

We now wish to investigate the possibility of a series representation of a function f near a *singularity*, i.e., a point where f is not analytic but which is the limit of points where f is analytic. After all, if (for example) the occurrence of a singularity is merely due to a vanishing denominator, might it not be possible to express the function as something like $A/(z - z_0)^p + g(z)$, where g is analytic and has a Taylor series around z_0? To be sure, not all singularities are of this type (recall $\operatorname{Log} z$ at $z_0 = 0$). However, if the function is analytic in an annulus surrounding one or more of its singularities (note that $\operatorname{Log} z$ does not have this property), we can display its "singular part" according to

THEOREM 14 *Let $f(z)$ be analytic in an open annulus $r < |z - z_0| < R$. Then $f(z)$ can be expressed there as the sum of two series*

$$f(z) = \sum_{j=0}^{\infty} a_j(z - z_0)^j + \sum_{j=1}^{\infty} a_{-j}(z - z_0)^{-j},$$

both series converging in the annulus, and converging uniformly in any closed subannulus $r < \rho_1 \leq |z - z_0| \leq \rho_2 < R$. The coefficients a_j are given by

$$(27) \qquad a_j = \frac{1}{2\pi i} \oint_C \frac{f(\xi)}{(\xi - z_0)^{j+1}} \, d\xi \qquad (j = 0, \pm 1, \pm 2, \ldots),$$

where C is any positively oriented simple closed contour lying in the annulus and containing z_0 in its interior.

Such an expansion, containing negative as well as positive powers of $(z - z_0)$, is called the *Laurent series* for f in this annulus. It is usually abbreviated

$$\sum_{j=-\infty}^{\infty} a_j(z - z_0)^j.$$

Notice that if f is analytic throughout the disk $|z - z_0| < R$, the coefficients in (27) with negative subscripts are zero, and the others reproduce the Taylor series for f.

Proof of Theorem 14   It suffices to prove uniform convergence in every closed subannulus, for this implies (pointwise) convergence in the open annulus.

First we show that for any z satisfying $r < \rho_1 \le |z - z_0| \le \rho_2 < R$ we have the representation

$$(28) \qquad f(z) = \frac{1}{2\pi i} \oint_{C_1} \frac{f(\xi)}{\xi - z}\, d\xi + \frac{1}{2\pi i} \oint_{C_2} \frac{f(\xi)}{\xi - z}\, d\xi,$$

where C_1 is the negatively oriented circle around z_0 of radius $R_1 = (r + \rho_1)/2$, and C_2 is the positively oriented circle around z_0 of radius $R_2 = (R + \rho_2)/2$; see Fig. 5.3. Indeed, Eq. (28) is just a slight varia-

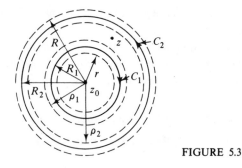

FIGURE 5.3

tion of the Cauchy integral formula in this case, as the following argument shows. Consider the contour Γ of Fig. 5.4(a); it is simple, closed, positively oriented, and contains z in its interior. Therefore

$$(29) \qquad f(z) = \frac{1}{2\pi i} \int_\Gamma \frac{f(\xi)}{\xi - z}\, d\xi.$$

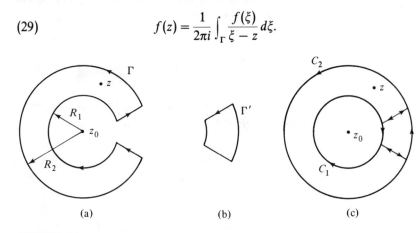

(a)   (b)   (c)

FIGURE 5.4

Let's think of Γ as a doughnut with a bite taken out of it, and let Γ′ denote the "bite" as in Fig. 5.4(b). Observe that

$$\frac{1}{2\pi i} \int_{\Gamma'} \frac{f(\xi)}{\xi - z} d\xi = 0$$

because the integrand is analytic inside and on Γ′. Consequently we can put the "bite" back into the doughnut and modify Eq. (29) to read

$$f(z) = \frac{1}{2\pi i} \int_{\Gamma + \Gamma'} \frac{f(\xi)}{\xi - z} d\xi,$$

where Γ + Γ′ is the path indicated in Fig. 5.4(c). But the integrals along the line segments in Fig. 5.4(c) cancel and, keeping track of the orientation, we arrive at Eq. (28).† Now we are ready to proceed with the derivation of the Laurent expansion.

Since z lies inside C_2, the integral over C_2 appearing in Eq. (28) is exactly like the integral which arose in the Taylor series theorem (Theorem 3); we treat it the same way, and find

$$\frac{1}{2\pi i} \oint_{C_2} \frac{f(\xi)}{\xi - z} d\xi = \sum_{j=0}^{n} a_j(z - z_0)^j + T_n(z),$$

where $T_n(z) \to 0$ uniformly as $n \to \infty$ for $|z - z_0| \le \rho_2$ and a_j is given by

$$(30) \qquad a_j = \frac{1}{2\pi i} \oint_{C_2} \frac{f(\xi)}{(\xi - z_0)^{j+1}} d\xi \qquad (j = 0, 1, 2, \ldots).$$

Hence

$$\frac{1}{2\pi i} \oint_{C_2} \frac{f(\xi)}{\xi - z} dz = \sum_{j=0}^{\infty} a_j(z - z_0)^j \qquad (|z - z_0| \le \rho_2).$$

We now turn to the integral around C_1 in Eq. (28). Since z lies outside C_1, we seek an expression for $1/(\xi - z)$ in powers of $(\xi - z_0)/(z - z_0)$; accordingly, we write

$$\frac{1}{\xi - z} = \frac{1}{(\xi - z_0) - (z - z_0)} = -\frac{1}{(z - z_0)} \frac{1}{1 - \dfrac{\xi - z_0}{z - z_0}}$$

$$= -\frac{1}{z - z_0} \left[1 + \frac{\xi - z_0}{z - z_0} + \frac{(\xi - z_0)^2}{(z - z_0)^2} + \cdots + \frac{(\xi - z_0)^m}{(z - z_0)^m} \right.$$

$$\left. + \frac{\dfrac{(\xi - z_0)^{m+1}}{(z - z_0)^{m+1}}}{1 - \dfrac{\xi - z_0}{z - z_0}} \right].$$

†Some readers may be able to use a deformation-of-contour argument to derive Eq. (28), but the doughnut analogy is probably easier to digest.

Inserting this into the integral, we find

$$\frac{1}{2\pi i} \oint_{C_1} \frac{f(\xi)}{\xi - z} \, d\xi = \sum_{j=1}^{m+1} a_{-j}(z - z_0)^{-j} + \mathcal{T}_m(z),$$

where

(31) $$a_{-j} = -\frac{1}{2\pi i} \oint_{C_1} \frac{f(\xi)}{(\xi - z_0)^{-j+1}} \, d\xi \qquad (j = 1, 2, 3, \dots)$$

(observe the exponent with care) and

$$\mathcal{T}_m(z) = \frac{1}{2\pi i} \oint_{C_1} \frac{f(\xi)}{\xi - z} \frac{(\xi - z_0)^{m+1}}{(z - z_0)^{m+1}} \, d\xi.$$

Now for ξ on C_1 we have $|\xi - z| \geq \rho_1 - R_1$, $|\xi - z_0| = R_1$, and $|z - z_0| \geq \rho_1$ (see Fig. 5.3). Thus

$$|\mathcal{T}_m(z)| \leq \frac{1}{2\pi} \cdot \max_{\xi \text{ on } C_1} |f(\xi)| \frac{1}{\rho_1 - R_1} \left(\frac{R_1}{\rho_1}\right)^{m+1} 2\pi R_1.$$

Since $R_1/\rho_1 < 1$, $\mathcal{T}_m(z) \to 0$ uniformly for $|z - z_0| \geq \rho_1$, and so

$$\frac{1}{2\pi i} \oint_{C_1} \frac{f(\xi)}{\xi - z} \, d\xi = \sum_{j=1}^{\infty} a_{-j}(z - z_0)^{-j} \qquad (|z - z_0| \geq \rho_1).$$

We have thus expressed both integrals in Eq. (28) as uniformly convergent series of the form mentioned in the theorem with the common region of convergence $\rho_1 \leq |z - z_0| \leq \rho_2$; we still have to verify formula (27) for the coefficients. If j is nonnegative, formula (30) applies, but the analysis of Chapter 4 justifies replacing the integral over C_2 by the integral over the contour C mentioned in the theorem, since the intervening region contains no singularities of $f(\xi)/(\xi - z_0)^{j+1}$; hence Eq. (27) holds for $j \geq 0$. Similarly, the integral over C_1 in formula (31) can be changed into an integral over C, incorporating the minus sign to account for the change in orientation. Hence (27) is verified for every j. ∎

Replacing $(z - z_0)$ with $1/(z - z_0)$ in Theorem 7, it is easily seen that any formal series of the form $\sum_{j=1}^{\infty} c_{-j}(z - z_0)^{-j}$ will converge *outside* some "circle of convergence" $|z - z_0| = r$ whose radius depends on the coefficients, with uniform convergence holding in each region $|z - z_0| \geq r' > r$. Thus termwise integration is justified by Theorem 8, and proceeding in a manner analogous to that of Sec. 5.3 we can prove

THEOREM 15   *Let $\sum_{j=0}^{\infty} c_j(z - z_0)^j$ and $\sum_{j=1}^{\infty} c_{-j}(z - z_0)^{-j}$ be any two series with the following properties:*

(i) $\sum_{j=0}^{\infty} c_j(z - z_0)^j$ *converges for* $|z - z_0| < R,$

(ii) $\sum_{j=1}^{\infty} c_{-j}(z - z_0)^{-j}$ *converges for* $|z - z_0| > r,$ *and*

(iii) $r < R.$

Then there is a function $f(z)$, analytic for $r < |z - z_0| < R$, whose Laurent series in this annulus is given by $\sum_{j=-\infty}^{\infty} c_j(z - z_0)^j$.

The proof is left to the Exercises (Prob. 9).

This theorem, like Theorem 7, justifies the use of other methods of finding Laurent series since it implies that *any* convergent series of the form $\sum_{j=-\infty}^{\infty} c_j(z - z_0)^j$, however obtained, must be the Laurent series of its sum function. In Examples 13 and 14 we shall derive Laurent expansions by making judicious use of the fact that the geometric series $\sum_{j=0}^{\infty} w^j$ converges to $(1 - w)^{-1}$ when $|w| < 1$.

EXAMPLE 13 Find the Laurent series for the function $(z^2 - 2z + 3)/(z - 2)$ in the region $|z - 1| > 1$.

SOLUTION Notice that this region is centered at $z_0 = 1$ and excludes the singularity at $z = 2$. First we manipulate $1/(z - 2)$ so that we can apply the geometric series result in the specified region:

$$\frac{1}{z - 2} = \frac{1}{(z - 1) - 1} = \frac{1}{z - 1} \cdot \frac{1}{1 - \dfrac{1}{z - 1}}.$$

Thus for $|1/(z - 1)| < 1$,

$$\frac{1}{z - 2} = \frac{1}{z - 1} \cdot \sum_{j=0}^{\infty} \frac{1}{(z - 1)^j}$$

$$= \frac{1}{z - 1} + \frac{1}{(z - 1)^2} + \frac{1}{(z - 1)^3} + \cdots.$$

Expressing the numerator $z^2 - 2z + 3$ in powers of $z - 1$, we find

$$z^2 - 2z + 3 = (z - 1)^2 + 0 \cdot (z - 1) + 2 = (z - 1)^2 + 2.$$

Therefore

$$\frac{z^2 - 2z + 3}{z - 2} = [(z - 1)^2 + 2] \cdot \left[\frac{1}{z - 1} + \frac{1}{(z - 1)^2} + \frac{1}{(z - 1)^3} + \cdots \right]$$

$$= \left[(z - 1) + 1 + \frac{1}{(z - 1)} + \frac{1}{(z - 1)^2} + \cdots \right] + \left[\frac{2}{(z - 1)} + \frac{2}{(z - 1)^2} + \cdots \right]$$

$$= (z - 1) + 1 + \sum_{j=1}^{\infty} \frac{3}{(z - 1)^j}. \quad \blacksquare$$

EXAMPLE 14 For the function

$$\frac{1}{(z - 1)(z - 2)},$$

find the Laurent series expansion in
(a) The region $|z| < 1$.
(b) The region $1 < |z| < 2$.
(c) The region $|z| > 2$.

SOLUTION   Using partial fractions, write

$$\frac{1}{(z-1)(z-2)} = \frac{1}{z-2} - \frac{1}{z-1}.$$

Now we proceed differently in each region in order to derive convergent series.

(a) For $|z| < 1$,

$$(32) \qquad \frac{1}{z-2} = -\frac{1}{2}\frac{1}{1-\frac{z}{2}} = -\frac{1}{2}\sum_{j=0}^{\infty}\left(\frac{z}{2}\right)^j = -\sum_{j=0}^{\infty}\frac{z^j}{2^{j+1}},$$

and

$$(33) \qquad \frac{1}{z-1} = -\frac{1}{1-z} = -\sum_{j=0}^{\infty}z^j.$$

Subtracting Eq. (33) from Eq. (32) gives

$$\frac{1}{(z-1)(z-2)} = \sum_{j=0}^{\infty}\left(-\frac{1}{2^{j+1}}+1\right)z^j = \frac{1}{2} + \frac{3}{4}z + \frac{7}{8}z^2 + \cdots$$

$$(|z| < 1).$$

(b) For $1 < |z| < 2$, Eq. (32) is still valid, but we have

$$(34) \qquad \frac{1}{z-1} = \frac{1}{z}\frac{1}{1-\frac{1}{z}} = \frac{1}{z}\sum_{j=0}^{\infty}\frac{1}{z^j} = \sum_{j=0}^{\infty}\frac{1}{z^{j+1}}.$$

Thus

$$\frac{1}{(z-1)(z-2)} = -\sum_{j=0}^{\infty}\frac{z^j}{2^{j+1}} - \sum_{j=0}^{\infty}\frac{1}{z^{j+1}} = \cdots - \frac{1}{z^2} - \frac{1}{z} - \frac{1}{2} - \frac{z}{4} - \cdots.$$

(c) For $|z| > 2$, Eq. (34) is valid and

$$\frac{1}{z-2} = \frac{1}{z}\frac{1}{1-\frac{2}{z}} = \frac{1}{z}\sum_{j=0}^{\infty}\left(\frac{2}{z}\right)^j = \sum_{j=0}^{\infty}\frac{2^j}{z^{j+1}}.$$

Hence

$$\frac{1}{(z-1)(z-2)} = \sum_{j=0}^{\infty}\frac{2^j-1}{z^{j+1}} = \frac{1}{z^2} + \frac{3}{z^3} + \frac{7}{z^4} + \cdots. \quad\blacksquare$$

EXAMPLE 15 Expand $e^{1/z}$ in a Laurent series around $z = 0$.

SOLUTION   As we already know,

$$e^w = 1 + w + \frac{w^2}{2!} + \frac{w^3}{3!} + \cdots$$

for all (finite) w. Thus if $z \neq 0$, we let $w = 1/z$ and find

(35) $$e^{1/z} = 1 + \frac{1}{z} + \frac{1}{2! z^2} + \frac{1}{3! z^3} + \cdots. \quad \blacksquare$$

EXAMPLE 16 What is the Laurent expansion around $z = 0$ for the function

$$f(z) = \begin{cases} \sin z & \text{if } z \neq 0, \\ 5 & \text{if } z = 0 \end{cases} \quad ?$$

SOLUTION   The alert reader, upon seeing this example, will undoubtedly accuse the authors of sophistry—and with good reason! The function f is simply defined "incorrectly" at $z = 0$ in order to make a point. We ask, however, that the audience bear with us, because this example will be useful in the next section.

Observe that f satisfies the hypothesis of Theorem 14, so it does have a Laurent series, valid for $|z| > 0$. But for such z we have [recall Eq. (14)]

(36) $$f(z) = \sin z = z - \frac{z^3}{3!} + \frac{z^5}{5!} - \cdots \quad (|z| > 0).$$

Therefore, Eq. (36) must be the Laurent expansion for f. $\blacksquare$

Exercises 5.5

1. Does the principal branch $\sqrt{z}$ have a Laurent series expansion in the domain $0 < |z|$?

2. Find the Laurent series for the function $1/(z + z^2)$ in each of the following domains.
   (a) $0 < |z| < 1$ (b) $1 < |z|$
   (c) $0 < |z + 1| < 1$ (d) $1 < |z + 1|$

3. Find the Laurent series for the function $z/(z + 1)(z - 2)$ in each of the following domains.
   (a) $|z| < 1$ (b) $1 < |z| < 2$ (c) $2 < |z|$

4. Find the Laurent series for $(\sin 2z)/z^3$ in $|z| > 0$.

5. Find the Laurent series for $(z + 1)/z(z - 4)^3$ in $0 < |z - 4| < 4$.

6. Find the Laurent series for $z^2 \cos(1/3z)$ in $|z| > 0$.

7. Obtain the first few terms of the Laurent series for each of the following functions in the specified domains.

(a) $\dfrac{e^{1/z}}{z^2 - 1}$ for $|z| > 1$ (b) $\dfrac{1}{e^z - 1}$ for $0 < |z| < 2\pi$

(c) $\csc z$ for $0 < |z| < \pi$ (d) $1/e^{(1-z)}$ for $1 < |z|$

8. Determine the annulus of convergence of the Laurent series

$$\sum_{j=-\infty}^{\infty} \frac{z^j}{2^{|j|}}.$$

9. Give a proof of Theorem 15.

10. Prove that the Laurent series expansion of the function

$$f(z) = e^{(\lambda/2)[z - (1/z)]}$$

in $|z| > 0$ is given by

$$\sum_{k=-\infty}^{\infty} J_k(\lambda)z^k,$$

where

$$J_k(\lambda) = (-1)^k J_{-k}(\lambda) = \frac{1}{2\pi} \int_0^{2\pi} \cos(k\theta - \lambda \sin \theta) \, d\theta.$$

The $J_k(\lambda)$ are known as *Bessel functions* of the first kind. [HINT: Use the integral formula (27) with $C: |z| = 1$.]

11. Obtain a general formula for the Laurent expansion of

$$f_n(z) = \frac{1}{(z - \alpha)^n} \qquad (n = 1, 2, \ldots)$$

which is valid for $|z| > |\alpha|$.

12. Prove that if $f(z)$ has a Laurent series expansion of the form $\sum_{j=0}^{\infty} a_j z^j$ in $0 < |z| < \rho$, then $\lim_{z \to 0} f(z)$ exists.

13. Let $f(z)$ be analytic in the annulus $r < |z - z_0| < R$ and bounded by M there. Prove that the coefficients a_j of the Laurent expansion of $f(z)$ in the annulus satisfy

$$|a_j| \le \frac{M}{R^j}, \qquad |a_{-j}| \le Mr^j \qquad (\text{for } j = 0, 1, 2, \ldots).$$

5.6 ZEROS AND SINGULARITIES

In this section we shall use the Laurent expansion to classify, in general terms, the behavior of an analytic function near its zeros and isolated singularities. A *zero* of a function f is a point z_0 where f is analytic and

$f(z_0) = 0$. An *isolated singularity* of f is a point z_0 such that f is analytic in some punctured disk $0 < |z - z_0| < R$ but not analytic at z_0 itself. For example, $\tan(\pi z/2)$ has a zero at each even integer and an isolated singularity at each odd integer.

We shall begin by examining the zeros of f.

DEFINITION 7 *A point z_0 is called a* **zero of order** *m for the function f if f is analytic at z_0 and f and its first $m - 1$ derivatives vanish at z_0, but $f^{(m)}(z_0) \neq 0$.*

In other words, we have

$$f(z_0) = f'(z_0) = f''(z_0) = \cdots = f^{(m-1)}(z_0) = 0 \neq f^{(m)}(z_0).$$

In this case the Taylor series for f around z_0 takes the form

$$f(z) = a_m(z - z_0)^m + a_{m+1}(z - z_0)^{m+1} + a_{m+2}(z - z_0)^{m+2} + \cdots$$

or

(37) $$f(z) = (z - z_0)^m[a_m + a_{m+1}(z - z_0) + a_{m+2}(z - z_0)^2 + \cdots],$$

where $a_m = f^{(m)}(z_0)/m! \neq 0$. The bracketed series in Eq. (37) clearly converges wherever the series for f does (at any particular point one is just a multiple of the other); hence it defines a function $g(z)$ analytic in a neighborhood of z_0, with $g(z_0) \neq 0$. Conversely, any function with a representation like Eq. (37) must have a zero of order m, so we deduce

THEOREM 16 *Let f be analytic at z_0. Then f has a zero of order m at z_0 if and only if f can be written as*

$$f(z) = (z - z_0)^m g(z),$$

where $g(z)$ is analytic at z_0 and $g(z_0) \neq 0$.

A zero of order 1 is sometimes called a *simple zero*. For instance, the zeros of the function $\sin z$, which occur (as we saw in Chapter 3) at integer multiples of π, are all simple (at such points the first derivative, $\cos z$, is nonzero).

An easy consequence of Theorem 16 is the following result which asserts that zeros of nonconstant analytic functions are isolated.

COROLLARY 3 *If f is an analytic function such that $f(z_0) = 0$, then either f is identically zero in a neighborhood of z_0 or there is a punctured disk about z_0 in which f has no zeros.*

Proof If all the Taylor coefficients $a_k = f^{(k)}(z_0)/k!$ of f at z_0 are zero, then f vanishes throughout a disk centered at z_0 by Theorem 3.

Otherwise, let $m(\geq 1)$ be the smallest subscript such that $a_m \neq 0$. Then, by Definition 7, f has a zero of order m at z_0, and so the representation $f(z) = (z - z_0)^m g(z)$ of Theorem 16 is valid. Since $g(z_0) \neq 0$ and g is continuous at z_0 (indeed, it is analytic there), there exists a disk $|z - z_0| < \delta$ throughout which g is nonzero. Consequently $f(z) \neq 0$ for $0 < |z - z_0| < \delta$. ∎

We now turn to the isolated singularities of f. We know that f has a Laurent expansion around any isolated singularity z_0;

$$(38) \qquad f(z) = \sum_{j=-\infty}^{\infty} a_j (z - z_0)^j,$$

for, say, $0 < |z - z_0| < R$. (The "r" of Theorem 14 is zero for an isolated singularity.) We can classify z_0 into one of three categories:

DEFINITION 8 *Let f have an isolated singularity at z_0, and let (38) be the Laurent expansion of f in $0 < |z - z_0| < R$. Then*

   (i) *If $a_j = 0$ for all $j < 0$, we say that z_0 is a **removable singularity** of f,*

   (ii) *If $a_{-m} \neq 0$ for some positive integer m but $a_j = 0$ for all $j < -m$, we say that z_0 is a **pole of order** m for f, and*

   (iii) *If $a_j \neq 0$ for an infinite number of negative values of j, we say that z_0 is an **essential singularity** of f.*

By examining separately each of these three types of isolated singularities we shall show that they can be distinguished by the qualitative behavior of $f(z)$ near the singularity. The resulting characterizations shall be summarized in the final theorem of this section.

When f has a removable singularity at z_0, its Laurent series takes the form

$$(39) \quad f(z) = a_0 + a_1(z - z_0) + a_2(z - z_0)^2 + \cdots \qquad (0 < |z - z_0| < R).$$

Example 16 of the previous section provides an illustration of this. Other examples of functions having removable singularities are

$$\frac{\sin z}{z} = \frac{1}{z}\left(z - \frac{z^3}{3!} + \frac{z^5}{5!} - \cdots\right) = 1 - \frac{z^2}{3!} + \frac{z^4}{5!} - \cdots \qquad (z_0 = 0),$$

$$\frac{\cos z - 1}{z} = \frac{1}{z}\left[\left(1 - \frac{z^2}{2!} + \frac{z^4}{4!} - \cdots\right) - 1\right] = -\frac{z}{2!} + \frac{z^3}{4!} - \cdots \qquad (z_0 = 0),$$

$$\frac{z^2 - 1}{z - 1} = z + 1 = 2 + (z - 1) + 0 + 0 + \cdots \qquad (z_0 = 1).$$

From (39) we can see that, except for the point z_0 itself, $f(z)$ is equal to a function $h(z)$ which is analytic at z_0. In other words, the only reason for the singularity is that $f(z)$ is undefined or defined "peculiarly" at z_0. Since the function $h(z)$ is analytic at z_0, it is obviously bounded in some neighborhood of z_0,† and so we have established

LEMMA 5   If f has a removable singularity at z_0, then

(i) *$f(z)$ is bounded in some punctured neighborhood of z_0,*

(ii) *$f(z)$ has a limit as z approaches z_0, and*

(iii) *$f(z)$ can be redefined at z_0 so that the new function is analytic at z_0.*

Conversely, if a function is bounded in some punctured neighborhood of an isolated singularity, that singularity is removable; see Prob. 13 for a direct proof.

Clearly removable singularities are not too important in the theory of analytic functions. But as we shall see in Lemmas 6 and 8, the concept is occasionally helpful in describing the properties of the other kinds of singularities.

The Laurent series for a function with a pole of order m looks like

$$(40) \qquad f(z) = \frac{a_{-m}}{(z - z_0)^m} + \frac{a_{-(m-1)}}{(z - z_0)^{m-1}} + \cdots + \frac{a_{-1}}{z - z_0}$$
$$+ a_0 + a_1(z - z_0) + \cdots$$
$$(a_{-m} \neq 0),$$

valid in some punctured neighborhood of z_0. For example,

$$\frac{e^z}{z^2} = \frac{1}{z^2}\left(1 + z + \frac{z^2}{2!} + \cdots\right) = \frac{1}{z^2} + \frac{1}{z} + \frac{1}{2!} + \frac{z}{3!} + \cdots$$

has a pole of order 2, and

$$\frac{\sin z}{z^5} = \frac{1}{z^5}\left(z - \frac{z^3}{3!} + \frac{z^5}{5!} - \cdots\right) = \frac{1}{z^4} - \frac{1}{3!\,z^2} + \frac{1}{5!} - \frac{z^2}{7!} + \cdots$$

has a pole of order 4, at $z = 0$.

A pole of order 1 is called a *simple pole*. Obviously $z = 0$ is a simple pole of the function $(\sin z)/z^2$.

From Eq. (40) we can deduce the following characterization of a pole:

LEMMA 6   If the function f has a pole of order m at z_0, then $|(z - z_0)^l f(z)| \to \infty$ as $z \to z_0$‡ for all integers $l < m$, while $(z - z_0)^m f(z)$ has a removable singularity at z_0. In particular, $|f(z)| \to \infty$ as z approaches a pole.

---

†That is, there exists a neighborhood of z_0 and a constant M such that $|h(z)| \leq M$ for all z in this neighborhood.

‡We remind the reader that the notation "$|h(z)| \to \infty$ as $z \to z_0$" means that $|h(z)|$ exceeds any given number for all z sufficiently near z_0.

Proof Equation (40) implies that in some punctured neighborhood of z_0

$$(z - z_0)^{m-1} f(z) = \frac{a_{-m}}{z - z_0} + a_{-m+1} + a_{-m+2}(z - z_0) + \cdots;$$

i.e.,

(41)
$$(z - z_0)^{m-1} f(z) = \frac{a_{-m}}{z - z_0} + h(z),$$

where $h(z)$ is defined by a convergent power series and is thus analytic at z_0. Since $a_{-m} \neq 0$, Eq. (41) shows that $|(z - z_0)^{m-1} f(z)| \to \infty$ as $z \to z_0$. It follows immediately that $|(z - z_0)^l f(z)| \to \infty$ for any $l < m$. Again from Eq. (40), we have

(42)
$$(z - z_0)^m f(z) = a_{-m} + a_{-m+1}(z - z_0) + \cdots,$$

and since there are no negative powers, the singularity of $(z - z_0)^m f(z)$ at z_0 is removable. ∎

Another transparent property of functions possessing a pole is

LEMMA 7 *If $f(z) = g(z)/(z - z_0)^m$ in a punctured neighborhood of z_0 and if $g(z)$ is analytic at z_0 with $g(z_0) \neq 0$, then z_0 is a pole of order m for f.*

Proof Write the Taylor series for $g(z)$:

$$g(z) = a_0 + a_1(z - z_0) + a_2(z - z_0)^2 + \cdots.$$

Then the Laurent series for f near z_0 must be

(43)
$$f(z) = \frac{g(z)}{(z - z_0)^m} = \frac{a_0}{(z - z_0)^m} + \frac{a_1}{(z - z_0)^{m-1}} + \cdots.$$

Since $a_0 = g(z_0) \neq 0$, Eq. (43) displays the predicted pole for f. ∎

EXAMPLE 17 Classify the singularity at $z = 1$ of the function $(\sin z)/(z^2 - 1)^2$.

SOLUTION Since

$$\frac{\sin z}{(z^2 - 1)^2} = \frac{(\sin z)/(z + 1)^2}{(z - 1)^2}$$

and the numerator is analytic and nonzero at $z = 1$, Lemma 7 implies that the function has a pole of order 2. ∎

EXAMPLE 18 Show that the only singularities of rational functions are removable singularities or poles.

SOLUTION   Recall that a rational function is the ratio of two polynomials, $P(z)/Q(z)$, and is analytic everywhere except at the zeros of $Q(z)$. If $Q(z)$ has a zero, say of order m, at z_0, then $Q(z) = (z - z_0)^m q(z)$, where $q(z)$ is a polynomial and $q(z_0) \neq 0$.

If $P(z_0) \neq 0$, we apply Lemma 7 to the expression

$$\frac{P(z)}{Q(z)} = \frac{1}{(z - z_0)^m} \frac{P(z)}{q(z)}$$

to deduce that $P(z)/Q(z)$ has a pole of order m. If, on the other hand, $P(z_0) = 0$, we can write $P(z) = (z - z_0)^n p(z)$, where n is the order of the zero at z_0 [we ignore the trivial case $P(z) \equiv 0$]; thus

$$\frac{P(z)}{Q(z)} = \frac{(z - z_0)^n}{(z - z_0)^m} \frac{p(z)}{q(z)},$$

and clearly $P(z)/Q(z)$ will have a pole if $n < m$ or a removable singularity if $n \geq m$.   ∎

The following lemma relating zeros and poles is easily derived using the above methods of analysis, so we simply state it here for reference purposes and assign the proof to the reader (Prob. 5).

LEMMA 8   If f has a zero of order m at z_0, then $1/f$ has a pole of order m at z_0. Conversely, if f has a pole of order m at z_0, then $1/f$ has a removable singularity at z_0, and if we define $1/f(z_0) = 0$, then $1/f$ has a zero of order m at z_0.

Some students may have felt that it is obvious that $|f(z)| \to \infty$ as z approaches a pole and that our painstaking analysis was unnecessary. They will probably be shocked to learn that such behavior does not occur as z approaches an essential singularity; instead we have

THEOREM 17   (Picard's Theorem) A function with an essential singularity assumes every complex number, with possibly one exception, as a value in any neighborhood of this singularity.

The proof of this theorem is beyond our text, but we invite the student to prove a somewhat weaker result, the *Casorati-Weierstrass Theorem*, in Prob. 14. We illustrate the Picard Theorem in

EXAMPLE 19   Verify Picard's result for $e^{1/z}$ near $z = 0$.

SOLUTION   (Observe first of all that $z = 0$ is an essential singularity; see Example 15.) Obviously $e^{1/z}$ is never zero. However, if $c \neq 0$, we can show that $e^{1/z}$ achieves the value c for $|z|$ less than any positive ε. To this end, recall that

$$\log c = \text{Log } |c| + i \text{ Arg } c + 2n\pi i \qquad (n = 0, \pm 1, \pm 2, \ldots)$$

(Sec. 3.2). By picking n sufficiently large, we can find a value w of $\log c$ such that $|w| > 1/\varepsilon$. Then let $z = 1/w$. We will have $|z| < \varepsilon$, and

$$e^{1/z} = e^w = e^{\log c} = c. \qquad \blacksquare$$

From the above results we observe that the three different kinds of isolated singularities produce qualitatively different behaviors near these points. *Thus boundedness indicates a removable singularity, approaching ∞ indicates a pole, and anything else must indicate an essential singularity.*

EXAMPLE 20   Classify the zeros and singularities of the function $\sin(1 - z^{-1})$.

SOLUTION   Since the zeros of $\sin w$ occur only when w is an integer multiple of π, the function $\sin(1 - z^{-1})$ has zeros when

$$1 - z^{-1} = n\pi,$$

i.e., at

$$z = \frac{1}{1 - n\pi} \qquad (n = 0, \pm 1, \pm 2, \ldots).$$

Furthermore, the zeros are simple because the derivative at these points is

$$\frac{d}{dz} \sin(1 - z^{-1}) \bigg|_{z=(1-n\pi)^{-1}} = \frac{1}{z^2} \cos(1 - z^{-1}) \bigg|_{z=(1-n\pi)^{-1}}$$

$$= (1 - n\pi)^2 \cos n\pi \neq 0.$$

The only singularity of $\sin(1 - z^{-1})$ appears at $z = 0$. Since 0 is the limit point of the sequence $(1 - n\pi)^{-1}$, $n = 1, 2, 3, \ldots$, we see that this function has a zero in every neighborhood of the origin. Consequently, $z = 0$ cannot be a pole. But this point is not a removable singularity either, because $\sin(1 - z^{-1})$ does not *approach* 0 as $z \to 0$; indeed, $\sin(1 - z_p^{-1}) = 1$ for the sequence of values $z_p = (1 - 2p\pi - \pi/2)^{-1}$, $p = 1, 2, 3, \ldots$. By elimination, then, we infer that $z = 0$ must be an essential singularity. $\blacksquare$

EXAMPLE 21   Classify the zeros and poles of the function $f(z) = (\tan z)/z$.

SOLUTION   Since $(\tan z)/z = (\sin z)/z \cos z$, the only possible zeros are those of $\sin z$; i.e., $z = n\pi$ $(n = 0, \pm 1, \pm 2, \ldots)$. However, $z = 0$ is, in fact, a singularity. Furthermore, the points $z = (n + \frac{1}{2})\pi$, which are the zeros of $\cos z$, are also singularities. We shall investigate these in turn.

If n is a nonzero integer, the reader should have no trouble convincing himself that $z = n\pi$ is a *simple* zero for the given function.

Near the point $z = 0$ we can write

$$\frac{\tan z}{z} = \frac{\sin z}{z \cos z} = \frac{1}{z \cos z}\left(z - \frac{z^3}{3!} + \frac{z^5}{5!} - \cdots\right)$$

$$= \frac{1}{\cos z}\left(1 - \frac{z^2}{3!} + \frac{z^4}{5!} - \cdots\right),$$

and we see that $(\tan z)/z \to 1$ as $z \to 0$. Hence the origin is a *removable singularity*.

Finally, since $\cos z$ has simple zeros at $z = (n + \frac{1}{2})\pi$ ($n = 0, \pm 1, \pm 2,$ ...), it is easy to see that $f(z)$ has simple poles at these points. ∎

We conclude this section by summarizing the various equivalent characterizations of the three types of isolated singularities. For economy of notation the statement of Theorem 18 utilizes the logician's symbol "$\Leftrightarrow$" to denote logical equivalence; it can be translated "if and only if."

THEOREM 18   If $f(z)$ has an isolated singularity at z_0, then the following equivalences hold:

(i) *z_0 is a removable singularity $\Leftrightarrow |f|$ is bounded near $z_0 \Leftrightarrow f(z)$ has a limit as $z \to z_0 \Leftrightarrow f$ can be redefined at z_0 so that f is analytic at z_0.*

(ii) *z_0 is a pole $\Leftrightarrow |f(z)| \to \infty$ as $z \to z_0 \Leftrightarrow f$ can be written $f(z) = g(z)/(z - z_0)^m$ for some integer $m > 0$ and some function $g(z)$ analytic at z_0 with $g(z_0) \neq 0$.*

(iii) *z_0 is an essential singularity $\Leftrightarrow |f(z)|$ neither is bounded near z_0 nor goes to infinity as $z \to z_0 \Leftrightarrow f(z)$ assumes every complex number, with possibly one exception, as a value in every neighborhood of z_0.*

Exercises 5.6

1. Find and classify the isolated singularities of each of the following functions.

(a) $\dfrac{z^3 + 1}{z^2(z + 1)}$ (b) $z^3 e^{1/z}$

(c) $\dfrac{\cos z}{z^2 + 1} + 4z$ (d) $\dfrac{1}{e^z - 1}$

(e) $\tan z$ (f) $\cos\left(1 - \dfrac{1}{z}\right)$

(g) $\dfrac{\sin(3z)}{z^2} - \dfrac{3}{z}$ (h) $\cot\left(\dfrac{1}{z}\right)$

2. What is the order of the pole of

$$f(z) = \frac{1}{(2 \cos z - 2 + z^2)^2}$$

at $z = 0$? [HINT: Work with $1/f(z)$.]

3. For each of the following, construct a function $f(z)$ analytic in the plane except for isolated singularities which satisfies the given conditions.

   (a) $f(z)$ has a zero of order 2 at $z = i$ and a pole of order 5 at $z = 2 - 3i$.

   (b) $f(z)$ has a simple zero at $z = 0$ and an essential singularity at $z = 1$.

   (c) $f(z)$ has a removable singularity at $z = 0$, a pole of order 6 at $z = 1$, and an essential singularity at $z = i$.

   (d) $f(z)$ has a pole of order 2 at $z = 1 + i$ and essential singularities at $z = 0$ and $z = 1$.

4. For each of the following, determine whether the statement made is always true or sometimes false.

   (a) If f and g have a pole at z_0, then $f + g$ has a pole at z_0.

   (b) If f has an essential singularity at z_0 and g has a pole at z_0, then $f + g$ has an essential singularity at z_0.

   (c) If $f(z)$ has a pole of order m at $z = 0$, then $f(z^2)$ has a pole of order $2m$ at $z = 0$.

   (d) If f has a pole at z_0 and g has an essential singularity at z_0, then the product $f \cdot g$ has a pole at z_0.

   (e) If f has a zero of order m at z_0 and g has a pole of order n, $n \leq m$, at z_0, then the product $f \cdot g$ has a removable singularity at z_0.

5. Give a proof of Lemma 8.

6. Prove that if $f(z)$ has a pole of order m at z_0, then $f'(z)$ has a pole of order $m + 1$ at z_0.

7. If $f(z)$ is analytic in $D: 0 < |z| \leq 1$, and $z^l \cdot f(z)$ is unbounded in D for every integer l, then what kind of singularity does $f(z)$ have at $z = 0$?

8. Verify Picard's Theorem for the function $\cos(1/z)$ at $z_0 = 0$.

9. Does there exist a function $f(z)$ having an essential singularity at z_0 which is bounded along some line segment emanating from z_0?

10. If the function $f(z)$ is analytic in a domain D and has zeros at the distinct points $z_1, z_2, \ldots, z_n$ of respective orders $m_1, m_2, \ldots, m_n$, then prove that there exists a function $g(z)$ analytic in D such that

$$f(z) = (z - z_1)^{m_1}(z - z_2)^{m_2} \cdots (z - z_n)^{m_n} g(z).$$

11. If f has a pole at z_0, show that $\operatorname{Re} f$ and $\operatorname{Im} f$ take on arbitrarily large positive as well as negative values in any punctured neighborhood of z_0.

12. Prove that if $f(z)$ has a pole of order m at z_0, then $g(z) \equiv f'(z)/f(z)$ has a simple pole at z_0. What is the coefficient of $(z - z_0)^{-1}$ in the Laurent expansion for $g(z)$?

13. Let $f(z)$ have an isolated singularity at z_0 and suppose that $f(z)$ is bounded in some punctured neighborhood of z_0. Prove directly from the integral formula (27) for the Laurent coefficients that $a_{-j} = 0$ for all $j = 1, 2, \ldots$; i.e., $f(z)$ must have a removable singularity at z_0.

14. Without appealing to Picard's Theorem, prove the *Casorati-Weierstrass Theorem*: If $f(z)$ has an essential singularity at z_0, then in any punctured neighborhood of z_0 the function $f(z)$ comes arbitrarily close to any specified complex number. [HINT: Let the specified number be c and assume to the contrary that $|f(z) - c| \geq \delta > 0$ in every small punctured neighborhood of z_0. Then, using Prob. 13, show that $f(z) - c$ [and hence $f(z)$ itself] must have either a pole or a removable singularity at z_0.]

15. Prove that if $f(z)$ has an essential singularity at z_0, then so does the function $e^{f(z)}$.

16. By completing each of the following steps prove

SCHWARZ'S LEMMA   If $f(z)$ is analytic in the unit disk U: $|z| < 1$ and satisfies the conditions

$$f(0) = 0, \text{ and } |f(z)| \leq 1 \text{ for all } z \text{ in } U,$$

then $|f(z)| \leq |z|$ for all z in U.

(a) Define $F(z) \equiv f(z)/z$, for $z \neq 0$, and $F(0) = f'(0)$. Show that F is analytic in U.

(b) Let $\xi(\neq 0)$ be any fixed point in U, and r be any real number which satisfies $|\xi| < r < 1$. Show by means of the Maximum-Modulus Principle that if C_r denotes the circle $|z| = r$, then

$$|F(\xi)| \leq \max_{z \text{ on } C_r} \frac{|f(z)|}{r} \leq \frac{1}{r}.$$

(c) Letting $r \to 1^-$ in part (b), deduce that $|f(\xi)| \le |\xi|$ for all ξ in U.

17. Let f be a function satisfying the conditions of Schwarz's Lemma (Prob. 16). Prove that if $|f(z_0)| = |z_0|$ for some nonzero z_0 in U, then f must be a function of the form $f(z) = e^{i\theta}z$ for some real θ. Show also that f must be of this form if $|f'(0)| = 1$.

5.7 THE POINT AT INFINITY

From our discussion of singularities in Sec. 5.6 we know that if a mapping is given by an analytic function possessing a pole, it carries points near that pole to indefinitely distant points. It must have occurred to the reader that one might define the value of f *at* the pole to be ∞. Before taking this plunge, however, we should be aware of all the ramifications. Let us look in detail at the behavior of $1/z$ near $z = 0$.

As $z \to 0$ along the positive real axis, $1/z$ goes to "plus infinity"; along the negative real axis, $1/z$ goes to "minus infinity"; and along the positive y-axis, $1/z$ goes to—what? "Minus i times infinity?" If we are to assign the symbol ∞ to $1/0$ we must realize that we are identifying all these "limits" as a single number; geometrically, we are speaking of *the point at infinity*, which can be reached, in a manner of speaking, by proceeding infinitely far along any direction in the complex plane.

It is somewhat enlightening to try to visualize the situation in the following manner: Consider the rays emanating from the origin in the complex plane to all be joined at their "ends," deforming the complex plane into something like an upside-down parachute with its lines tied together. In fact, such an image motivates the *stereographic projection* depicted in Fig. 5.5 wherein the xy-plane is mapped onto the surface of the three-dimensional unit sphere which has the xy-plane as its equatorial plane. The mapping is defined as follows: Starting from a point P in the plane, we draw a line through the north pole of the sphere. Then the point P is mapped to the point P', where this line intersects the sphere, as illustrated in Fig. 5.5.

Thus the open disk $|z| < 1$ is mapped onto the southern hemisphere, the circle $|z| = 1$ onto the equator, and the exterior $|z| > 1$ onto the northern hemisphere excluding the north pole. Notice that all distant points get mapped onto "arctic zones" near the north pole; the latter is, in fact, the actual limit of such points, in the topology of the sphere. It corresponds to the "point at ∞."

With this model in mind, we officially append "∞" to the set of complex numbers, calling the resulting collection the *extended complex plane*. A neighborhood of ∞ is any set of the form $|z| > M$, and a

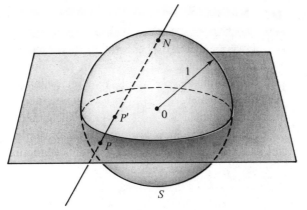

FIGURE 5.5

sequence of points z_n $(n = 1, 2, 3, \ldots)$ approaches ∞ if $|z_n|$ can be made arbitrarily large by taking n large.

Consequently we shall write $f(z_0) = \infty$ when $|f(z)|$ increases without bound as $z \to z_0$† and shall write $f(\infty) = w_0$ when $f(z) \to w_0$ as $z \to \infty$. For example, if

(44) $$f(z) = \frac{2z + 1}{z - 1},$$

then $f(1) = \infty$ and $f(\infty) = 2$.

Observe that for $f(z) = 2z + 1$, we have $f(\infty) = \infty$.

It is convenient in some applications to carry this notion still further and speak of functions which are "analytic at ∞." The analyticity properties of f at ∞ are classified by first performing the mapping $w = 1/z$, which maps the point at infinity to the origin, and then examining the behavior of the composite function $g(w) \equiv f(1/w)$ at the origin $w = 0$. Thus we say

(i) $f(z)$ is analytic at ∞ if $f(1/w)$ is analytic (or has a removable singularity) at $w = 0$,‡

(ii) $f(z)$ has a pole of order m at ∞ if $f(1/w)$ has a pole of order m at $w = 0$, and

(iii) $f(z)$ has an essential singularity at ∞ if $f(1/w)$ has an essential singularity at $w = 0$.

From Theorem 18, we can interpret these conditions for a function analytic outside some disk as follows:

(i') $f(z)$ is analytic at ∞ if $|f(z)|$ is bounded for sufficiently large $|z|$,

†Technically, $f(z_0) = \infty$ if for any $M > 0$ there is a $\delta > 0$ such that $0 < |z - z_0| < \delta$ implies that $|f(z)| > M$.

‡Some authors also allow the possibility of a removable singularity at ∞, but we feel that nothing is gained by this generality.

(*ii'*) $f(z)$ has a pole at ∞ if $f(z) \to \infty$ as $z \to \infty$, and

(*iii'*) $f(z)$ has an essential singularity at ∞ if $|f(z)|$ neither is bounded for large $|z|$ nor goes to infinity as $z \to \infty$.

EXAMPLE 22 Classify the behavior at ∞ of the functions $z^2 + 2$, $(iz + 1)/(z - 1)$, and $\sin z$.

SOLUTION Obviously $f(z) = z^2 + 2$ has a pole at ∞. The pole is of order 2, because

$$f\left(\frac{1}{w}\right) = \frac{1}{w^2} + 2$$

has a pole of order 2 at $w = 0$.

Since

$$\frac{iz + 1}{z - 1} \to i \quad \text{as} \quad z \to \infty,$$

this function is certainly analytic at ∞.

Finally, $\sin z$ has no limit as $z \to \infty$, even for real z (it oscillates). Hence ∞ must be an essential singularity.† ∎

EXAMPLE 23 Find all the functions f which are analytic everywhere in the extended complex plane.

SOLUTION Since f is analytic at ∞, it is bounded for, say, $|z| > M$. By continuity, f is also bounded for $|z| \le M$. Consequently, f is a bounded entire function. Hence f is constant, by Liouville's Theorem. ∎

EXAMPLE 24 Classify all the functions which are everywhere analytic in the extended complex plane except for a pole at one point.

SOLUTION If $f(z)$ has a pole, say, of order m at some finite point z_0, then the Laurent series for f

$$(45) \quad f(z) = \frac{a_{-m}}{(z - z_0)^m} + \frac{a_{-m+1}}{(z - z_0)^{m-1}} + \cdots + \frac{a_{-1}}{z - z_0} + \sum_{n=0}^{\infty} a_n(z - z_0)^n$$

converges for all $z \ne z_0$. Moreover, since we are assuming that $z_0 \ne \infty$, the function f must be analytic, and hence bounded, at ∞. From Eq. (45), then, we see that the *entire* function defined by the series $\sum_{n=0}^{\infty} a_n(z - z_0)^n$

†Alternatively this can be seen directly from the Laurent expansion for $\sin(1/w)$ about $w = 0$.

is also bounded at ∞; thus it must be constant, i.e., equal to a_0. Therefore, the most general form for such a function is

(46) $$f(z) = \frac{a_{-m}}{(z - z_0)^m} + \frac{a_{-m+1}}{(z - z_0)^{m-1}} + \cdots + \frac{a_{-1}}{z - z_0} + a_0 .$$

If the pole occurs at $z = \infty$, then $f(1/w)$ has a pole at the origin and can be expressed in the form

(47) $$f\left(\frac{1}{w}\right) = \frac{a_{-m}}{w^m} + \frac{a_{-m+1}}{w^{m-1}} + \cdots + \frac{a_{-1}}{w} + \sum_{n=0}^{\infty} a_n w^n.$$

Since $f(z)$ is bounded near $z = 0$, $f(1/w)$ is bounded for large $|w|$, and, as before, we conclude that $a_n = 0$ for $n > 0$. Hence Eq. (47) becomes

(48) $$f(z) = a_{-m} z^m + a_{-m+1} z^{m-1} + \cdots + a_{-1} z + a_0;$$

i.e., $f(z)$ is a *polynomial* in z.

Equations (46) and (48) categorize the totality of all functions possessing one pole in the extended complex plane. ∎

We note in passing that the theory of *Fuchsian equations* is based upon considerations of singularities in the extended complex plane, and these have been extremely helpful in relating many of the so-called "special functions" which arise in mathematical physics; Ref. [3] discusses this application.

Exercises 5.7

1. Classify the behavior at ∞ for each of the following functions.

(a) e^z (b) $\cosh z$ (c) $\dfrac{z - 1}{z + 1}$

(d) $\dfrac{z}{z^3 + i}$ (e) $\dfrac{z^3 + i}{z}$ (f) $e^{\sinh z}$

(g) $\dfrac{\sin z}{z^2}$ (h) $\dfrac{1}{\sin z}$ (i) $e^{\tan 1/z}$

2. State Picard's Theorem (Sec. 5.6) for functions with an essential singularity at ∞. Verify for e^z.

3. Prove that if $f(z)$ is analytic at ∞, then it has a series expansion of the form

$$f(z) = \sum_{n=0}^{\infty} \frac{a_n}{z^n}$$

converging uniformly outside some disk.

4. Construct the series mentioned in Prob. 3 for the following functions.

(a) $\dfrac{z-1}{z+1}$ (b) $\dfrac{z^2}{z^2+1}$ (c) $\dfrac{1}{z^3-i}$

5. What is the meaning of the statement "$f(z)$ has a zero of order m at ∞"? What is the order of the zero at ∞ if $f(z)$ is a rational function of the form $P(z)/Q(z)$ with deg $P <$ deg Q?

6. Let the point $z = x + iy$ correspond to the point (x_1, x_2, x_3) on the sphere $x_1^2 + x_2^2 + x_3^2 = 1$ under the stereographic projection. Show that

$$x_1 = \frac{2x}{|z|^2+1}, \qquad x_2 = \frac{2y}{|z|^2+1}, \qquad x_3 = \frac{|z|^2-1}{|z|^2+1}.$$

[HINT: The three points $(0, 0, 1)$, (x_1, x_2, x_3), and $(x, y, 0)$ must be collinear.]

7. Prove that the image under the stereographic projection of a circle or line in the z-plane is a circle on the sphere $x_1^2 + x_2^2 + x_3^2 = 1$. [HINT: Show that the image is the intersection of the sphere with a plane $Ax_1 + Bx_2 + Cx_3 = D$.]

*5.8 ANALYTIC CONTINUATION

Analytic continuation is a very simple concept which is difficult to formulate in abstract terms. So by way of motivation, we open our discussion with a parable.

A mathematician wishes to investigate the function $f(z)$ defined by the power series

(49) $$f(z) = \sum_{j=1}^{\infty} \frac{(-1)^{j-1}(z-1)^j}{j};$$

he does not recognize that this is a Taylor series for Log z. By, say, the Ratio Test he determines that the circle of convergence turns out to be $C: |z-1| = 1$ (reaching, of course, to the singular point $z = 0$). Thus he knows the values of this analytic function inside C; to be realistic, we should say he can compute, to arbitrary accuracy, the values of the function (and its derivatives) inside the circle. In particular, he knows the function and its derivatives at the point z_1 in Fig. 5.6.

Now he forms the power series

(50) $$\sum_{j=0}^{\infty} \frac{f^{(j)}(z_1)}{j!}(z-z_1)^j.$$

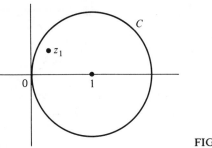

FIGURE 5.6

Of course, this is a Taylor series for f and will converge to f at least inside the small circle C' of Fig. 5.7. But when he actually calculates the circle of convergence, he finds it to be the larger circle C'', reaching to the origin (which *we* know is still the nearest singularity of Log z). He can thus extend the *domain of definition* of the analytic function f. Such a procedure is called *analytic continuation*. To be precise, we formulate

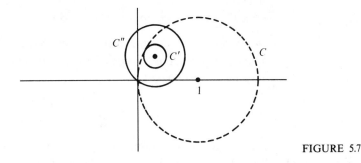

FIGURE 5.7

DEFINITION 9 Suppose that f is analytic in a domain D_1 and that g is analytic in a domain D_2. Then we say that g is a **direct analytic continuation** *of f to D_2 if $D_1 \cap D_2$ is nonempty and $f(z) = g(z)$ for all z in $D_1 \cap D_2$.* (The adjective "direct" is used here to distinguish this concept from another kind of analytic continuation to be discussed below.)

In the parable the first domain D_1 is the interior of C, the second domain D_2 is the interior of C'', and g is the sum of the power series (50). We have not actually proved that $f = g$ over the whole lens-shaped domain $D_1 \cap D_2$; equality has only been established inside C'. So we shall use the following theorem to show that (50) is a bona fide direct analytic continuation of $f(z)$ to D_2.

THEOREM 19 If $F(z)$ is analytic in a domain D and vanishes on some open disk contained in D, then it vanishes throughout D.

Proof By hypothesis, D contains a disk $\mathcal{N}$, say with center z_0, such that $F(z) = 0$ for all z in $\mathcal{N}$. Now let's assume, contrary to the

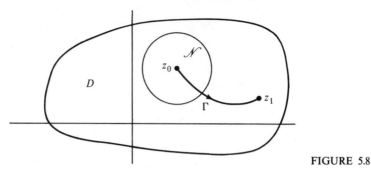

FIGURE 5.8

conclusion of the theorem, that there is some point z_1 in D such that $F(z_1) \neq 0$, and let Γ be a path in D joining z_0 to z_1. (See Fig. 5.8.) As we move along Γ from z_0, at first we only observe points where $F(z) = 0$. Eventually, however, we must encounter a point w with the properties

(*i*) For all z on Γ preceding w, $F(z) = 0$, and

(*ii*) There are points z on Γ arbitrarily close to w such that $F(z) \neq 0$. (Recall a similar situation in Sec. 4.6.)

First we observe that condition (*i*) implies that the derivative of $F(z)$ is zero on the portion of Γ *preceding w*, because at any point z on this part of the curve we can evaluate

$$F'(z) = \lim_{\xi \to z} \frac{F(\xi) - F(z)}{\xi - z}$$

by letting ξ approach z along this portion of Γ, where $F(\xi) = F(z) = 0$. Hence the limit is zero. Continuing in this fashion, we express

$$F''(z) = \lim_{\substack{\xi \to z \\ \xi \text{ on } \Gamma}} \frac{F'(\xi) - F'(z)}{\xi - z} = \lim_{\substack{\xi \to z \\ \xi \text{ on } \Gamma}} 0 = 0,$$

and so on, to conclude that *all derivatives of $F(z)$ vanish on the portion of Γ preceding w*. By continuity, then, F and all its derivatives also vanish at w, which implies that the Taylor coefficients for F about $z = w$ are identically zero; i.e., F must vanish in some *disk* around w. But this contradicts condition (*ii*), so our assumption that F is not identically zero must be wrong. ∎

COROLLARY 4   *If f and g are analytic in a domain D and $f = g$ in some disk contained in D, then $f = g$ throughout D.*

Proof   Simply apply Theorem 19 to the difference $F \equiv f - g$. ∎

Returning to our discussion of the situation depicted in Fig. 5.7 we can now conclude from the fact that the series (49) and (50) agree inside the circle C', that they must agree inside the lens-shaped domain formed by the intersection of the interiors of C and C''. This, then, implies that (50) is a direct analytic continuation of $f(z)$ from the interior of C to the interior of C'', in strict accordance with Definition 9.

It is worthwhile at this point to list some salient observations about direct analytic continuation. The first two are so trivial that we shall delete the proofs:

THEOREM 20 *If f is analytic in a domain D_1 and g is a direct analytic continuation of f to the domain D_2, then the function*

$$(51) \qquad\qquad F(z) \equiv \begin{cases} f(z) & (\text{for } z \text{ in } D_1), \\ g(z) & (\text{for } z \text{ in } D_2), \end{cases}$$

is single-valued and analytic on $D_1 \cup D_2$.

THEOREM 21 *If f is analytic in a domain D_1 and D_2 is a domain such that $D_1 \cap D_2$ is nonempty, then the direct analytic continuation of f to D_2, if it exists, is unique.*

Theorem 19 and its corollary can be generalized as follows:

THEOREM 22 *Suppose that f is analytic in a domain D and that $\{z_n\}$ is an infinite sequence of distinct points converging to a point z_0 in D. Suppose, moreover, that $f(z_n) = 0$ for each $n = 1, 2, \dots$. Then $f(z) \equiv 0$ throughout D.*

Proof By continuity, z_0 is a zero of f. However, it is not an isolated zero because every punctured disk about z_0 contains points of the sequence $\{z_n\}$. Consequently, by Corollary 3, Sec. 5.6, f must be identically zero in some neighborhood of z_0. Hence Theorem 19 implies $f(z) \equiv 0$ throughout D. ∎

COROLLARY 5 *If f and g are analytic functions in a domain D and $f(z_n) = g(z_n)$ for an infinite sequence of distinct points $\{z_n\}$ converging to a point z_0 in D, then $f \equiv g$ throughout D.*

Proof Again, consider $f - g$. ∎

Often this corollary is used to extend a known equality from a curve to a domain, as in

EXAMPLE 25 Prove, by direct analytic continuation, that $\sin^2 z + \cos^2 z = 1$ for all z.

SOLUTION We know from elementary trigonometry that the equality is true for real z. In other words, the two entire functions $f(z) \equiv$

$\sin^2 z + \cos^2 z$ and $g(z) \equiv 1$ agree on the real axis. Corollary 5 thus extends the equality to the whole plane. ▮

Now let's turn to a related topic, the concept of *analytic continuation along a curve*. The situation is as follows (refer to Fig. 5.9): $f(z)$ is analytic in a domain D, z_1 is a point in D, and γ is some path connecting z_1 to a point z^*.

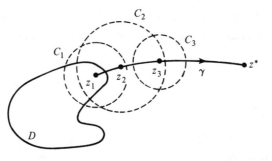

FIGURE 5.9

We expand $f(z)$ in a Taylor series around z_1; the resulting power series converges to a function $f_1(z)$ inside, say, the circle C_1. Staying inside C_1, we proceed along γ until we come to some point z_2, and then we expand $f_1(z)$ around z_2. This series converges to some analytic function $f_2(z)$ inside the circle C_2. Next we move further along γ, now staying inside C_2, to some point z_3, and expand around z_3. And so on. If this process eventually produces a circle of convergence, say C_n, enclosing the portion of γ between z_n and z^*, we say that the scheme derived from the sequence of points $\{z_1, z_2, \ldots, z_n, z^*\}$ and the corresponding functions $\{f, f_1, f_2, \ldots, f_n\}$ constitutes an *analytic continuation of $f(z)$ to z^* along the curve γ*. The value of this analytic continuation at z^* is, of course, $f_n(z^*)$.

(We remark that there can be situations in which f_1 is not a *direct* analytic continuation of f; Prob. 6 illustrates this possibility.)

Let's see what would happen if our aforementioned colleague investigating the series $\sum_{j=1}^{\infty} (-1)^{j-1}(z-1)^j/j$ sought to establish an analytic continuation along the curve γ in Fig. 5.10. He expands around z_1 (as before), computing the function $f_1(z)$, which we know to equal Log z.

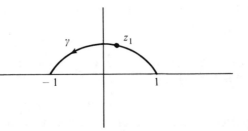

FIGURE 5.10

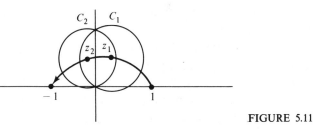

FIGURE 5.11

inside C_1. Next he expands around z_2, deriving the function $f_2(z)$ inside the circle C_2; see Fig. 5.11.

But now the domain of the analytic function f_2 extends beyond the negative real axis, so we cannot identify f_2 with Log z over the whole interior of C_2. However, $f_2(z)$ is an analytic function, defined on the interior of C_2, whose value and derivatives coincide with those of Log z at z_2; hence $f_2(z)$ must agree with some appropriately chosen *branch* of log z whose branch cut does not intersect C_2. For example, the branch given by Log $|z| + i$ arg z with $0 \le$ arg $z < 2\pi$ matches Log z at z_2; hence it agrees with $f_2(z)$ inside C_2.

Finally our tireless mathematician expands around the point z_3 and derives the function $f_3(z)$, analytic inside C_3 (Fig. 5.12). Once again we see that $f_3(z) =$ Log $|z| + i$ arg z, $0 \le$ arg $z < 2\pi$, inside C_3; in particular, $f_3(-1) = \pi i$. In short, the mathematician has analytically continued the power series $f(z) = \sum_{j=1}^{\infty} (-1)^{j-1}(z-1)^j/j$ along the curve γ, and the value of the continuation at -1 is πi.

FIGURE 5.12   FIGURE 5.13

It is instructive to study the result of continuing this same function f along the curve γ' in Fig. 5.13. In this case, the scheme might consist of the points $\{+1, \xi_1, \xi_2, \xi_3, -1\}$ and the functions $\{f, g_1, g_2, g_3\}$. The power series computed around the point ξ_2 would sum to the function $g_2(z)$, whose derivatives would agree with those of Log z at ξ_2 but whose domain of analyticity (the interior of C_2') would enclose a portion of the negative real axis. Thus $g_2(z)$ and, in fact, $g_3(z)$ would agree with the branch $\text{Log } |z| + i \arg z$, $-2\pi \le \arg z < 0$. In particular, $g_3(-1) = -\pi i$.

Summarizing, we have seen that an analytic continuation of $f(z)$ along γ gives the value πi at $z = -1$, but an analytic continuation along γ' gives the value $-\pi i$. We conclude that, in general, the value obtained by analytic continuation along a curve may depend on the curve itself, not merely on its terminal point.

The alert reader has probably surmised by now that the source of this anomaly is the singularity of all the branches of $\log z$ at the origin. This is in fact the case, and the following result, known as the *Monodromy Theorem*, can be considered a vindication of the procedure of analytic continuation along curves:

THEOREM 23 (Monodromy Theorem) *Let $f(z)$ be analytic in a domain D, and suppose that γ and γ' are two directed smooth curves connecting the point z_1 in D to some point z^*. Suppose further that there is some domain D' with the following properties:*

(i) *The loop $\Gamma = \{\gamma, -\gamma'\}$ lies in D' and can be continuously deformed† to a point in D', and*

(ii) *$f(z)$ can be analytically continued along any smooth curve in D'.*

Then the value at z^* of the analytic continuation of f along γ agrees with the value of its continuation along γ'.

Thus, for instance, if we continue $\text{Log } z$ from $z = +1$ to $z = -1$ along *any* curve lying in the upper half-plane, we shall arrive at the value πi, while the value of the continuation along any curve in the lower half-plane will be $-\pi i$.

The proof of the Monodromy Theorem involves some topological constructions with which our readers may be unfamiliar, so we shall delete it (see Ref. [4]). It should be noted, however, that one consequence of the theorem is the fact (which we have tacitly assumed) that the analytic continuation of a function along a particular curve does not depend on the sequence of points $\{z_1, z_2, \ldots, z_n, z^*\}$ used in the scheme—again we direct the reader to the references for a rigorous proof.

EXAMPLE 26 Consider the function $f(z)$ defined for Re $z > 0$ as the principal branch of $z^{1/2}$, i.e., that branch which takes the value $+1$ when $z = 1$. What is the value obtained by continuation of f along a simple closed positively oriented curve γ beginning at $z = 1$ and terminating at the same point, $z = 1$?

†Compare Sec. 4.4a.

SOLUTION   We must consider three possibilities: Either the origin lies on γ, outside γ, or inside γ.

Case 1: The origin lies on γ. Then analytic continuation along γ is not possible; there is no scheme of points and power series which can pass through the "barrier" $z = 0$, since this singularity will be excluded from every circle of convergence of a Taylor series for $z^{1/2}$. See Fig. 5.14.

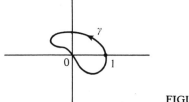

FIGURE 5.14

Case 2: The origin lies outside γ (see Fig. 5.15). Then the conditions of the Monodromy Theorem are satisfied with the curve γ' consisting of the single point $z = 1$, and the domain D' given by the entire plane with the origin deleted. The continuation along γ' results, of course, in the value $+1$.

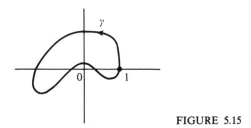

FIGURE 5.15

Case 3: The origin lies inside γ. We can apply the Monodromy Theorem, again identifying D' as the punctured plane but now taking γ' to be the unit circle $|z| = 1$, positively oriented (see Fig. 5.16). To get the continuation along γ' we use De Moivre's formula for $z^{1/2}$. Then it is clear that if γ' is parametrized by $z = e^{i\theta}, 0 \leq \theta \leq 2\pi$, the value of the continuation of $z^{1/2}$ at $z = e^{i\theta}$ is $e^{i\theta/2}$; this gives $e^{i2\pi/2} = -1$ at the terminal point

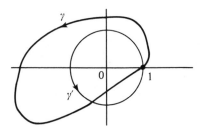

FIGURE 5.16

$z = 1$. Thus we learn that analytic continuation around a *closed* curve may yield functional values which are different from the original ones! ∎

We conclude this section by remarking that there exist functions analytic in a domain D which cannot be analytically continued to *any* point outside D. In such a case the boundary of D is called a *natural boundary*. This situation is illustrated in Prob. 9.

Exercises 5.8

1. Prove that if $f(z)$ is analytic and agrees with a polynomial $\sum_{j=0}^{n} a_j x^j$ for $z = x$ on a segment of the real axis, then $f(z) = \sum_{j=0}^{n} a_j z^j$ everywhere.

2. Prove that if f is analytic in a deleted neighborhood of $z = 0$ and if $f(1/n) = 0$ for all $n = \pm 1, \pm 2, \ldots$, then either f is identically zero or f has an essential singularity at $z = 0$.

3. Given that $f(z)$ is analytic at $z = 0$ and that $f(1/n) = 1/n^2$, $n = 1, 2, \ldots$, find $f(z)$.

4. Does there exist a function $f(z)$, not identically zero, which is analytic in the open disk D: $|z| < 1$ and vanishes at infinitely many points in D?

5. Let $f(z) = \sum_{j=0}^{\infty} z^j$, $|z| < 1$. For what values of α ($|\alpha| < 1$) does the Taylor expansion of $f(z)$ about $z = \alpha$ yield a direct analytic continuation of $f(z)$ to a disk extending outside $|z| < 1$?

6. Show that when a function f is analytically continued along a curve (as depicted in Fig. 5.9), the first function $f_1(z)$ generated by the power series expansion around the initial point z_1 of Γ need not be a direct analytic continuation of f. [HINT: Take $f(z) = \operatorname{Log} z$, with D and z_1 as depicted in Fig. 5.17.]

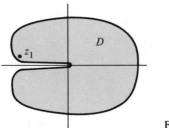

FIGURE 5.17

7. By summing the series, show that $-\sum_{j=0}^{\infty} (2 - z)^j$ is an analytic continuation, along some curve, of $\sum_{j=0}^{\infty} z^j$.

8. For each of the following functions, choose a branch which is analytic in the circle $|z - 2| < 1$. Then analytically continue this branch along the curve γ indicated in Fig. 5.18. Do the new functional values agree with the old?

   (a) $3z^{2/3}$ (b) $\sin 5z$

   (c) $(e^z)^{1/3}$ (d) $\sin(z^{1/2})$

   (e) $(z^{1/2})^2$ (f) $(z^2)^{1/2}$

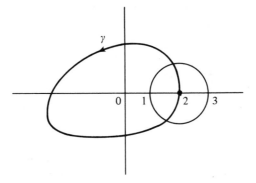

FIGURE 5.18

9. Show that the unit circle $|z| = 1$ is a natural boundary for the function $f(z) = \sum_{j=1}^{\infty} z^{j!}$, $|z| < 1$. [HINT: Argue that $|f(re^{i\theta\pi})| \to \infty$ as $r \to 1^-$ for any rational number θ.]

10. Show that $|z| = 1$ is a natural boundary for the function $g(z) = \sum_{j=1}^{\infty} z^{j!}/j!$, although the series converges for $|z| = 1$. [HINT: Relate $g(z)$ to the function $f(z)$ of Prob. 9.]

11. Show that $|z| = 1$ is a natural boundary for $\sum_{j=0}^{\infty} z^{2^j}$.

12. The *Gamma function* $\Gamma(z)$ is defined for Re $z > 0$ by the integral

$$\Gamma(z) = \int_0^{\infty} e^{-t} t^{z-1} \, dt.$$

   (a) Show that $\Gamma(z + 1) = z\Gamma(z)$ for Re $z > 0$.

   (b) In most advanced texts it is shown that $\Gamma(z)$ is analytic in the right half-plane. Assuming this, argue that the functional equation in part (a) can be used to analytically continue $\Gamma(z)$ to the entire plane, except for the negative integers $z = -n$, $n = 1, 2, \ldots$.

   (c) Show that $\Gamma(n) = (n - 1)!$ for positive integers n.

13. The *Schwarz Reflection Principle* provides a formula for analytic continuation across a straight-line segment under certain circumstances. Its simplest form is stated as follows: Suppose that $f(z) = u(x, y) + iv(x, y)$ is analytic in a simply connected domain D which lies in the upper half-plane and which has a segment γ of the real axis as part of its boundary (see Fig. 5.19). Suppose furthermore

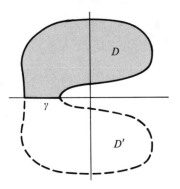

FIGURE 5.19

that $v(x, y) \to 0$ as (x, y) approaches γ and that $u(x, y)$ also takes continuous limiting values on γ, denoted by $U(x)$ [so that $U(x)$ is continuous on γ]. Then the function f can be analytically continued across γ into the domain D', which is the reflection of D in the real axis. Specifically, the function $F(z)$ defined by

$$F(z) = \begin{cases} u(x, y) + iv(x, y) & \text{(for } z = x + iy \text{ in } D\text{)}, \\ U(x) & \text{(for } z = x \text{ on } \gamma\text{)}, \\ u(x, -y) - iv(x, -y) & \text{(for } z = x + iy \text{ in } D'\text{)} \end{cases}$$

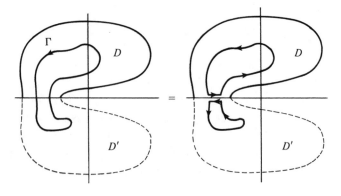

FIGURE 5.20

is analytic in the domain $D \cup \gamma \cup D'$. Justify this principle based upon the following observations.

(a) $F(z)$ satisfies the Cauchy-Riemann equations in D'.

(b) $F(z)$ is continuous in $D \cup \gamma \cup D'$.

(c) Morera's Theorem can be applied to $F(z)$, if the contour of integration Γ is decomposed as illustrated in Fig. 5.20.

14. State and prove a generalization of the Schwarz Reflection Principle to the case where the boundary of D contains an arbitrary line segment upon which the limiting values of $f(z)$ all lie on some straight line (as would be the case, for example, if these limiting values were all real).

15. State and prove two reflection principles for harmonic functions $\phi(x, y)$, based upon the Schwarz Reflection Principle. One should cover the case when $\phi(x, y) \to 0$ on the real axis, and the other should apply when $\partial\phi/\partial y \to 0$ on the real axis.

SUMMARY

The principal achievement of this chapter was to establish an equivalence (roughly speaking) between analytic functions and convergent power series. The equivalence is as follows: Any function f can be expressed as a power series around any point z_0 at which f is analytic, and the series converges uniformly in every closed disk (centered at z_0) which excludes the singularities of f. This series is known as the Taylor series and has the form

$$\sum_{j=0}^{\infty} \frac{f^{(j)}(z_0)}{j!} (z - z_0)^j.$$

On the other hand, any power series $\sum_{j=0}^{\infty} a_j(z - z_0)^j$ converging in a disk $|z - z_0| < R$ sums to an analytic function and, in fact, is the Taylor series of this function.

Power series can be added, integrated, and differentiated termwise, as can any uniformly convergent sequence of analytic functions. Moreover, power series can be multiplied like polynomials.

If a function f fails to be analytic at certain points but is analytic in an annulus surrounding or excluding these points, it can be expanded in a Laurent series,

$$f(z) = \sum_{j=-\infty}^{\infty} a_j(z - z_0)^j,$$

converging uniformly in any closed subannulus; the nonanalyticity is reflected in the appearance of negative exponents in the expansion. In fact, if z_0 is an isolated singularity of the function f, the Laurent series can be used to classify z_0 into one of three categories: a removable singularity (f bounded near z_0), a pole ($|f| \to \infty$ at z_0), or an essential singularity (neither of the above). Similarly, the Taylor series allows one to classify the order of an isolated zero of f.

The proof of the Taylor and Laurent expansions, and the actual computation of many Taylor and Laurent series, is facilitated by the use of the geometric series $\sum_{j=0}^{\infty} w^j$, which converges uniformly to $(1-w)^{-1}$ on any closed disk of the form $|w| \le \rho < 1$.

Finally, we have seen how the power series expansions lead one to the possibility of extending the domain of definition of f as an analytic function. Analytic continuation, however, is a subtle process in that it may result in multiple-valuedness.

SUGGESTED READING

More detailed treatments of some of the topics of this chapter can be found in the following references:

Theory of Series

[1] DIENES, P. *The Taylor Series*. Dover Publications, Inc., New York, 1957.

[2] KNOPP, K. *Infinite Sequences and Series*. Dover Publications, Inc., New York, 1956.

Fuchsian Equations

[3] BIRKHOFF, G., and ROTA, G. C. *Ordinary Differential Equations*, 2nd ed. Xerox College Publishing, Lexington, Mass., 1969.

Analytic Continuation and the Reflection Principle

[4] HILLE, E. *Analytic Function Theory*, Vol. II. Ginn/Blaisdell, Waltham, Mass., 1962.

[5] NEHARI, Z. *Conformal Mapping*. McGraw-Hill Book Company, New York, 1952.

6

Residue Theory

We have already seen how the theory of contour integration lends great insight into the properties of analytic functions. In this chapter we shall explore another dividend of this theory, namely its usefulness in evaluating certain *real* integrals. We shall begin by presenting a technique for evaluating contour integrals which is known as *residue theory*.

6.1 THE RESIDUE THEOREM

Let us consider the problem of evaluating the integral

$$\int_\Gamma f(z)\, dz,$$

where Γ is a simple closed positively oriented contour, and $f(z)$ is analytic on and inside Γ *except* for a single isolated singularity, z_0, lying interior to Γ. As we know, the function $f(z)$ has a Laurent series expansion

(1)
$$f(z) = \sum_{j=-\infty}^{\infty} a_j(z - z_0)^j,$$

converging in some punctured neighborhood of z_0; in particular Eq. (1) is valid for all z on the small positively oriented circle C indicated in Fig. 6.1. By the methods of Sec. 4.4, integration over Γ can be converted to integration over C without changing the integral:

$$\int_\Gamma f(z)\, dz = \int_C f(z)\, dz.$$

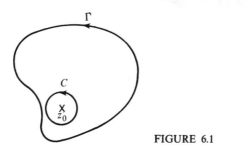

FIGURE 6.1

This last integral can be computed by termwise integration of the series (1) along C. For all $j \neq -1$ the integral is zero, and for $j = -1$ we obtain the value $2\pi i a_{-1}$. Consequently we have

(2)
$$\int_\Gamma f(z)\, dz = 2\pi i a_{-1}.$$

(Compare this with the formula for a_{-1} given in Theorem 14 of Chapter 5.)

Thus the constant a_{-1} plays an important role in contour integration. Accordingly, we adopt the following terminology:

DEFINITION 1 *If f has an isolated singularity at the point z_0, then the coefficient a_{-1} of $1/(z - z_0)$ in the Laurent expansion for f around z_0 is called the **residue of f** at z_0 and is denoted by*

$$\text{Res}(f; z_0) \quad \text{or} \quad \text{Res}(z_0).$$

EXAMPLE 1 Find the residue at $z = 0$ of the function $f(z) = ze^{3/z}$ and compute

$$\oint_{|z|=4} ze^{3/z}\, dz.$$

SOLUTION   Since e^w has the Taylor expansion

$$e^w = \sum_{j=0}^{\infty} \frac{w^j}{j!} \qquad \text{(for all } w\text{)},$$

the Laurent expansion for $ze^{3/z}$ around $z = 0$ is given by

$$ze^{3/z} = z \sum_{j=0}^{\infty} \frac{1}{j!}\left(\frac{3}{z}\right)^j = z + 3 + \frac{3^2}{2!\,z} + \frac{3^3}{3!\,z^2} + \cdots.$$

Hence

$$\text{Res}(0) = \frac{3^2}{2!} = \frac{9}{2},$$

and since $z = 0$ is the only singularity inside $|z| = 4$, we have, by formula (2),

$$\oint_{|z|=4} ze^{3/z}\, dz = 2\pi i \cdot \frac{9}{2} = 9\pi i. \qquad ∎$$

Now if f has a *removable* singularity at z_0, all the coefficients of the negative powers of $(z - z_0)$ in its Laurent expansion are zero, and so, in particular, the residue at z_0 is zero. Furthermore, if f has a *pole* at z_0, we shall see that its residue there can be computed from a formula. Suppose first that z_0 is a simple pole, i.e., a pole of order 1. Then for z near z_0 we have

$$f(z) = \frac{a_{-1}}{z - z_0} + a_0 + a_1(z - z_0) + a_2(z - z_0)^2 + \cdots,$$

and so

$$(z - z_0)f(z) = a_{-1} + (z - z_0)[a_0 + a_1(z - z_0) + a_2(z - z_0)^2 + \cdots].$$

By taking the limit as $z \to z_0$ we deduce that

$$\lim_{z \to z_0} (z - z_0)f(z) = a_{-1} + 0.$$

Hence *at a simple pole*

(3) $$\text{Res}(f; z_0) = \lim_{z \to z_0} (z - z_0)f(z).$$

For example, the function $f(z) = e^z/z(z + 1)$ has simple poles at $z = 0$ and $z = -1$; therefore

$$\text{Res}(f; 0) = \lim_{z \to 0} zf(z) = \lim_{z \to 0} \frac{e^z}{z + 1} = 1,$$

and

$$\text{Res}(f; -1) = \lim_{z \to -1} (z + 1)f(z) = \lim_{z \to -1} \frac{e^z}{z} = -e^{-1}.$$

Another consequence of formula (3) is illustrated in

EXAMPLE 2 Let $f(z) = P(z)/Q(z)$, where the functions $P(z)$ and $Q(z)$ are both analytic at z_0, and Q has a simple zero at z_0, while $P(z_0) \neq 0$. Prove that

$$\text{Res}(f; z_0) = \frac{P(z_0)}{Q'(z_0)}.$$

SOLUTION Obviously f has a simple pole at z_0, so we can apply formula (3). Using the fact that $Q(z_0) = 0$ we see directly that

$$\text{Res}(f; z_0) = \lim_{z \to z_0} (z - z_0)\frac{P(z)}{Q(z)} = \lim_{z \to z_0} \frac{P(z)}{\left[\dfrac{Q(z) - Q(z_0)}{z - z_0}\right]} = \frac{P(z_0)}{Q'(z_0)}. \qquad \blacksquare$$

EXAMPLE 3 Compute the residue at each singularity of $f(z) = \cot z$.

SOLUTION Since $\cot z = \cos z/\sin z$, the singularities of this function are simple poles occurring at the points $z = n\pi, n = 0, \pm 1, \pm 2, \dots$. Utilizing Example 2 with $P(z) = \cos z$, $Q(z) = \sin z$, the residues at these points are given by

$$\text{Res}(\cot z; n\pi) = \frac{\cos z}{(\sin z)'}\bigg|_{z = n\pi} = \frac{\cos n\pi}{\cos n\pi} = 1. \qquad \blacksquare$$

To obtain the general formula for the residue at a pole of order m we need some method of picking out the coefficient a_{-1} from the Laurent expansion. The reader should encounter no difficulty in following the derivation of

THEOREM 1 *If f has a pole of order m at z_0, then*

$$(4) \qquad \text{Res}(f; z_0) = \lim_{z \to z_0} \frac{1}{(m - 1)!}\frac{d^{m-1}}{dz^{m-1}}[(z - z_0)^m f(z)].$$

Proof Starting with the Laurent expansion for f around z_0,

$$f(z) = \frac{a_{-m}}{(z - z_0)^m} + \cdots + \frac{a_{-2}}{(z - z_0)^2} + \frac{a_{-1}}{z - z_0} + a_0 + a_1(z - z_0) + \cdots,$$

we multiply by $(z - z_0)^m$,

$$(z - z_0)^m f(z) = a_{-m} + \cdots + a_{-2}(z - z_0)^{m-2} + a_{-1}(z - z_0)^{m-1}$$
$$+ a_0(z - z_0)^m + a_1(z - z_0)^{m+1} + \cdots,$$

and differentiate $m - 1$ times to derive

$$\frac{d^{m-1}}{dz^{m-1}}[(z - z_0)^m f(z)] = (m - 1)!\, a_{-1} + m!\, a_0(z - z_0)$$
$$+ \frac{(m + 1)!}{2}\, a_1(z - z_0)^2 + \cdots.$$

Hence

$$\lim_{z \to z_0} \frac{d^{m-1}}{dz^{m-1}} [(z - z_0)^m f(z)] = (m - 1)! a_{-1} + 0,$$

which is equivalent to Eq. (4). ∎

EXAMPLE 4 Compute the residues at the singularities of

$$f(z) = \frac{\cos z}{z^2(z - \pi)^3}.$$

SOLUTION The above function has a pole of order 2 at $z = 0$ and a pole of order 3 at $z = \pi$. Applying formula (4) we find

$$\operatorname{Res}(0) = \lim_{z \to 0} \frac{1}{1!} \frac{d}{dz} [z^2 f(z)] = \lim_{z \to 0} \frac{d}{dz} \left[\frac{\cos z}{(z - \pi)^3} \right]$$

$$= \lim_{z \to 0} \left[\frac{-(z - \pi) \sin z - 3 \cos z}{(z - \pi)^4} \right] = \frac{-3}{\pi^4},$$

$$\operatorname{Res}(\pi) = \lim_{z \to \pi} \frac{1}{2!} \frac{d^2}{dz^2} [(z - \pi)^3 f(z)] = \lim_{z \to \pi} \frac{1}{2} \frac{d^2}{dz^2} \left[\frac{\cos z}{z^2} \right]$$

$$= \lim_{z \to \pi} \frac{1}{2} \left[\frac{(6 - z^2) \cos z + 4z \sin z}{z^4} \right] = \frac{-(6 - \pi^2)}{2\pi^4}. ∎$$

We have already seen how to compute the integral $\int_\Gamma f(z) \, dz$ when $f(z)$ has only one singularity inside Γ. Let's now turn to the more general case where Γ is a simple closed positively oriented contour and $f(z)$ is analytic inside and on Γ except for a finite number of isolated singularities at the points $z_1, z_2, \ldots, z_n$ inside Γ (see Fig. 6.2). Notice that by the methods of Sec. 4.4 we can express the integral along Γ in terms of the integrals around the circles C_j in Fig. 6.3:

$$\int_\Gamma f(z) \, dz = \sum_{j=1}^{n} \int_{C_j} f(z) \, dz.$$

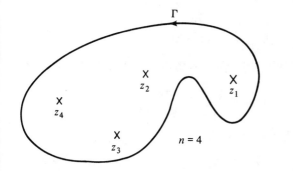

FIGURE 6.2

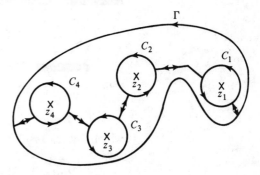

FIGURE 6.3

However, because z_j is the only singularity of f inside C_j, we know that

$$\int_{C_j} f(z)\, dz = 2\pi i \, \text{Res}(z_j).$$

Hence we have established the important

THEOREM 2 (Cauchy's Residue Theorem) *If Γ is a simple closed positively oriented contour and f is analytic inside and on Γ except at the points $z_1, z_2, \ldots, z_n$ inside Γ, then*

(5)
$$\int_{\Gamma} f(z)\, dz = 2\pi i \sum_{j=1}^{n} \text{Res}(z_j).$$

EXAMPLE 5 Evaluate

$$\oint_{|z|=2} \frac{1 - 2z}{z(z - 1)(z - 3)}\, dz.$$

SOLUTION The function $f(z) = (1 - 2z)/z(z - 1)(z - 3)$ has simple poles at $z = 0$, $z = 1$, and $z = 3$. However, only the first two of these points lie inside Γ: $|z| = 2$. Thus by the residue theorem

$$\oint_{|z|=2} f(z)\, dz = 2\pi i[\text{Res}(0) + \text{Res}(1)],$$

and since

$$\text{Res}(0) = \lim_{z \to 0} z f(z) = \lim_{z \to 0} \frac{1 - 2z}{(z - 1)(z - 3)} = \frac{1}{3},$$

$$\text{Res}(1) = \lim_{z \to 1} (z - 1) f(z) = \lim_{z \to 1} \frac{1 - 2z}{z(z - 3)} = \frac{1}{2},$$

we obtain

$$\oint_{|z|=2} f(z)\, dz = 2\pi i\left(\frac{1}{3} + \frac{1}{2}\right) = \frac{5\pi i}{3}. \qquad \blacksquare$$

EXAMPLE 6 Compute

$$\oint_{|z|=5} \left[ze^{3/z} + \frac{\cos z}{z^2(z - \pi)^3} \right] dz.$$

SOLUTION The given integral can obviously be expressed as the sum

$$\oint_{|z|=5} ze^{3/z}\, dz + \oint_{|z|=5} \frac{\cos z}{z^2(z - \pi)^3}\, dz,$$

which, by the residue theorem, equals

$$2\pi i\left[\text{Res}(ze^{3/z}; 0) + \text{Res}\left(\frac{\cos z}{z^2(z - \pi)^3}; 0\right) + \text{Res}\left(\frac{\cos z}{z^2(z - \pi)^3}; \pi\right) \right].$$

These residues were computed in Examples 1 and 4; the desired answer is therefore

$$2\pi i \left[\frac{9}{2} - \frac{3}{\pi^4} - \frac{(6 - \pi^2)}{2\pi^4} \right]. \quad \blacksquare$$

Exercises 6.1

1. Determine all the isolated singularities of each of the following functions and compute the residue at each singularity.

   (a) $\dfrac{e^{3z}}{z - 2}$ (b) $\dfrac{z + 1}{z^2 - 3z + 2}$

   (c) $\dfrac{\cos z}{z^2}$ (d) $\left(\dfrac{z - 1}{z + 1}\right)^3$

   (e) $\dfrac{e^z}{z(z + 1)^3}$ (f) $\sin\left(\dfrac{1}{3z}\right)$

   (g) $\tan z$ (h) $\dfrac{z - 1}{\sin z}$

2. Find the residue at $z = 1$ for the function $z^2/(1 - \sqrt{z})$, where $\sqrt{z}$ denotes the principal branch.

3. Evaluate each of the following integrals by means of the Cauchy Residue Theorem.

   (a) $\displaystyle\oint_{|z| = 5} \frac{\sin z}{z^2 - 4} \, dz$ (b) $\displaystyle\oint_{|z| = 3} \frac{e^z}{z(z - 2)^3} \, dz$

   (c) $\displaystyle\oint_{|z| = 2\pi} \tan z \, dz$ (d) $\displaystyle\oint_{|z| = 3} \frac{e^{iz}}{z^2(z - 2)(z + 5i)} \, dz$

   (e) $\displaystyle\oint_{|z| = 1} \frac{1}{z^2 \sin z} \, dz$ (f) $\displaystyle\oint_{|z| = 3} \frac{3z + 2}{z^4 + 1} \, dz$

   (g) $\displaystyle\oint_{|z| = 8} \frac{1}{z^2 + z + 1} \, dz$

4. Evaluate

$$\oint_{|z| = 1} e^{1/z} \sin(1/z) \, dz.$$

5. Explain why Cauchy's Integral Formula can be regarded as a special case of the residue theorem.

6. Does there exist a function f having a simple pole at z_0 with $\text{Res}(f; z_0) = 0$? How about a function with a pole of order 2 at z_0 and $\text{Res}(f; z_0) = 0$?

7. Let f have an isolated singularity at z_0 (f analytic in a punctured neighborhood of z_0). Show that the residue of the derivative f' at z_0 is equal to zero.

8. Suppose that f is analytic and has a zero of order m at the point z_0. Show that the function $g(z) = f'(z)/f(z)$ has a simple pole at z_0 with $\text{Res}(g; z_0) = m$.

6.2 TRIGONOMETRIC INTEGRALS OVER $[0, 2\pi]$

Our goal here is to apply the residue theorem to evaluate real integrals of the form

(6)
$$\int_0^{2\pi} U(\cos\theta, \sin\theta)\, d\theta,$$

where $U(\cos\theta, \sin\theta)$ is a rational function (with real coefficients) of $\cos\theta$ and $\sin\theta$, and is finite over $[0, 2\pi]$. An example of such an integral is

$$\int_0^{2\pi} \frac{\sin^2\theta}{a + b\cos\theta}\, d\theta \qquad (0 < b < a).$$

We shall show that (6) can be identified as the parametrized form of a contour integral, $\int_C F(z)\, dz$, of some complex function F around the positively oriented unit circle $C: |z| = 1$. To establish this identification we parametrize C by

$$z = e^{i\theta} \qquad (0 \le \theta \le 2\pi).$$

For such z we have

$$\frac{1}{z} = \frac{1}{e^{i\theta}} = e^{-i\theta},$$

and since

$$\cos\theta = \frac{e^{i\theta} + e^{-i\theta}}{2}, \qquad \sin\theta = \frac{e^{i\theta} - e^{-i\theta}}{2i},$$

we have the identities

(7)
$$\cos\theta = \frac{1}{2}\left(z + \frac{1}{z}\right), \qquad \sin\theta = \frac{1}{2i}\left(z - \frac{1}{z}\right).$$

Furthermore, when integrating along C,

$$dz = ie^{i\theta} \, d\theta = iz \, d\theta,$$

so that

(8)
$$d\theta = \frac{dz}{iz}.$$

Making the substitutions (7) and (8) in the integral (6), we see that

(9)
$$\int_0^{2\pi} U(\cos\theta, \sin\theta) \, d\theta = \int_C F(z) \, dz,$$

where the new integrand F is

$$F(z) \equiv U\left[\frac{1}{2}\left(z + \frac{1}{z}\right), \frac{1}{2i}\left(z - \frac{1}{z}\right)\right] \cdot \frac{1}{iz};$$

integration over $[0, 2\pi]$ has thus been replaced by integration around C.

Because of the form of U, the function F must be a rational function of z. Hence it has only removable singularities (which can be ignored in evaluating integrals) or poles. Consequently, by the residue theorem, our trigonometric integral equals $2\pi i$ times the sum of the residues at those poles of F which lie inside C.

The procedure is illustrated in

EXAMPLE 7   Evaluate

$$I = \int_0^{2\pi} \frac{\sin^2\theta}{5 + 4\cos\theta} \, d\theta.$$

SOLUTION   First observe that the denominator, $5 + 4\cos\theta$, is never zero, so the integrand is finite over $[0, 2\pi]$. Performing the substitutions (7) and (8) for $\cos\theta$, $\sin\theta$, and $d\theta$, we obtain

$$I = \int_C \frac{\left[\frac{1}{2i}\left(z - \frac{1}{z}\right)\right]^2}{5 + 4\left[\frac{1}{2}\left(z + \frac{1}{z}\right)\right]} \frac{dz}{iz},$$

which after some algebra reduces to

$$I = -\frac{1}{4i}\int_C \frac{(z^2 - 1)^2}{z^2(2z^2 + 5z + 2)} \, dz.$$

Clearly the integrand

$$g(z) \equiv \frac{(z^2 - 1)^2}{z^2(2z^2 + 5z + 2)} = \frac{(z^2 - 1)^2}{2z^2(z + \frac{1}{2})(z + 2)}$$

has simple poles at $z = -\frac{1}{2}$ and $z = -2$, and has a pole of order 2 at the origin. However, only $-\frac{1}{2}$ and 0 lie inside the unit circle C, so that

$$I = -\frac{1}{4i} \cdot 2\pi i \left[\text{Res}\left(g; -\frac{1}{2}\right) + \text{Res}(g; 0) \right].$$

Utilizing the formulas of the preceding section we find

$$\text{Res}\left(g; -\frac{1}{2}\right) = \lim_{z \to -1/2} \left(z + \frac{1}{2}\right) g(z) = \lim_{z \to -1/2} \frac{(z^2 - 1)^2}{2z^2(z + 2)} = \frac{3}{4},$$

and

$$\text{Res}(g; 0) = \lim_{z \to 0} \frac{1}{1!} \frac{d}{dz} [z^2 g(z)] = \lim_{z \to 0} \frac{d}{dz} \left[\frac{(z^2 - 1)^2}{2z^2 + 5z + 2} \right]$$

$$= \frac{(2z^2 + 5z + 2) \cdot 2(z^2 - 1)2z - (z^2 - 1)^2(4z + 5)}{(2z^2 + 5z + 2)^2} \bigg|_{z=0}$$

$$= \frac{-5}{4}.$$

Hence

$$I = \frac{-1}{4i} 2\pi i \left[\frac{3}{4} - \frac{5}{4} \right] = \frac{\pi}{4}. \quad \blacksquare$$

EXAMPLE 8 Evaluate

$$I = \int_0^\pi \frac{d\theta}{2 - \cos \theta}.$$

SOLUTION The catch here is that the integral is taken over $[0, \pi]$ instead of $[0, 2\pi]$. However it is easy to see that, since $\cos \theta = \cos(2\pi - \theta)$,

$$\int_0^\pi \frac{d\theta}{2 - \cos \theta} = \int_\pi^{2\pi} \frac{d\theta}{2 - \cos \theta},$$

and therefore

$$\int_0^{2\pi} \frac{d\theta}{2 - \cos \theta} = 2I.$$

Substituting for $\cos \theta$ and $d\theta$ we have

(10) $$2I = \int_C \frac{1}{2 - \frac{1}{2}\left(z + \frac{1}{z}\right)} \cdot \frac{dz}{iz} = -\frac{2}{i} \int_C \frac{dz}{z^2 - 4z + 1}.$$

By the quadratic formula the zeros of the denominator are

$$z_1 \equiv 2 - \sqrt{3} \quad \text{and} \quad z_2 \equiv 2 + \sqrt{3},$$

and so the integrand

$$g(z) = \frac{1}{z^2 - 4z + 1} = \frac{1}{(z - z_1)(z - z_2)}$$

has simple poles at these points. But only z_1 lies inside C, and the residue there is given by

$$\text{Res}(g; z_1) = \lim_{z \to z_1} (z - z_1)g(z) = \lim_{z \to z_1} \frac{1}{(z - z_2)}$$

$$= \frac{1}{z_1 - z_2} = -\frac{1}{2\sqrt{3}}.$$

Hence from Eq. (10)

$$2I = -\frac{2}{i} \cdot 2\pi i \left(-\frac{1}{2\sqrt{3}} \right) = \frac{2\pi}{\sqrt{3}},$$

or

$$I = \frac{\pi}{\sqrt{3}}. \quad \blacksquare$$

Exercises 6.2

1. Evaluate $\int_0^{2\pi} \frac{d\theta}{2 + \sin \theta}$.

2. Evaluate $\int_0^{\pi} \frac{8 \, d\theta}{5 + 2 \cos \theta}$.

3. Evaluate $\int_{-\pi}^{\pi} \frac{d\theta}{1 + \sin^2 \theta}$.

4. Evaluate $\int_0^{\pi} \frac{d\theta}{(3 + 2 \cos \theta)^2}$.

Using the method of residues, verify each of the following.

5. $\int_0^{2\pi} \frac{d\theta}{1 + a \cos \theta} = \frac{2\pi}{\sqrt{1 - a^2}}, \quad a^2 < 1.$

6. $\int_0^{2\pi} \frac{\sin^2 \theta}{a + b \cos \theta} \, d\theta = \frac{2\pi}{b^2} (a - \sqrt{a^2 - b^2}), \quad a > |b| > 0.$

7. $\int_0^{\pi} \frac{d\theta}{(a + \sin^2 \theta)^2} = \frac{\pi(2a + 1)}{2\sqrt{(a^2 + a)^3}}, \quad a > 0.$

8. $\displaystyle\int_0^{2\pi} \frac{d\theta}{a^2 \sin^2 \theta + b^2 \cos^2 \theta} = \frac{2\pi}{ab},\quad a, b > 0.$

9. $\displaystyle\int_0^{2\pi} (\cos \theta)^{2n}\, d\theta = \frac{\pi \cdot (2n)!}{2^{2n-1}(n!)^2},\quad n = 1, 2, \ldots .$

10. $\displaystyle\int_0^{2\pi} e^{\cos \theta} \cos(n\theta - \sin \theta)\, d\theta = \frac{2\pi}{n!},\quad n = 1, 2, \ldots .$

11. $\displaystyle\int_0^{\pi} \tan(\theta + ia)\, d\theta = \pi i \cdot \operatorname{sign} a,\quad a \text{ real and nonzero.}$

6.3 IMPROPER INTEGRALS OF CERTAIN FUNCTIONS OVER $(-\infty, \infty)$

If $f(x)$ is a function continuous on the nonnegative real axis $0 \le x < \infty$, then the improper integral of f over $[0, \infty)$ is defined by

$$(11) \qquad \int_0^{\infty} f(x)\, dx \equiv \lim_{b \to \infty} \int_0^{b} f(x)\, dx,$$

provided that this limit exists.† For example,

$$\int_0^{\infty} e^{-2x}\, dx = \lim_{b \to \infty} \int_0^{b} e^{-2x}\, dx = \lim_{b \to \infty} \frac{-e^{-2x}}{2}\Big|_0^{b}$$

$$= \lim_{b \to \infty} \left[\frac{-e^{-2b}}{2} + \frac{1}{2} \right] = \frac{1}{2}.$$

Similarly, when $f(x)$ is continuous on $(-\infty, 0]$, we set

$$(12) \qquad \int_{-\infty}^{0} f(x)\, dx \equiv \lim_{c \to -\infty} \int_c^{0} f(x)\, dx.$$

If it turns out that both of the limits (11) and (12) exist for a function f continuous on the whole real line, then f is said to be *integrable over* $(-\infty, \infty)$, and we write

$$\int_{-\infty}^{\infty} f(x)\, dx \equiv \lim_{c \to -\infty} \int_c^{0} f(x)\, dx + \lim_{b \to \infty} \int_0^{b} f(x)\, dx$$

$$= \int_{-\infty}^{0} f(x)\, dx + \int_0^{\infty} f(x)\, dx.$$

†More generally, $\int_a^{\infty} f(x)\, dx \equiv \lim_{b \to \infty} \int_a^{b} f(x)\, dx$ if this limit exists.

249

Section 6.3
Improper Integrals
of Certain
Functions Over
$(-\infty, \infty)$

In such a case the value of the improper integral over $(-\infty, \infty)$ can be computed by taking a single limit, namely,

$$\int_{-\infty}^{\infty} f(x)\, dx = \lim_{\rho \to \infty} \int_{-\rho}^{\rho} f(x)\, dx.$$

However, we caution the reader that this last limit may exist even for certain *nonintegrable* functions f. Indeed, consider $f(x) = x$. This function is not integrable over $(-\infty, \infty)$ because the limit

$$\lim_{b \to \infty} \int_{0}^{b} x\, dx = \lim_{b \to \infty} \frac{x^2}{2}\Big|_0^b = \lim_{b \to \infty} \frac{b^2}{2}$$

does not exist (as a finite number). However,

$$\lim_{\rho \to \infty} \int_{-\rho}^{\rho} x\, dx = \lim_{\rho \to \infty} \frac{x^2}{2}\Big|_{-\rho}^{\rho} = \lim_{\rho \to \infty} 0 = 0.$$

For this reason we introduce the following terminology: Given *any* function f continuous on $(-\infty, \infty)$ the limit

$$\lim_{\rho \to \infty} \int_{-\rho}^{\rho} f(x)\, dx$$

(if it exists) is called the *Cauchy principal value* of the integral of f over $(-\infty, \infty)$, and we write

$$\text{p.v.} \int_{-\infty}^{\infty} f(x)\, dx \equiv \lim_{\rho \to \infty} \int_{-\rho}^{\rho} f(x)\, dx.$$

(Thus, in particular,

$$\text{p.v.} \int_{-\infty}^{\infty} x\, dx = 0.)$$

We reiterate that whenever the improper integral $\int_{-\infty}^{\infty} f(x)\, dx$ exists, it must equal its principal value.

We shall now show how the theory of residues can be used to compute "p.v. integrals" for certain functions f.

EXAMPLE 9   Evaluate

$$I = \text{p.v.} \int_{-\infty}^{\infty} \frac{dx}{x^4 + 4} \left(= \lim_{\rho \to \infty} \int_{-\rho}^{\rho} \frac{dx}{x^4 + 4} \right).$$

SOLUTION   As a first step, we recognize that the integral I_ρ defined by

$$I_\rho \equiv \int_{-\rho}^{\rho} \frac{dx}{x^4 + 4}$$

can be interpreted as a contour integral of an analytic function; in fact

$$I_\rho = \int_{\gamma_\rho} \frac{dz}{z^4 + 4},$$

where γ_ρ is the directed segment of the real axis from $-\rho$ to $+\rho$. Now the key to using residue theory to find I lies in constructing (for each sufficiently large value of ρ) a simple *closed* contour Γ_ρ such that γ_ρ is one of its components, i.e., $\Gamma_\rho = (\gamma_\rho, \gamma'_\rho)$, and such that the integral of $1/(z^4 + 4)$ along the other component γ'_ρ is somehow known. For then we will have

$$\int_{\Gamma_\rho} \frac{dz}{z^4 + 4} = I_\rho + \int_{\gamma'_\rho} \frac{dz}{z^4 + 4},$$

and if Γ_ρ is positively oriented, the residue theorem yields

$$2\pi i \cdot \sum (\text{residues inside } \Gamma_\rho) = I_\rho + \int_{\gamma'_\rho} \frac{dz}{z^4 + 4}.$$

Consequently, the integral I is evaluated by

(13) $\quad I = \lim_{\rho \to \infty} I_\rho = \lim_{\rho \to \infty} 2\pi i \sum (\text{residues inside } \Gamma_\rho) - \lim_{\rho \to \infty} \int_{\gamma'_\rho} \frac{dz}{z^4 + 4},$

provided the limits on the right exist. Actually, we see from Eq. (13) that it is only the *limiting value* of the integrals over γ'_ρ which must be known in order to apply residue theory.

Now among the many curves which "close the contour γ_ρ," i.e., which start at $z = \rho$ and terminate at $z = -\rho$, how are we to find a suitable curve γ'_ρ? Observe that the integrand, $1/(z^4 + 4)$, is quite small in modulus when $|z|$ is large. This suggests that if we choose our curves γ'_ρ far enough away from the origin, the integrals over them might well be negligible, i.e., approach zero as $\rho \to \infty$. Thus an obvious thing to try for γ'_ρ is the half-circle C_ρ^+ parametrized by

(14) $\qquad\qquad C_\rho^+ : z = \rho e^{it} \qquad (0 \le t \le \pi)$

(see Fig. 6.4). To see if this works, we note that $|z| = \rho$ on C_ρ^+, so that by the triangle inequality

$$\left| \frac{1}{z^4 + 4} \right| \le \frac{1}{|z|^4 - 4} = \frac{1}{\rho^4 - 4} \qquad (\text{for } \rho^4 > 4),$$

and hence

$$\left| \int_{C_\rho^+} \frac{dz}{z^4 + 4} \right| \le \frac{1}{\rho^4 - 4} \cdot \pi\rho,$$

which certainly does go to zero as $\rho \to \infty$.

251

Section 6.3
*Improper Integrals
of Certain
Functions Over*
$(-\infty, \infty)$

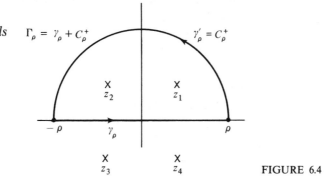

FIGURE 6.4

So now all we have to do in Eq. (13) is to evaluate the appropriate residues. First we locate the singularities of $1/(z^4 + 4)$. These occur at the zeros of $z^4 + 4$, i.e., at

$$z_1 = 1 + i, \qquad z_2 = -1 + i, \qquad z_3 = -1 - i, \qquad z_4 = 1 - i,$$

and the function

$$\frac{1}{z^4 + 4} = \frac{1}{(z - z_1)(z - z_2)(z - z_3)(z - z_4)}$$

has simple poles at these points. Since z_3 and z_4 lie in the lower half-plane, they are always excluded from the interior of the semicircular contour Γ_ρ of Fig. 6.4, but z_1 and z_2 lie inside Γ_ρ for every $\rho > \sqrt{2}$. Thus, for such ρ,

$$\int_{\Gamma_\rho} \frac{dz}{z^4 + 4} = 2\pi i [\text{Res}(z_1) + \text{Res}(z_2)]$$

$$= 2\pi i \left(\lim_{z \to z_1} \frac{z - z_1}{z^4 + 4} + \lim_{z \to z_2} \frac{z - z_2}{z^4 + 4} \right)$$

$$= 2\pi i \left[\frac{1}{(z_1 - z_2)(z_1 - z_3)(z_1 - z_4)} + \frac{1}{(z_2 - z_1)(z_2 - z_3)(z_2 - z_4)} \right]$$

$$= 2\pi i \left[\frac{1}{2(2 + 2i)2i} + \frac{1}{(-2)(2i)(-2 + 2i)} \right]$$

$$= 2\pi i \left[\frac{-1 - i}{16} + \frac{1 - i}{16} \right] = \frac{\pi}{4}.$$

Putting this all together in Eq. (13) (with $\gamma'_\rho = C_\rho^+$) we have

$$I = \lim_{\rho \to \infty} \frac{\pi}{4} - \lim_{\rho \to \infty} \int_{C_\rho^+} \frac{dz}{z^4 + 4} = \frac{\pi}{4} - 0 = \frac{\pi}{4}. \qquad \blacksquare$$

The technique of using expanding semicircular contours Γ_ρ can readily be applied to a general class of integrands f. Indeed, the success of the procedure illustrated in Example 9 depends only on the following two conditions:

(*i*) f is analytic on and above the real axis except for a *finite* number of isolated singularities in the open upper half-plane Im $z > 0$ (this ensures that for ρ sufficiently large, all the singularities in the upper half-plane will lie inside the contour Γ_ρ of Fig. 6.4), and

(*ii*) $\lim_{\rho \to \infty} \int_{C_\rho^+} f(z)\, dz = 0$.

Whenever these conditions are satisfied, the value of the integral

$$\text{p.v.} \int_{-\infty}^{\infty} f(x)\, dx$$

is given by $2\pi i$ times the sum of the residues of f at the singularities in the upper half-plane. (Of course, the lower half-plane can be used whenever analogous conditions hold there; see Prob. 8.)

A class of rational functions having property (*ii*) is given in

LEMMA 1 *If $f(z) = P(z)/Q(z)$ is the quotient of two polynomials such that*

(15) $$\text{degree } Q \geq 2 + \text{degree } P,$$

then

(16) $$\lim_{\rho \to \infty} \int_{C_\rho^+} f(z)\, dz = 0,$$

where C_ρ^+ is the upper half-circle of radius ρ defined in Eq. (14).

Proof For large $|z|$, inequality (15) implies that

$$|f(z)| \leq \frac{K}{|z|^2},$$

where K is some constant (recall Lemma 1 of Sec. 4.6); consequently

$$\left| \int_{C_\rho^+} f(z)\, dz \right| \leq \frac{K}{\rho^2} \cdot \pi\rho = \frac{K\pi}{\rho} \to 0 \quad \text{as} \quad \rho \to \infty. \quad \blacksquare$$

We remark that the same proof shows that Eq. (16) remains valid if integration along C_ρ^+ is replaced by integration along the lower half-circle C_ρ^-.

253

Section 6.3
*Improper Integrals
of Certain
Functions Over*
$(-\infty, \infty)$

EXAMPLE 10 Compute

$$I = \text{p.v.} \int_{-\infty}^{\infty} \frac{x^2}{(x^2 + 1)^2}\, dx.$$

SOLUTION   Since the integrand has no singularities on the real axis and has numerator degree 2 and denominator degree 4, the expanding semicircular contour method is justified thanks to Lemma 1. On writing

$$f(z) \equiv \frac{z^2}{(z^2 + 1)^2} = \frac{z^2}{(z - i)^2(z + i)^2},$$

we see that poles occur at $z = \pm i$. Thus, for any $\rho > 1$, the integral along the closed contour Γ_ρ in Fig. 6.5 is given by

$$\int_{\Gamma_\rho} f(z)\, dz = 2\pi i\ \text{Res}(f;\ +i).$$

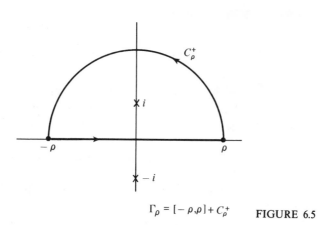

$$\Gamma_\rho = [-\rho,\rho] + C_\rho^+ \qquad \textsf{FIGURE 6.5}$$

Since $+i$ is a second-order pole, we have from the residue formula (4)

$$\text{Res}(f;\ +i) = \lim_{z \to i} \frac{1}{1!}\frac{d}{dz}[(z - i)^2 f(z)] = \lim_{z \to i} \frac{d}{dz}\left[\frac{z^2}{(z + i)^2}\right]$$

$$= \lim_{z \to i}\left[\frac{(z + i)2z - 2z^2}{(z + i)^3}\right] = \frac{1}{4i}.$$

Therefore

(17) $$\int_{\Gamma_\rho} f(z)\, dz = 2\pi i \cdot \frac{1}{4i} = \frac{\pi}{2} \qquad \text{(for all } \rho > 1).$$

On the other hand,

$$\int_{\Gamma_\rho} f(z)\, dz = \int_{-\rho}^{\rho} f(x)\, dx + \int_{C_\rho^+} f(z)\, dz,$$

and so on taking the limit as $\rho \to \infty$ in this last equation we deduce from Eq. (17) and Lemma 1 that

$$\frac{\pi}{2} = \lim_{\rho \to \infty} \int_{-\rho}^{\rho} f(x) \, dx + 0.$$

Hence

$$\frac{\pi}{2} = \text{p.v.} \int_{-\infty}^{\infty} \frac{x^2}{(x^2 + 1)^2} \, dx. \qquad \blacksquare$$

As we shall see in the next section, semicircular contours are also useful in evaluating certain integrals involving trigonometric functions. However, there are situations which call for other types of contours, as in the following example.

EXAMPLE 11   Compute

$$I = \text{p.v.} \int_{-\infty}^{\infty} \frac{e^{ax}}{1 + e^x} \, dx, \qquad \text{for } 0 < a < 1.$$

SOLUTION   Observe that the function $e^{az}/(1 + e^z)$ has an infinite number of singularities in both the upper and lower half-planes; these occur at the points

$$z = (2n + 1)\pi i \qquad (n = 0, \pm 1, \pm 2, \ldots).$$

Hence if we employ expanding semicircles, the contribution due to residues will result in an infinite series. A much better choice is the rectangular contour Γ_ρ in Fig. 6.6. First let's show how we can deal with the integrals over γ_2, γ_3, and γ_4 when ρ is large.

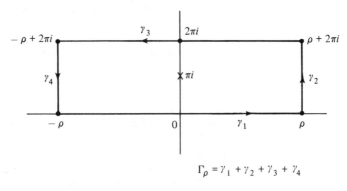

$$\Gamma_\rho = \gamma_1 + \gamma_2 + \gamma_3 + \gamma_4$$

FIGURE 6.6

255

Section 6.3
Improper Integrals
of Certain
Functions Over
$(-\infty, \infty)$

For γ_2: $z = \rho + it$, $0 \le t \le 2\pi$, we have

$$\left| \int_{\gamma_2} \frac{e^{az}}{1 + e^z} \, dz \right| = \left| \int_0^{2\pi} \frac{e^{a(\rho + it)}}{1 + e^{\rho + it}} i \, dt \right|$$

$$\le \frac{e^{a\rho}}{e^{\rho} - 1} \cdot 2\pi,$$

which goes to zero as $\rho \to \infty$ since $a < 1$.

Similarly, on γ_4: $z = -\rho + i(2\pi - t)$, $0 \le t \le 2\pi$, we have

$$\left| \int_{\gamma_4} \frac{e^{az}}{1 + e^z} \, dz \right| = \left| \int_0^{2\pi} \frac{e^{a[-\rho + i(2\pi - t)]}}{1 + e^{-\rho + i(2\pi - t)}} (-i) \, dt \right|$$

$$\le \frac{e^{-a\rho}}{1 - e^{-\rho}} \cdot 2\pi,$$

again approaching zero as $\rho \to \infty$ since $a > 0$.

For the integral over the horizontal segment γ_3: $z = -t + 2\pi i$, $-\rho \le t \le \rho$, we obtain

$$\int_{\gamma_3} \frac{e^{az}}{1 + e^z} \, dz = \int_{-\rho}^{\rho} \frac{e^{a(-t + 2\pi i)}}{1 + e^{-t + 2\pi i}} (-1) \, dt.$$

Making the change of variables $x = -t$, $dx = -dt$, this becomes

$$\int_{\gamma_3} \frac{e^{az}}{1 + e^z} \, dz = \int_{\rho}^{-\rho} \frac{e^{ax} \cdot e^{a2\pi i}}{1 + e^x \cdot 1} \, dx$$

$$= -e^{a2\pi i} \int_{-\rho}^{\rho} \frac{e^{ax}}{1 + e^x} \, dx,$$

which is just $-e^{a2\pi i}$ times the integral over γ_1. In other words, on taking the limit as $\rho \to \infty$ we have

$$(18) \qquad \lim_{\rho \to \infty} \int_{\Gamma_\rho} \frac{e^{az}}{1 + e^z} \, dz = (1 - e^{a2\pi i}) \text{ p.v.} \int_{-\infty}^{\infty} \frac{e^{ax}}{1 + e^x} \, dx.$$

Now we use residue theory to evaluate the contour integral in Eq. (18). For each $\rho > 0$, the function $e^{az}/(1 + e^z)$ is analytic inside and on Γ_ρ except for a simple pole at $z = \pi i$, the residue there being given by

$$(19) \qquad \text{Res}(\pi i) = \left. \frac{e^{az}}{\dfrac{d}{dz}(1 + e^z)} \right|_{z = \pi i} = \frac{e^{a\pi i}}{e^{\pi i}} = -e^{a\pi i}$$

(recall Example 2). Consequently, putting Eqs. (18) and (19) together we obtain

$$\text{p.v.} \int_{-\infty}^{\infty} \frac{e^{ax}}{1 + e^x} \, dx = \frac{1}{1 - e^{a2\pi i}} \cdot (2\pi i)(-e^{a\pi i}),$$

which can be reduced to $\pi/\sin a\pi$. ∎

Exercises 6.3

Verify the integral formulas in Problems 1–3 with the aid of residues.

1. p.v. $\displaystyle\int_{-\infty}^{\infty} \frac{dx}{x^2 + 2x + 2} = \pi.$

2. p.v. $\displaystyle\int_{-\infty}^{\infty} \frac{x^2}{(x^2 + 9)^2}\, dx = \frac{\pi}{6}.$

3. $\displaystyle\int_{0}^{\infty} \frac{x^2 + 1}{x^4 + 1}\, dx = \frac{\pi}{\sqrt{2}}.$

Use residues to find the values of the integrals in Problems 4–7.

4. p.v. $\displaystyle\int_{-\infty}^{\infty} \frac{dx}{(x^2 + 1)(x^2 + 4)}.$

$\pi/6$

5. p.v. $\displaystyle\int_{-\infty}^{\infty} \frac{x}{(x^2 + 4x + 13)^2}\, dx.$

$\dfrac{3\pi\sqrt{2}}{16}$

6. $\displaystyle\int_{0}^{\infty} \frac{x^2}{(x^2 + 1)(x^2 + 4)}\, dx.$

7. $\displaystyle\int_{0}^{\infty} \frac{x^6}{(x^4 + 1)^2}\, dx.$

8. Show that if $f(z) = P(z)/Q(z)$ is the quotient of two polynomials such that deg $Q \geq 2 + $ deg P, where Q has no real zeros, then

$$\text{p.v.} \int_{-\infty}^{\infty} f(x)\, dx = -2\pi i \cdot \sum [\text{residues of } f(z) \text{ at}$$
$$\text{the poles in the lower half-plane}].$$

9. Show that

$$\text{p.v.} \int_{-\infty}^{\infty} \frac{e^{2x}}{\cosh(\pi x)}\, dx = \sec 1$$

by integrating $e^{2z}/\cosh(\pi z)$ around rectangles with vertices at $z = \pm\rho,\ \rho + i,\ -\rho + i.$

10. Given that $\int_{0}^{\infty} e^{-x^2}\, dx = \sqrt{\pi}/2$, integrate e^{-z^2} around a rectangle with vertices at $z = \pm\rho,\ \pm\rho + bi$, where $b > 0$, and let $\rho \to \infty$ to prove that

$$\int_{0}^{\infty} e^{-x^2} \cos(2bx)\, dx = \frac{\sqrt{\pi}}{2} e^{-b^2}.$$

11. Compute

$$\int_{0}^{\infty} \frac{dx}{x^3 + 1}$$

257

Section 6.3
Improper Integrals
of Certain
Functions Over
$(-\infty, \infty)$

by integrating $1/(z^3 + 1)$ around the boundary of the circular sector $S_\rho : \{z = re^{i\theta} : 0 \leq \theta \leq 2\pi/3, \, 0 \leq r \leq \rho\}$ and letting $\rho \to \infty$.

Summation of Series

12. Let $f(z)$ be a rational function of the form $P(z)/Q(z)$, where deg $Q \geq 2 + $ deg P. Assume that no poles of $f(z)$ occur at the integer points $z = 0, \pm 1, \pm 2, \dots$. Complete each of the following steps to establish the summation formula

$$\sum_{k=-\infty}^{\infty} f(k) = -\{\text{sum of the residues of } \pi f(z) \cot(\pi z) \text{ at the poles of } f(z)\}.$$

(a) Show that for the function $g(z) \equiv \pi f(z) \cot(\pi z)$, we have $\text{Res}(g; k) = f(k), \, k = 0, \pm 1, \pm 2, \dots$.

(b) Let Γ_N be the boundary of the square with vertices at $(N + \frac{1}{2})(1 + i), \, (N + \frac{1}{2})(-1 + i), \, (N + \frac{1}{2})(-1 - i), \, (N + \frac{1}{2})(1 - i)$, taken in that order, where N is a positive integer. Show that there is a constant M independent of N such that $|\pi \cot(\pi z)| \leq M$ for all z on Γ_N.

(c) Prove that

$$\lim_{N \to +\infty} \int_{\Gamma_N} \pi f(z) \cot(\pi z) \, dz = 0,$$

where Γ_N is as defined above.

(d) Use the residue theorem and parts (a) and (c) to show that

$$\lim_{N \to +\infty} \sum_{k=-N}^{N} f(k) = -\{\text{sum of the residues of } \pi f(z) \cot(\pi z) \text{ at the poles of } f(z)\}.$$

13. Using the summation formula in Prob. 12, verify that

(a) $\displaystyle\sum_{k=-\infty}^{\infty} \frac{1}{k^2 + 1} = \pi \coth(\pi)$. [HINT: Take $f(z) = 1/(z^2 + 1)$.]

(b) $\displaystyle\sum_{k=-\infty}^{\infty} \frac{1}{(k - \frac{1}{2})^2} = \pi^2$.

(c) $\displaystyle\sum_{k=1}^{\infty} \frac{1}{k^2} = \frac{\pi^2}{6}$. [HINT: The formula in Prob. 12 needs to be modified to compensate for the pole of $f(z) = 1/z^2$ at $z = 0$.]

6.4 IMPROPER INTEGRALS INVOLVING TRIGONOMETRIC FUNCTIONS

Our purpose here is to use residue theory to evaluate integrals of the general forms

$$\text{p.v.} \int_{-\infty}^{\infty} \frac{P(x)}{Q(x)} \cos mx \, dx, \qquad \text{p.v.} \int_{-\infty}^{\infty} \frac{P(x)}{Q(x)} \sin mx \, dx,$$

where m is real and $P(x)/Q(x)$ denotes a certain rational function continuous on $(-\infty, \infty)$. As we shall show in the following example, the semicircular contour technique of the previous section can be applied, but some modifications are necessary.

EXAMPLE 12 Compute

$$I = \text{p.v.} \int_{-\infty}^{\infty} \frac{\cos 3x}{x^2 + 4} \, dx.$$

SOLUTION In utilizing semicircular contours our first inclination is to deal with the complex function

(20)
$$\frac{\cos 3z}{z^2 + 4}.$$

However, with this choice for $f(z)$ we are doomed to failure because the modulus of (20) does not go to zero in either the upper or lower half-plane. Indeed, when $z = \pm \rho i$ we have

$$\left| \frac{\cos 3z}{z^2 + 4} \right| = \frac{e^{-3\rho} + e^{3\rho}}{2 |-\rho^2 + 4|},$$

which becomes infinite as $\rho \to \infty$.

To circumvent this difficulty we notice that since $\cos 3x$ is the real part of e^{3ix}, we have

$$I = \text{Re}(I_0), \text{ where } I_0 \equiv \text{p.v.} \int_{-\infty}^{\infty} \frac{e^{3ix}}{x^2 + 4} \, dx.$$

Now if we deal with the function

$$f(z) \equiv \frac{e^{3iz}}{z^2 + 4},$$

we encounter singularities at $z = \pm 2i$, and because

$$|f(z)| = |f(x + iy)| = \frac{|e^{3ix} \cdot e^{-3y}|}{|z^2 + 4|} = \frac{e^{-3y}}{|z^2 + 4|},$$

we have in the *upper* half-plane $(y \geq 0)$

$$|f(z)| \leq \frac{1}{|z^2 + 4|}.$$

259

Section 6.4
Improper Integrals
Involving
Trigonometric
Functions

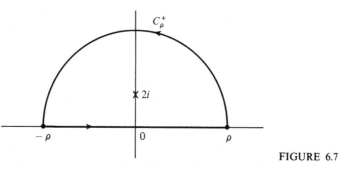

FIGURE 6.7

Thus for any $\rho > 2$, the integral over the upper half-circle C_ρ^+ in Fig. 6.7 is bounded by

$$\left| \int_{C_\rho^+} f(z)\, dz \right| \le \frac{\pi\rho}{\rho^2 - 4},$$

and so it goes to zero as $\rho \to \infty$. Furthermore, since $+2i$ is the only singularity in the upper half-plane, we have for $\rho > 2$

$$\int_{-\rho}^{\rho} f(x)\, dx + \int_{C_\rho^+} f(z)\, dz = 2\pi i\ \text{Res}(f\,;2i).$$

Hence on taking the limit as $\rho \to \infty$ we get

$$\text{p.v.} \int_{-\infty}^{\infty} \frac{e^{3ix}}{x^2 + 4}\, dx + 0 = 2\pi i\ \text{Res}(f;2i).$$

But

$$\text{Res}(f;2i) = \lim_{z \to 2i} (z - 2i) f(z) = \lim_{z \to 2i} \frac{e^{3iz}}{z + 2i} = \frac{e^{-6}}{4i}.$$

Thus

$$I_0 = 2\pi i \cdot \frac{e^{-6}}{4i} = \frac{\pi}{2e^6},$$

and finally

$$I = \text{Re}(I_0) = \text{Re}\left(\frac{\pi}{2e^6}\right) = \frac{\pi}{2e^6}. \quad\blacksquare$$

The technique of Example 12 can be used to evaluate any integral of the form

(21) $$\text{p.v.} \int_{-\infty}^{\infty} e^{imx} \frac{P(x)}{Q(x)}\, dx \qquad (m > 0),$$

where P and Q are polynomials, Q has no real zeros, and the degree of Q exceeds that of P by at least 2. Indeed, the function $e^{imz} P(z)/Q(z)$ has only a finite number of singularities and, for large ρ, its integral over C_ρ^+ is bounded by $1 \cdot (K/\rho^2) \cdot \pi\rho$, which goes to zero as $\rho \to \infty$. However, in applications it is sometimes necessary to evaluate integrals such as (21) where the degree of Q is just *one* higher than that of P. For example, consider

$$(22) \qquad \text{p.v.} \int_{-\infty}^{\infty} \frac{e^{ix}x}{1 + x^2} \, dx.$$

If we estimate the integral of $e^{iz}z/(1 + z^2)$ over C_ρ^+ as before, we find that it is bounded by $1 \cdot (K/\rho) \cdot \pi\rho = K\pi$, for some constant K. Since this does not go to zero, it is by no means obvious that the semicircular contour method will work in evaluating (22).

Surprisingly, it turns out that the integrals of such a function over C_ρ^+ *do* go to zero as $\rho \to \infty$; however, a much finer integral estimate is needed to show this. The next two lemmas fill this need.

LEMMA 2 *Suppose that $f(t)$ and $M(t)$ are continuous functions on the real interval $a \le t \le b$, with f complex and M real-valued. If $|f(t)| \le M(t)$ on this interval, then*

$$(23) \qquad \left| \int_a^b f(t) \, dt \right| \le \int_a^b M(t) \, dt.$$

Proof This result is probably quite obvious to the reader; for a rigorous proof, however, we must revert to the integration theory presented in Chapter 4. Choose an arbitrary subdivision of the interval $[a, b]$, say,

$$a = \tau_0 < \tau_1 < \cdots < \tau_n = b,$$

and form a Riemann sum for f:

$$\sum_{k=1}^n f(c_k) \, \Delta\tau_k .$$

Since $|f(t)| \le M(t)$ for all t, we have by the triangle inequality

$$\left| \sum_{k=1}^n f(c_k) \, \Delta\tau_k \right| \le \sum_{k=1}^n M(c_k) \, \Delta\tau_k ,$$

and we note that the right-hand side is a Riemann sum for M over $[a, b]$. Because the inequality holds for every partition of $[a, b]$, (23) follows by letting the mesh tend to zero. ∎

As an immediate consequence of this lemma we deduce that for any continuous $f(t)$,

$$(24) \qquad \left| \int_a^b f(t) \, dt \right| \le \int_a^b |f(t)| \, dt.$$

261

Section 6.4
Improper Integrals
Involving
Trigonometric
Functions

LEMMA 3 (Jordan Lemma) *If $m > 0$ and P/Q is the quotient of two polynomials such that*

(25)
$$\text{degree } Q \geq 1 + \text{degree } P,$$

then

(26)
$$\lim_{\rho \to \infty} \int_{C_\rho^+} e^{imz} \frac{P(z)}{Q(z)} \, dz = 0,$$

where C_ρ^+ is the upper half-circle of radius ρ.

Proof Parametrizing C_ρ^+ we have

$$\int_{C_\rho^+} e^{imz} \frac{P(z)}{Q(z)} \, dz = \int_0^\pi g(t) \, dt,$$

where

$$g(t) \equiv e^{im(\rho e^{it})} \frac{P(\rho e^{it})}{Q(\rho e^{it})} \rho i e^{it}.$$

Now

$$\left| e^{im(\rho e^{it})} \right| = \left| e^{im\rho \cos t - m\rho \sin t} \right| = e^{-m\rho \sin t}.$$

Furthermore, from (25) we know that there is some constant K such that

$$\left| \frac{P(\rho e^{it})}{Q(\rho e^{it})} \right| \leq \frac{K}{\rho} \qquad \text{(for } \rho \text{ large)}.$$

Thus

$$|g(t)| \leq e^{-m\rho \sin t} \cdot \frac{K}{\rho} \cdot \rho = K e^{-m\rho \sin t},$$

and so by Lemma 2

(27)
$$\left| \int_{C_\rho^+} e^{imz} \frac{P(z)}{Q(z)} \, dz \right| = \left| \int_0^\pi g(t) \, dt \right| \leq K \int_0^\pi e^{-m\rho \sin t} \, dt.$$

To estimate the right-hand integral we first observe that the function $e^{-m\rho \sin t}$ on $[0, \pi]$ is symmetric about $t = \pi/2$. Consequently

$$\int_0^\pi e^{-m\rho \sin t} \, dt = 2 \int_0^{\pi/2} e^{-m\rho \sin t} \, dt.$$

Furthermore, if we consider the graph of $\sin t$, we notice that it is concave downward on $[0, \pi/2]$; thus it lies above the dashed line in Fig. 6.8. In other words,

$$\sin t \geq \frac{2}{\pi} t, \qquad \text{for } 0 \leq t \leq \frac{\pi}{2}.$$

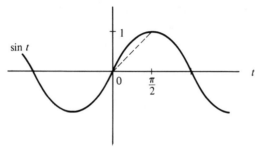

FIGURE 6.8

When $m\rho > 0$, this last inequality implies that

$$e^{-m\rho \sin t} \le e^{-m\rho 2t/\pi} \quad \text{on} \quad [0, \pi/2],$$

and so

$$\int_0^\pi e^{-m\rho \sin t}\, dt = 2\int_0^{\pi/2} e^{-m\rho \sin t}\, dt \le 2\int_0^{\pi/2} e^{-m\rho 2t/\pi}\, dt$$

$$= 2\left(\frac{-\pi}{m\rho 2}\right)[e^{-m\rho} - 1] < \frac{\pi}{m\rho}.$$

Hence from (27) we see that

$$\left|\int_{C_\rho^+} e^{imz}\frac{P(z)}{Q(z)}\, dz\right| \le K \cdot \frac{\pi}{m\rho};$$

consequently, the integral goes to zero as $\rho \to \infty$. ∎

We remark that when m is *negative* the function $e^{imz}P(z)/Q(z)$ is not bounded in the upper half-plane. However, under assumption (25), we do have

$$\lim_{\rho \to \infty} \int_{C_\rho^-} e^{imz}\frac{P(z)}{Q(z)}\, dz = 0 \qquad (m < 0),$$

where C_ρ^- is the lower half-circle of radius ρ. To prove this result we need only make the change of variable $w = -z$ in (26).

EXAMPLE 13   Evaluate

$$\text{p.v.} \int_{-\infty}^\infty \frac{x \sin x}{1 + x^2}\, dx.$$

SOLUTION   We shall first compute the integral

$$I_0 \equiv \text{p.v.} \int_{-\infty}^\infty \frac{xe^{ix}}{1 + x^2}\, dx$$

263

Section 6.4
*Improper Integrals
Involving
Trigonometric
Functions*

and then take its imaginary part as our answer. Since the hypotheses of the Jordan Lemma are satisfied when $m = 1$ and $P/Q = x/(1 + x^2)$, the integral I_0 equals $2\pi i$ times the sum of the residues of $ze^{iz}/(1 + z^2)$ in the upper half-plane. Writing

$$\frac{ze^{iz}}{1 + z^2} = \frac{ze^{iz}}{(z + i)(z - i)},$$

we find for the residue at $z = +i$ the value

$$\lim_{z \to i} \frac{ze^{iz}}{z + i} = \frac{ie^{i2}}{i + i} = \frac{e^{-1}}{2}.$$

Consequently

$$I_0 = 2\pi i \cdot \frac{e^{-1}}{2},$$

and so

$$\text{p.v.} \int_{-\infty}^{\infty} \frac{x \sin x}{1 + x^2} \, dx = \text{Im}(I_0) = \text{Im}\left(\frac{2\pi i e^{-1}}{2}\right) = \frac{\pi}{e}. \qquad \blacksquare$$

EXAMPLE 14   Evaluate

$$I = \text{p.v.} \int_{-\infty}^{\infty} \frac{\sin x}{x + i} \, dx.$$

SOLUTION   One may be tempted to say that this equals the imaginary part of

$$\text{p.v.} \int_{-\infty}^{\infty} \frac{e^{ix}}{x + i} \, dx.$$

This is wrong! (Why?) Moreover, we can't use $(\sin z)/(z + i)$ either, because it is unbounded in both the upper and lower half-planes. So let's try the substitution

$$\sin x = \frac{e^{ix} - e^{-ix}}{2i},$$

which leads to the representation

(28) $$I = \frac{1}{2i}\left(\text{p.v.} \int_{-\infty}^{\infty} \frac{e^{ix}}{x + i} \, dx - \text{p.v.} \int_{-\infty}^{\infty} \frac{e^{-ix}}{x + i} \, dx\right).$$

Now we deal with each integral separately.

For

$$I_1 \equiv \text{p.v.} \int_{-\infty}^{\infty} \frac{e^{ix}}{x + i} \, dx$$

we close the contour $[-\rho, \rho]$ with the half-circle C_ρ^+ in the upper half-plane. Then, by Lemma 3, we have

$$\lim_{\rho \to \infty} \int_{C_\rho^+} \frac{e^{iz}}{z + i} \, dz = 0,$$

and since the only singularity of the integrand is in the *lower* half-plane at $z = -i$, we deduce that $I_1 = 0$.

Now the second integral

$$I_2 \equiv \text{p.v.} \int_{-\infty}^{\infty} \frac{e^{-ix}}{x + i} \, dx$$

involves the function e^{-iz}, which is unbounded in the upper half-plane, so we close the contour $[-\rho, \rho]$ in the lower half-plane with the semicircle $C_\rho^- : z = \rho e^{-it}$, $0 \le t \le \pi$ (see Fig. 6.9). Then by the analogue of the Jordan Lemma for the case when $m < 0$ we deduce that

$$\lim_{\rho \to \infty} \int_{C_\rho^-} \frac{e^{-iz}}{z + i} \, dz = 0.$$

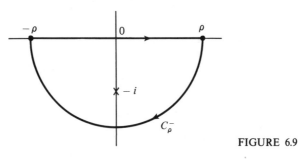

FIGURE 6.9

Observing that the closed contour in Fig. 6.9 is negatively oriented, we obtain

$$I_2 = -2\pi i \, \text{Res}\left(\frac{e^{-iz}}{z + i} ; -i \right)$$

$$= -2\pi i \lim_{z \to -i} e^{-iz} = -2\pi i \cdot e^{-1}.$$

Consequently, from Eq. (28),

$$I = \frac{1}{2i}[I_1 - I_2] = \frac{1}{2i}[0 + 2\pi i e^{-1}] = \frac{\pi}{e}. \quad \blacksquare$$

Exercises 6.4

Using the method of residues, verify the integral formulas in Problems 1–3.

1. p.v. $\displaystyle\int_{-\infty}^{\infty} \frac{\cos(2x)}{x^2 + 1}\, dx = \frac{\pi}{e^2}$.

2. p.v. $\displaystyle\int_{-\infty}^{\infty} \frac{x \sin x}{x^2 - 2x + 10}\, dx = \frac{\pi}{3e^3}(3\cos 1 + \sin 1)$.

3. $\displaystyle\int_0^{\infty} \frac{\cos x}{(x^2 + 1)^2}\, dx = \frac{\pi}{2e}$.

Compute each of the integrals in Problems 4–9.

4. p.v. $\displaystyle\int_{-\infty}^{\infty} \frac{x \sin(3x)}{x^4 + 4}\, dx$.

5. p.v. $\displaystyle\int_{-\infty}^{\infty} \frac{e^{3ix}}{x - 2i}\, dx$.

$\frac{\pi}{5}\left(\cos(1)\,e^{-2}\right)$

6. p.v. $\displaystyle\int_{-\infty}^{\infty} \frac{\cos x}{(x^2 + 1)(x^2 + 4)}\, dx$.

7. p.v. $\displaystyle\int_{-\infty}^{\infty} \frac{e^{-2ix}}{x^2 + 4}\, dx$.

8. p.v. $\displaystyle\int_{-\infty}^{\infty} \frac{\cos(2x)}{x - 3i}\, dx$.

9. $\displaystyle\int_0^{\infty} \frac{x^3 \sin(2x)}{(x^2 + 1)^2}\, dx$.

10. Give conditions under which the following formula is valid:

$$\text{p.v.} \int_{-\infty}^{\infty} e^{imx} \frac{P(x)}{Q(x)}\, dx = 2\pi i \cdot \sum \text{[residues of } e^{imz} P(z)/Q(z) \text{ at poles in the upper half-plane]}.$$

11. Given that $\int_0^{\infty} e^{-x^2}\, dx = \sqrt{\pi}/2$, integrate e^{iz^2} around the boundary of the circular sector $S_\rho : \{z = re^{i\theta} : 0 \le \theta \le \pi/4, 0 \le r \le \rho\}$, and let $\rho \to +\infty$ to prove that

$$\int_0^{\infty} e^{ix^2}\, dx = \frac{\sqrt{2\pi}}{4}(1 + i).$$

6.5 INDENTED CONTOURS

In the preceding sections the integrands f were assumed to be defined and continuous over the whole interval of integration. We turn now to the problem of evaluating special integrals where $|f(x)| \to \infty$ as x approaches certain finite points. Our first step is to give precise meaning to the integrals of f.

Let $f(x)$ be continuous on $[a, b]$ except at the point c, $a < c < b$. Then the *improper integrals* of f over the intervals $[a, c]$, $[c, b]$, and $[a, b]$ are defined by

$$\int_a^c f(x)\, dx \equiv \lim_{r \to 0^+} \int_a^{c-r} f(x)\, dx,$$

$$\int_c^b f(x)\, dx \equiv \lim_{s \to 0^+} \int_{c+s}^b f(x)\, dx,$$

and

(29) $$\int_a^b f(x)\, dx \equiv \lim_{r \to 0^+} \int_a^{c-r} f(x)\, dx + \lim_{s \to 0^+} \int_{c+s}^b f(x)\, dx,$$

provided the appropriate limit(s) exists. For example,

$$\int_0^1 \frac{1}{\sqrt{x}}\, dx = \lim_{s \to 0^+} \int_s^1 \frac{1}{\sqrt{x}}\, dx = \lim_{s \to 0^+} 2\sqrt{x}\,\Big|_s^1$$

$$= \lim_{s \to 0^+} [2 - 2\sqrt{s}] = 2.$$

Even if the limits in (29) fail to exist, the single limit

$$\lim_{r \to 0^+} \left(\int_a^{c-r} f(x)\, dx + \int_{c+r}^b f(x)\, dx \right)$$

may exist, and again we adopt the principal value notation; i.e.,

$$\text{p.v.} \int_a^b f(x)\, dx \equiv \lim_{r \to 0^+} \left(\int_a^{c-r} f(x)\, dx + \int_{c+r}^b f(x)\, dx \right).$$

As an illustration we have

$$\text{p.v.} \int_1^4 \frac{dx}{x - 2} = \lim_{r \to 0^+} \left(\int_1^{2-r} \frac{dx}{x - 2} + \int_{2+r}^4 \frac{dx}{x - 2} \right)$$

$$= \lim_{r \to 0^+} \left(\text{Log}\,|x - 2|\,\Big|_1^{2-r} + \text{Log}\,|x - 2|\,\Big|_{2+r}^4 \right)$$

$$= \lim_{r \to 0^+} (\text{Log}\, r + \text{Log}\, 2 - \text{Log}\, r) = \text{Log}\, 2.$$

For this example it is easy to see that the improper integral

$$\int_1^4 \frac{dx}{x - 2}$$

does not exist. However, if an improper integral does exist, it must equal the p.v. integral.

When the function $f(x)$ is continuous on the whole real line except at c, the principal value of its integral over $(-\infty, \infty)$ is defined by

$$(30) \qquad \text{p.v.} \int_{-\infty}^{\infty} f(x)\, dx \equiv \lim_{\substack{\rho \to \infty \\ r \to 0^+}} \left[\int_{-\rho}^{c-r} f(x)\, dx + \int_{c+r}^{\rho} f(x)\, dx \right],$$

provided the limit exists as $\rho \to \infty$ and $r \to 0^+$ independently.† In the case of several discontinuities occurring at points $x = c_i$ we extend the definition of the p.v. integral over $(-\infty, \infty)$ in a natural way; namely, we remove a small symmetric interval $(c_i - r_i, c_i + r_i)$ about each c_i and then take the limit of the integral as the variables $r_i \to 0^+$ and $\rho \to \infty$, independently. (See Fig. 6.10.)

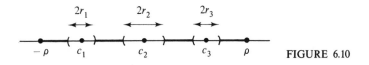

FIGURE 6.10

Residue theory is useful in evaluating certain integrals of the form (30) when the integrand, considered as a function of z, has a simple pole at the exceptional point c. Assuming this to be the case, we must consider the integrals of f along $[-\rho, c - r]$ and $[c + r, \rho]$, and to utilize residue theory we must form some closed contour which contains these segments. In the last two sections we discussed suitable ways to join $+\rho$ to $-\rho$. But now we also need to join $c - r$ to $c + r$. In so doing we cannot proceed along the real axis, for such a segment would pass through the singularity at c. Instead, we detour around c by forming, for example, the half-circle S_r indicated in Fig. 6.11. Eventually we will let r tend to zero, and so it will be necessary to determine the limit

$$\lim_{r \to 0^+} \int_{S_r} f(z)\, dz.$$

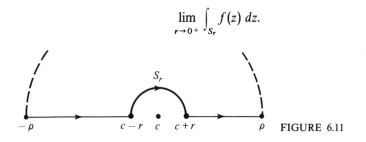

FIGURE 6.11

This is handled by the following lemma, which deals not only with half-circles, but with arbitrary circular arcs:

---

†More precisely, we say that the limit in (30) exists and equals L if for every $\varepsilon > 0$ there exist positive constants M and δ such that for $\rho > M$ and $0 < r < \delta$ the bracketed expression in (30) is within ε of L.

LEMMA 4   *If f has a simple pole at $z = c$ and T_r is the circular arc of Fig. 6.12 defined by*

$$(31) \qquad\qquad T_r: z = c + re^{i\theta} \qquad (\theta_1 \le \theta \le \theta_2),$$

then

$$(32) \qquad\qquad \lim_{r \to 0^+} \int_{T_r} f(z)\, dz = i(\theta_2 - \theta_1)\text{Res}(f; c).$$

Consequently, for the clockwise oriented half-circle S_r of Fig. 6.11 we have

$$(33) \qquad\qquad \lim_{r \to 0^+} \int_{S_r} f(z)\, dz = -i\pi\, \text{Res}(f; c).$$

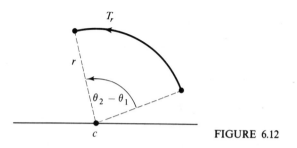

FIGURE 6.12

Proof   Since f has a simple pole at c, its Laurent expansion has the form

$$f(z) = \frac{a_{-1}}{z - c} + \sum_{k=0}^{\infty} a_k(z - c)^k,$$

valid in some punctured neighborhood of c, say $0 < |z - c| < R$. Thus if $0 < r < R$, we can write

$$(34) \qquad\qquad \int_{T_r} f(z)\, dz = a_{-1} \int_{T_r} \frac{dz}{z - c} + \int_{T_r} g(z)\, dz,$$

where

$$g(z) \equiv \sum_{k=0}^{\infty} a_k(z - c)^k.$$

Now g is analytic at c (why?), and hence is bounded in some neighborhood of this point; i.e.,

$$|g(z)| \le M, \qquad \text{for } |z - c| < R_1.$$

Consequently, for $0 < r < R_1$, we have

$$\left| \int_{T_r} g(z)\, dz \right| \le M l(T_r) = M(\theta_2 - \theta_1)r,$$

and the last term goes to zero as $r \to 0^+$. Therefore,

$$\lim_{r \to 0^+} \int_{T_r} g(z) \, dz = 0.$$

To deal with the integral of $1/(z - c)$ we use the parametrization (31) to derive

$$\int_{T_r} \frac{dz}{z - c} = \int_{\theta_1}^{\theta_2} \frac{1}{re^{i\theta}} rie^{i\theta} \, d\theta = i \int_{\theta_1}^{\theta_2} d\theta = i(\theta_2 - \theta_1),$$

the value being independent of r. Hence from Eq. (34) we obtain

$$\lim_{r \to 0^+} \int_{T_r} f(z) \, dz = a_{-1} i(\theta_2 - \theta_1) + 0 = \text{Res}(f; c) \, i(\theta_2 - \theta_1),$$

which is the desired limit (32).

In particular, when the T_r are counterclockwise-oriented *half*-circles, we get the limiting value $i\pi \, \text{Res}(f; c)$, and thus for the oppositely oriented half-circles S_r in Fig. 6.11 we get minus this value. ∎

EXAMPLE 15   Evaluate

$$I \equiv \text{p.v.} \int_{-\infty}^{\infty} \frac{e^{ix}}{x} \, dx.$$

SOLUTION   First notice that the integrand is continuous except at $x = 0$. Hence

$$I = \lim_{\substack{\rho \to \infty \\ r \to 0^+}} \left(\int_{-\rho}^{-r} \frac{e^{ix}}{x} \, dx + \int_{r}^{\rho} \frac{e^{ix}}{x} \, dx \right).$$

Now we introduce the complex function

$$f(z) \equiv \frac{e^{iz}}{z},$$

which has a simple pole at the origin but is elsewhere analytic. Next we must form a closed contour containing the segments $[-\rho, -r]$ and $[r, \rho]$. Observing that the Jordan Lemma applies to $f(z)$ we join $+\rho$ to $-\rho$ by the half-circle C_ρ^+ in the upper half-plane. In joining $-r$ to r we indent around the origin by using a half-circle S_r. This yields the closed contour of Fig. 6.13. Now since e^{iz}/z has no singularities inside the closed contour, we have

$$\left(\int_{-\rho}^{-r} + \int_{S_r} + \int_{r}^{\rho} + \int_{C_\rho^+} \right) \frac{e^{iz}}{z} \, dz = 0;$$

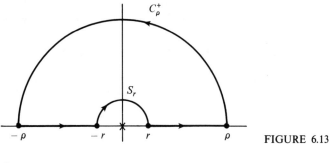

FIGURE 6.13

i.e.,

$$(35) \qquad \int_{-\rho}^{-r} \frac{e^{ix}}{x} \, dx + \int_{r}^{\rho} \frac{e^{ix}}{x} \, dx = -\int_{S_r} \frac{e^{iz}}{z} \, dz - \int_{C_\rho^+} \frac{e^{iz}}{z} \, dz.$$

By the Jordan Lemma,

$$\lim_{\rho \to \infty} \int_{C_\rho^+} \frac{e^{iz}}{z} \, dz = 0,$$

and by Eq. (33) of Lemma 4

$$\lim_{r \to 0^+} \int_{S_r} \frac{e^{iz}}{z} \, dz = -i\pi \, \text{Res}(0)$$

$$= -i\pi \lim_{z \to 0} z \cdot \frac{e^{iz}}{z} = -i\pi.$$

Thus from Eq. (35) we obtain

$$(36) \qquad \text{p.v.} \int_{-\infty}^{\infty} \frac{e^{ix}}{x} \, dx = -(-i\pi) - 0 = i\pi. \qquad \blacksquare$$

EXAMPLE 16 Find

$$\int_0^\infty \frac{\sin x}{x} \, dx = \lim_{\substack{\rho \to \infty \\ r \to 0^+}} \int_r^\rho \frac{\sin x}{x} \, dx \ .$$

SOLUTION Observe that the integrand $g(x) = (\sin x)/x$ is an even function of x; i.e., $g(-x) = g(x)$ for all x. Hence

$$2 \int_0^\infty \frac{\sin x}{x} \, dx = \text{p.v.} \int_{-\infty}^{\infty} \frac{\sin x}{x} \, dx.$$

Furthermore, the right-hand integral is the imaginary part of the integral of e^{ix}/x over $(-\infty, \infty)$ and so, by Example 15, it equals $\text{Im}(i\pi) = \pi$. Thus

$$\int_0^\infty \frac{\sin x}{x} \, dx = \frac{\pi}{2}. \qquad \blacksquare$$

We remark that as another consequence of Example 15 we have

$$\text{p.v.} \int_{-\infty}^{\infty} \frac{\cos x}{x}\, dx = \text{Re}(i\pi) = 0,$$

but this is scarcely surprising because the integrand $h(x) = (\cos x)/x$ is an odd function of x; i.e., $h(-x) = -h(x)$ for all x.

EXAMPLE 17   Compute

$$\text{p.v.} \int_{-\infty}^{\infty} \frac{xe^{2ix}}{x^2 - 1}\, dx.$$

SOLUTION   Here the integrand is discontinuous at two real points, $x = \pm 1$. Thus we need to find

$$\lim_{\substack{\rho \to \infty \\ r_1, r_2 \to 0+}} \left(\int_{-\rho}^{-1-r_1} + \int_{-1+r_1}^{1-r_2} + \int_{1+r_2}^{\rho} \right) \frac{xe^{2ix}}{x^2 - 1}\, dx.$$

For this purpose we work with

$$f(z) \equiv \frac{ze^{2iz}}{z^2 - 1}$$

and indent around each of its simple poles, as indicated in Fig. 6.14. Then since $f(z)$ is analytic inside the closed contour, we obtain

$$(37) \quad \left(\int_{-\rho}^{-1-r_1} + \int_{-1+r_1}^{1-r_2} + \int_{1+r_2}^{\rho} \right) \frac{xe^{2ix}}{x^2 - 1}\, dx + J_{r_1} + J_{r_2} + J_\rho = 0,$$

where J_{r_1}, J_{r_2}, J_ρ are the integrals of $f(z)$ over $S_{r_1}, S_{r_2}, C_\rho^+$, respectively. Now by the Jordan Lemma we have

$$\lim_{\rho \to \infty} J_\rho = 0,$$

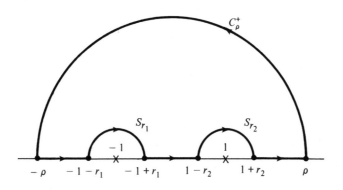

FIGURE 6.14

and from Eq. (33) of Lemma 4

$$\lim_{r_1 \to 0^+} J_{r_1} = -i\pi \, \text{Res}(-1) = -i\pi \lim_{z \to -1} (z+1) f(z)$$

$$= -i\pi \lim_{z \to -1} \frac{ze^{2iz}}{z-1} = \frac{-i\pi e^{-2i}}{2},$$

and

$$\lim_{r_2 \to 0^+} J_{r_2} = -i\pi \, \text{Res}(1) = -i\pi \lim_{z \to 1} (z-1) f(z)$$

$$= -i\pi \lim_{z \to 1} \frac{ze^{2iz}}{z+1} = \frac{-i\pi e^{2i}}{2}.$$

Hence on taking the limits in Eq. (37) we get

$$\text{p.v.} \int_{-\infty}^{\infty} \frac{xe^{2ix}}{x^2 - 1} \, dx = \frac{i\pi e^{-2i}}{2} + \frac{i\pi e^{2i}}{2} - 0 = i\pi \cos 2. \quad \blacksquare$$

Exercises 6.5

1. Compute each of the following limits along the given circular arcs.

   (a) $\displaystyle \lim_{r \to 0^+} \int_{T_r} \frac{2z^2 + 1}{z} \, dz$, where $T_r: z = re^{i\theta}, \, 0 \le \theta \le \dfrac{\pi}{2}$

   (b) $\displaystyle \lim_{r \to 0^+} \int_{\Gamma_r} \frac{e^{3iz}}{z^2 - 1} \, dz$, where $\Gamma_r: z = 1 + re^{i\theta}, \, \dfrac{\pi}{4} \le \theta \le \pi$

   (c) $\displaystyle \lim_{r \to 0^+} \int_{\gamma_r} \frac{\text{Log } z}{z - 1} \, dz$, where $\gamma_r: z = 1 + re^{-i\theta}, \, \pi \le \theta \le 2\pi$

   (d) $\displaystyle \lim_{r \to 0^+} \int_{S_r} \frac{e^z - 1}{z^2} \, dz$, where $S_r: z = re^{-i\theta}, \, \pi \le \theta \le 2\pi$

Using the technique of residues, verify each of the integral formulas in Problems 2–4.

2. p.v. $\displaystyle \int_{-\infty}^{\infty} \frac{e^{2ix}}{x+1} \, dx = \pi i e^{-2i}$.

3. p.v. $\displaystyle \int_{-\infty}^{\infty} \frac{e^{ix}}{(x-1)(x-2)} \, dx = \pi i (e^{2i} - e^i)$.

4. $\displaystyle \int_0^{\infty} \frac{\sin(2x)}{x(x^2+1)^2} \, dx = \pi \left(\frac{1}{2} - \frac{1}{e^2} \right)$.

5. Compute $\displaystyle \int_0^{\infty} \frac{\cos x - 1}{x^2} \, dx$.

6. Compute p.v. $\displaystyle\int_{-\infty}^{\infty} \frac{\sin x}{(x^2 + 4)(x - 1)}\, dx.$

7. Compute p.v. $\displaystyle\int_{-\infty}^{\infty} \frac{x \cos x}{x^2 - 3x + 2}\, dx.$

8. Compute p.v. $\displaystyle\int_{-\infty}^{\infty} \frac{\cos(2x)}{x^3 + 1}\, dx.$

9. Prove that for $a > 0$ and $b > 0$

$$\int_0^{\infty} \frac{\sin(ax)}{x(x^2 + b^2)}\, dx = \frac{\pi}{2b^2}\left(1 - e^{-ab}\right).$$

10. Prove that

$$\int_0^{\infty} \frac{\sin^2 x}{x^2}\, dx = \frac{\pi}{2}.$$

[HINT: $\sin^2 x = \frac{1}{2}(1 - \cos 2x) = \frac{1}{2}\,\text{Re}(1 - e^{2ix}).$]

11. Compute p.v. $\displaystyle\int_{-\infty}^{\infty} \frac{\sin^3 x}{x^3}\, dx.$

$$\left[\text{HINT: } \sin^3 x = \text{Im}\left(\frac{3e^{ix}}{4} - \frac{e^{3ix}}{4} - \frac{1}{2}\right).\right]$$

12. Compute p.v. $\displaystyle\int_{-\infty}^{\infty} \frac{e^{ax}}{e^x - 1}\, dx$ for $0 < a < 1.$

[HINT: Indent the contour of Fig. 6.6 around the points $z = 0$ and $z = 2\pi i.$]

6.6 INTEGRALS INVOLVING MULTIPLE-VALUED FUNCTIONS

In attempting to apply residue theory to compute an integral of $f(x)$, it may turn out that the complex function $f(z)$ is multiple-valued. If this happens, we need to modify our procedure by taking into account not only isolated singularities but also branch points and branch cuts. In fact we may find it necessary to integrate along a branch cut, so we turn first to a discussion of this technique.

To be specific let α denote a real number, but not an integer, and let $f(z)$ be the branch of z^α obtained by restricting the argument of z to lie between 0 and 2π; i.e.,

(38) $\qquad f(z) = e^{\alpha(\text{Log } r + i\theta)}, \qquad \text{where } z = re^{i\theta},\ 0 < \theta < 2\pi.$

As shown in Chapter 3 this function is analytic in the plane except along its branch cut, the nonnegative real axis. In fact, as z approaches a point x (>0) on the cut from the upper half-plane, θ goes to zero, and

$$(39) \qquad f(z) \to e^{\alpha\,\text{Log}\,x} \equiv x^{\alpha}$$

(x^{α} being the principal value as in calculus); while if z approaches x from the lower half-plane, θ tends to 2π, and so

$$(40) \qquad f(z) \to e^{\alpha(\text{Log}\,x + i2\pi)} = x^{\alpha} \cdot e^{2\pi i \alpha}.$$

In this sense we visualize f as being equal to x^{α} on the "upper side" of the cut, and being equal to $x^{\alpha} \cdot e^{2\pi i \alpha}$ on the "lower side."

Now if we are to integrate $f(z)$ along the cut, we avoid ambiguity by placing a direction arrow either above or below the cut to indicate which values of f are to be used. For example, the integrals of f along the segments γ_1 and γ_2 of Fig. 6.15 are given by

$$(41) \qquad \int_{\gamma_1} f(z)\,dz = \int_{\varepsilon}^{\rho} x^{\alpha}\,dx, \qquad \int_{\gamma_2} f(z)\,dz = -\int_{\varepsilon}^{\rho} x^{\alpha} \cdot e^{2\pi i \alpha}\,dx.$$

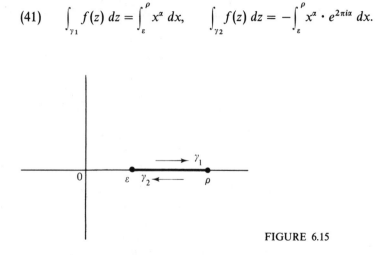

FIGURE 6.15

These same remarks apply to arbitrary functions with branch cuts—the values of the function on each side of the cut being determined by continuity from that side. We shall now illustrate how the residue theorem applies to such functions.

EXAMPLE 18   Let α be a real number (not an integer) and let $R(z)$ be a rational function having no poles on the closed contour Γ of Fig. 6.16. Prove that if $f(z)$ is the branch of $z^{\alpha}R(z)$ obtained by restricting the argument of z to be between 0 and 2π, i.e.,

$$f(z) = \begin{cases} e^{\alpha(\text{Log}\,r + i\theta)} \cdot R(re^{i\theta}), & \text{for } z = re^{i\theta}, \, 0 < \theta < 2\pi, \\ x^{\alpha}R(x), & \text{for } z = x \text{ when integrating on } \gamma_1, \\ x^{\alpha}e^{2\pi i \alpha}R(x), & \text{for } z = x \text{ when integrating on } \gamma_2, \end{cases}$$

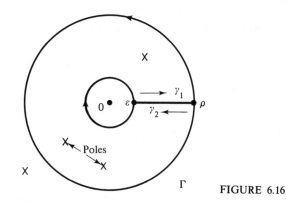

FIGURE 6.16

then

(42) $\int_{\Gamma} f(z)\, dz = 2\pi i \cdot \sum [\text{residues of } f(z) \text{ at the poles inside } \Gamma].\dagger$

SOLUTION   Notice that the residue theorem cannot be directly applied here because the integrand is multiple-valued on the portion $[\varepsilon, \rho]$ of the branch cut. (Notice also that the integrals along γ_1 and γ_2 do *not* cancel here.) To circumvent this difficulty we introduce a segment, indicated by the dashed line in Fig. 6.17, which joins the inner circle to the outer one and does not pass through any poles of $R(z)$. This creates two positively oriented closed contours Γ_1 and Γ_2 (as shown in Fig. 6.17) such that

(43) $\int_{\Gamma} f(z)\, dz = \int_{\Gamma_1} f(z)\, dz + \int_{\Gamma_2} f(z)\, dz$

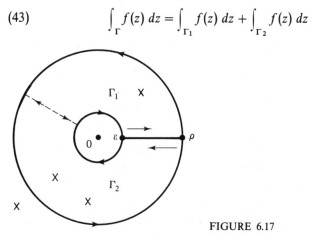

FIGURE 6.17

(the integrals along the dashed segment cancel). Now on Γ_1 we *can* apply the residue theorem because $f(z)$ agrees with a function analytic on this contour; indeed, for z on or inside Γ_1

$$f(z) = e^{\alpha \, \text{Log} \, z} R(z)$$

† To be precise, by the *inside* of Γ we mean the set of those points between the two circles but not lying on the segment $[\varepsilon, \rho]$.

(the *principal branch* Log z having its branch cut along the negative real axis). Hence

(44) $\displaystyle\int_{\Gamma_1} f(z)\,dz = 2\pi i \cdot \sum (\text{residues of } f \text{ at poles inside } \Gamma_1).$

Similarly, on Γ_2 the function $f(z)$ again agrees with an analytic function; for instance, for $z = re^{i\theta}$ on Γ_2

$$f(z) = e^{\alpha(\text{Log } r + i\theta)}R(re^{i\theta}), \quad \frac{\pi}{2} < \theta < \frac{5\pi}{2}$$

(with cut along the positive imaginary axis). Consequently

(45) $\displaystyle\int_{\Gamma_2} f(z)\,dz = 2\pi i \cdot \sum (\text{residues of } f \text{ at poles inside } \Gamma_2).$

Therefore, since every pole inside Γ lies either inside Γ_1 or inside Γ_2, adding Eqs. (44) and (45) gives the desired equation (42). ∎

Armed with the above application of the residue theorem we now can tackle problems of integrating certain functions involving fractional powers of x.

EXAMPLE 19   Compute

$$I \equiv \int_0^\infty \frac{x^{\lambda-1}}{x+4}\,dx, \qquad \text{where } 0 < \lambda < 1,$$

and $x^{\lambda-1}$ denotes the principal value for $x > 0$.

SOLUTION   Observe that we are required here to find

$$I = \lim_{\substack{\rho \to \infty \\ \varepsilon \to 0^+}} \int_\varepsilon^\rho \frac{x^{\lambda-1}}{x+4}\,dx,$$

and for this purpose we take the branch of $z^{\lambda-1}/(z+4)$ defined by

$$f(z) = \frac{e^{(\lambda-1)(\text{Log } r + i\theta)}}{re^{i\theta}+4}, \qquad \text{for } z = re^{i\theta}, 0 < \theta < 2\pi,$$

which has the nonnegative real axis as its branch cut. Then, according to our convention,

$$f(x) = \frac{x^{\lambda-1}}{x+4}$$

for $x > 0$ on the upper side of the cut, and

$$f(x) = \frac{x^{\lambda-1}e^{2\pi i(\lambda-1)}}{x+4} = \frac{x^{\lambda-1}e^{2\pi i\lambda}}{x+4}$$

for $x > 0$ on the lower side. Now we need to form a closed contour containing the segment $[\varepsilon, \rho]$, and in so doing we must take into account the branch point at the origin as well as the pole at $z = -4$. Consider then the closed contour of Fig. 6.18, where ε is small enough and ρ is large enough so that the pole at -4 lies inside the contour. Then for such ε and ρ we have, by Example 18,

$$(46) \qquad \left(\int_{\Gamma_\varepsilon} + \int_{C_\rho} + \int_{\gamma_1} + \int_{\gamma_2} \right) f(z)\, dz = 2\pi i\, \mathrm{Res}(f;\, -4).$$

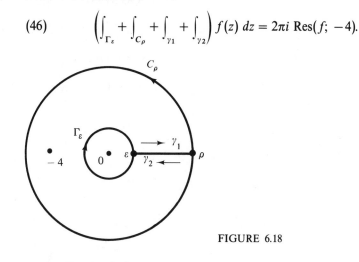

FIGURE 6.18

As discussed above,

$$\int_{\gamma_1} f(z)\, dz + \int_{\gamma_2} f(z)\, dz = \int_\varepsilon^\rho \frac{x^{\lambda-1}}{x+4}\, dx - \int_\varepsilon^\rho \frac{x^{\lambda-1} e^{2\pi i\lambda}}{x+4}\, dx$$

$$= (1 - e^{2\pi i\lambda}) \int_\varepsilon^\rho \frac{x^{\lambda-1}}{x+4}\, dx,$$

and so we can identify

$$\lim_{\substack{\rho \to \infty \\ \varepsilon \to 0^+}} \left(\int_{\gamma_1} f(z)\, dz + \int_{\gamma_2} f(z)\, dz \right) = (1 - e^{2\pi i\lambda}) I.$$

Furthermore, on the circle of radius ρ we have

$$|f(z)| = \frac{|z|^{\lambda-1}}{|z+4|} \leq \frac{\rho^{\lambda-1}}{\rho-4},$$

which yields the estimate

$$\left| \int_{C_\rho} f(z)\, dz \right| \leq \frac{\rho^{\lambda-1}}{\rho-4} \cdot 2\pi\rho.$$

Consequently, since $\lambda - 1 < 0$, the integral over C_ρ tends to zero as $\rho \to \infty$. Similarly, on the inner circle of radius ε we have

$$|f(z)| \leq \frac{\varepsilon^{\lambda-1}}{4-\varepsilon},$$

which implies that

$$\left| \int_{\Gamma_\varepsilon} f(z)\, dz \right| \le \frac{\varepsilon^{\lambda-1}}{4-\varepsilon} \cdot 2\pi\varepsilon = \frac{2\pi\varepsilon^\lambda}{4-\varepsilon}.$$

As $\varepsilon \to 0^+$ this also goes to zero (remember that $\lambda > 0$).

Hence on taking the limit as $\rho \to \infty$ and $\varepsilon \to 0^+$ in Eq. (46) we obtain

$$(47) \qquad\qquad 0 + 0 + (1 - e^{2\pi i \lambda})I = 2\pi i \, \text{Res}(f; -4).$$

Finally, since $z = -4$ is a simple pole of f,

$$\text{Res}(f; -4) = \lim_{z \to -4} (z+4)f(z) = \lim_{\substack{r \to 4 \\ \theta \to \pi}} e^{(\lambda-1)(\text{Log } r + i\theta)}$$

$$= e^{(\lambda-1)(\text{Log } 4 + i\pi)} = -4^{\lambda-1} e^{i\pi\lambda}.$$

Therefore from Eq. (47) we get

$$I = \frac{-2\pi i \cdot 4^{\lambda-1} e^{i\pi\lambda}}{1 - e^{2\pi i \lambda}} = \frac{-2\pi i \cdot 4^{\lambda-1}}{e^{-i\pi\lambda} - e^{i\pi\lambda}},$$

which can be written in the form

$$I = \frac{\pi 4^{\lambda-1}}{\sin(\pi\lambda)}. \qquad \blacksquare$$

An additional complication arises in

EXAMPLE 20 Compute

$$I \equiv \text{p.v.} \int_0^\infty \frac{x^{\lambda-1}}{x-4}\, dx, \qquad \text{where } 0 < \lambda < 1.$$

SOLUTION There is a significant difference between this and the preceding example; here we have a singularity at $x = +4$ which lies *on* the interval of integration. Thus we must compute

$$I = \lim_{\substack{\rho \to \infty \\ \varepsilon,\, \delta \to 0^+}} \left(\int_\varepsilon^{4-\delta} + \int_{4+\delta}^\rho \right) \frac{x^{\lambda-1}}{x-4}\, dx.$$

To do this we modify the approach of the preceding problem by indenting around the singularity. Choosing the branch

$$f(z) = \frac{e^{(\lambda-1)(\text{Log } r + i\theta)}}{re^{i\theta} - 4}, \qquad \text{for } z = re^{i\theta},\ 0 < \theta < 2\pi,$$

we form the contour of Fig. 6.19. Since $f(z)$ has no singularities "inside" the closed contour, the integral over the latter must be zero. Utilizing the

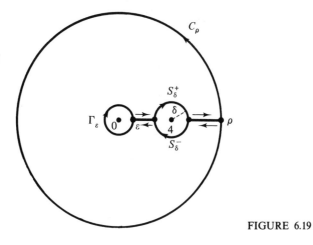

FIGURE 6.19

different definitions for f on the upper and lower sides of the branch cut, we can write this as

$$(48) \qquad (1 - e^{2\pi i \lambda})\left(\int_{\varepsilon}^{4-\delta} + \int_{4+\delta}^{\rho}\right)\frac{x^{\lambda-1}}{x-4}\,dx$$

$$+ \left(\int_{\Gamma_\varepsilon} + \int_{S_\delta^+} + \int_{S_\delta^-} + \int_{C_\rho}\right)f(z)\,dz = 0,$$

where the contours are as indicated in Fig. 6.19.

Now it follows by the same reasoning used in Example 19 that

$$(49) \qquad \lim_{\varepsilon \to 0+}\int_{\Gamma_\varepsilon} f(z)\,dz = 0 \quad \text{and} \quad \lim_{\rho \to \infty}\int_{C_\rho} f(z)\,dz = 0.$$

To compute the limits as $\delta \to 0^+$ of the integrals over S_δ^+ and S_δ^- we apply the result of the preceding section concerning the behavior of integrals near simple poles. On the upper half-circles around $z = 4$, the function f agrees with the principal branch

$$f_1(z) \equiv \frac{e^{(\lambda-1)\,\text{Log}\,z}}{z-4},$$

which is *analytic* on the positive real axis except for its simple pole at $z = 4$. Hence by Lemma 4 of Sec. 6.5

$$(50) \qquad \lim_{\delta \to 0+}\int_{S_\delta^+} f(z)\,dz = -i\pi\,\text{Res}(f_1;4) = -i\pi \lim_{z \to 4} e^{(\lambda-1)\text{Log}\,z}$$

$$= -i\pi 4^{\lambda-1}.$$

However, on the lower half-circles $f(z)$ equals $e^{2\pi i \lambda}$ times $f_1(z)$, and so

$$(51) \qquad \lim_{\delta \to 0+}\int_{S_\delta^-} f(z)\,dz = -i\pi 4^{\lambda-1}e^{2\pi i \lambda}.$$

Finally, on taking the limit as $\rho \to \infty$, $\varepsilon \to 0^+$, and $\delta \to 0^+$ in Eq. (48), we deduce from (49), (50), and (51) that

$$(1 - e^{2\pi i\lambda})I + 0 - i\pi 4^{\lambda-1} - i\pi 4^{\lambda-1}e^{2\pi i\lambda} + 0 = 0,$$

or, equivalently,

$$I = i\pi 4^{\lambda-1}\frac{(1 + e^{2\pi i\lambda})}{(1 - e^{2\pi i\lambda})}.$$

This can also be expressed as

$$I = -\pi 4^{\lambda-1}\cot(\pi\lambda). \quad \blacksquare$$

Exercises 6.6

Use residue theory to verify each of the integral formulas in Problems 1–7.

1. $\displaystyle\int_0^\infty \frac{\sqrt{x}}{x^2 + 1}\,dx = \frac{\pi}{\sqrt{2}}.$

2. $\displaystyle\int_0^\infty \frac{x^{\alpha-1}}{x + 1}\,dx = \frac{\pi}{\sin(\pi\alpha)},\ 0 < \alpha < 1.$

3. $\displaystyle\int_0^\infty \frac{x^\alpha}{(x + 9)^2}\,dx = \frac{9^{\alpha-1}\pi\alpha}{\sin(\pi\alpha)},\ -1 < \alpha < 1,\ \alpha \neq 0.$

4. $\displaystyle\int_0^\infty \frac{x^\alpha}{(x^2 + 1)^2}\,dx = \frac{\pi(1 - \alpha)}{4\cos(\alpha\pi/2)},\ -1 < \alpha < 3,\ \alpha \neq 1.$

5. $\displaystyle\int_0^\infty \frac{x^{\alpha-1}}{x^2 + x + 1}\,dx = \frac{2\pi}{\sqrt{3}}\cos\left(\frac{2\alpha\pi + \pi}{6}\right)\csc(\alpha\pi),\ 0 < \alpha < 2,\ \alpha \neq 1.$

6. p.v. $\displaystyle\int_0^\infty \frac{x^\alpha}{x^2 - 1}\,dx = \frac{\pi}{2\sin(\pi\alpha)}[1 - \cos(\pi\alpha)],\ -1 < \alpha < 1,\ \alpha \neq 0.$

7. $\displaystyle\int_0^\infty \frac{x^\alpha}{1 + 2x\cos\phi + x^2}\,dx = \frac{\pi}{\sin(\pi\alpha)}\frac{\sin(\phi\alpha)}{\sin\phi},$

$$-1 < \alpha < 1,\ \alpha \neq 0,\ -\pi < \phi < \pi,\ \phi \neq 0.$$

8. Prove that

$$\text{p.v.}\int_{-\infty}^\infty \frac{\text{Log}\,|x|}{x^2 + 4}\,dx = \frac{\pi}{2}\text{Log }2.$$

[HINT: Integrate $(\text{Log }z)/(z^2 + 4)$ around a semicircular contour indented at the origin (see Fig. 6.13) and note that $(\rho\,\text{Log }\rho)/(\rho^2 - 4) \to 0$ as $\rho \to +\infty$ or as $\rho \to 0^+$.]

9. Prove that

$$\int_0^\infty \frac{\text{Log } x}{x^2 + 1}\, dx = 0.$$

[HINT: See Prob. 8.]

10. Prove that

$$\int_0^\infty \frac{\text{Log } x}{(x^2 + 1)^2}\, dx = -\frac{\pi}{4}.$$

[HINT: See Prob. 8.]

11. Prove that

$$\int_0^\infty x^{\alpha - 1} \sin x\, dx = \sin\left(\frac{\pi\alpha}{2}\right) \cdot \Gamma(\alpha) \qquad (0 < \alpha < 1),$$

where $\Gamma(\alpha) \equiv \int_0^\infty e^{-x} x^{\alpha - 1}\, dx$ is the *Gamma function*. [HINT: Integrate $e^{-z} z^{\alpha - 1}$ around a quarter-circle indented at the origin.]

*6.7 THE ARGUMENT PRINCIPLE AND ROUCHÉ'S THEOREM

In this section we shall use Cauchy's Residue Theorem to derive two theoretical results which have important practical applications. These results pertain to functions all of whose isolated singularities are poles. Such functions are given a special name in

DEFINITION 2   A function f is said to be **meromorphic** *in a domain D if at every point of D it is either analytic or has a pole.*

In particular, we regard the analytic functions on D as being special cases of meromorphic functions. The rational functions are examples of functions which are meromorphic in the whole plane.

Suppose now that we are given a function f which is analytic and nonzero at each point of a simple closed contour C and is meromorphic inside C. Under these conditions it can be shown that f has at most a *finite* number of poles inside C. The proof of this depends on two facts: first, that the only singularities of f are *isolated* singularities (poles), and, second, that every infinite sequence of points inside C has a subsequence which converges to some point on or inside C. (The last fact is proved in advanced calculus texts under the name: Bolzano-Weierstrass Theorem.) Hence if f had an infinite number of poles inside C, some subsequence of them would converge to a point which must be a singularity, but not an isolated singularity, of f. By contradiction, then, the number of poles must be finite.

Applying this same argument to the reciprocal function $1/f$, which has poles at the zeros of f, it follows that f can also have at most a finite number of *zeros* inside C.

In counting the number of zeros or poles of a function it is common practice to include the multiplicity. For example, let $C: |z| = 4$ and take

(52)
$$f(z) = \frac{(z-8)^2 z^3}{(z-5)^4 (z+2)^2 (z-1)^5}.$$

Then the number $N_p(f)$ of poles of f *inside* C is to be interpreted as

$$N_p(f) \equiv \sum_{\text{poles inside } C} (\text{order of each pole})$$

$$= (\text{order of pole at } z = -2) + (\text{order of pole at } z = 1)$$

$$= 2 + 5 = 7,$$

while the number $N_0(f)$ of its zeros inside C equals

$$N_0(f) \equiv (\text{order of the zero at } z = 0) = 3.$$

We now can state the first of our main results, which relates the contour integral of the logarithmic derivative of f to the number of zeros and poles.

THEOREM 3 (Argument Principle) *If f is analytic and nonzero at each point of a simple closed positively oriented contour C and is meromorphic inside C, then*

(53)
$$\frac{1}{2\pi i} \int_C \frac{f'(z)}{f(z)} dz = N_0(f) - N_p(f),$$

where $N_0(f)$ and $N_p(f)$ are, respectively, the number of zeros and poles of f inside C (multiplicity included).

In particular, for the function f defined in Eq. (52) the theorem asserts that

$$\frac{1}{2\pi i} \oint_{|z| = 4} \frac{f'(z)}{f(z)} dz = 3 - 7 = -4.$$

Proof of Theorem 3 The strategy is quite straightforward; we locate the singularities and compute the residues of the integrand

$$G(z) \equiv \frac{f'(z)}{f(z)}.$$

Notice that this function is analytic at each point *on* C because f is analytic and nonzero there. Inside C the singularities of G occur at those points where f has a zero or a pole.

Consider first a point z_0 inside C which is a zero of f of order m. Then we know f can be written in the form

$$f(z) = (z - z_0)^m \cdot h(z),$$

where $h(z)$ is analytic and not zero at $z = z_0$ (recall Theorem 16 in Chapter 5). Hence in some punctured neighborhood of z_0 we compute that

$$G(z) = \frac{f'(z)}{f(z)} = \frac{m}{z - z_0} + \frac{h'(z)}{h(z)}.$$

Since the function h'/h is analytic at z_0, the above representation shows that G has a simple pole at z_0 with residue equal to m.

On the other hand, if f has a pole of order k at z_p, then

$$f(z) = \frac{H(z)}{(z - z_p)^k},$$

where $H(z)$ is analytic at z_p and $H(z_p) \neq 0$. This time we derive that in a punctured neighborhood of z_p

$$G(z) = \frac{f'(z)}{f(z)} = \frac{-k}{z - z_p} + \frac{H'(z)}{H(z)},$$

and since H'/H is analytic at z_p, we find that G has a simple pole at z_p with residue equal to *minus k*.

Finally, by the residue theorem, the integral of G around C must equal $2\pi i$ times the sum of the residues at the singularities inside C. This, from the above deliberations, equals $2\pi i$ times the sum of the orders of the zeros of f inside C plus the sum of the negatives of the orders of the poles of f inside C. That is,

$$\int_C G(z)\, dz = \int_C \frac{f'(z)}{f(z)}\, dz = 2\pi i[N_0(f) - N_p(f)],$$

which is the same as Eq. (53). ∎

Of course if the function f of Theorem 3 has no poles inside C, then $N_p(f) = 0$, and we have

COROLLARY 1 *If f is analytic inside and on a simple closed positively oriented contour C and if f is nonzero on C, then*

$$\frac{1}{2\pi i} \int_C \frac{f'(z)}{f(z)}\, dz = N_0(f),$$

where $N_0(f)$ is the number of zeros of f inside C (multiplicity included).

Concerning Theorem 3 we remark that the name "Argument Principle" is due to the fact that the left-hand side of Eq. (53) can be interpreted as a change in the argument of $f(z)$. Indeed, if we integrate f'/f along a sufficiently small arc of C from, say, z_1 to z_2, we get

$$\int_{z_1}^{z_2} \frac{f'(z)}{f(z)}\, dz = \log f(z)\Big|_{z_1}^{z_2}$$

for any branch of $\log f(z)$ which is analytic in a disk containing this portion of C. Hence if we apply the stepwise process of analytic continuation to the antiderivative $\log f(z)$, we ultimately obtain

$$\frac{1}{2\pi i} \int_C \frac{f'(z)}{f(z)}\, dz = \frac{1}{2\pi i} \Delta_C \log f(z),$$

where $\Delta_C \log f(z)$ denotes the change in the value of

$$\log f(z) = \mathrm{Log}\,|f(z)| + i\, \arg f(z)$$

after completing one full circuit around C. But since the real function $\mathrm{Log}\,|f(z)|$ returns to its original value, the net change is in the *argument* of f; i.e.,

$$\frac{1}{2\pi i} \int_C \frac{f'(z)}{f(z)}\, dz = \frac{1}{2\pi} \Delta_C \arg f(z).$$

Consequently the conclusion of the Argument Principle can be written in the form

(54)
$$\frac{1}{2\pi} \Delta_C \arg f(z) = N_0(f) - N_p(f).$$

There is yet another way to express the variation in the argument of f along C; it involves the *image curve* $f(C)$. This is simply the image (in the w-plane) of the curve C under the mapping $w = f(z)$; i.e., if C is parametrized by $z = z(t)$, $a \le t \le b$, then $f(C)$ is the curve parametrized by

(55)
$$w = f(z(t)) \qquad (a \le t \le b).$$

Obviously the image curve is closed, but unlike C it need not be simple or positively oriented.

Now if we sketch the image curve as in Fig. 6.20, it is easy to follow the net change in the argument of $f(z)$. Every time $f(z(t))$ encircles the origin $w = 0$ in the positive (counterclockwise) direction, $\arg f$ increases by 2π, while it decreases by 2π for a negative circuit. Hence, since $f(C)$ is closed, $\Delta_C \arg f(z)$ equals 2π multiplied by the net number of times $f(C)$ winds around $w = 0$ *in the positive sense*, i.e., counterclockwise minus clockwise.[†]

†Some authors call this the *winding number* of $f(C)$ about the origin.

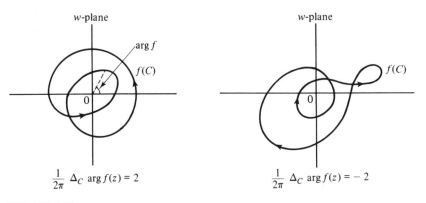

$$\frac{1}{2\pi} \Delta_C \arg f(z) = 2 \qquad \frac{1}{2\pi} \Delta_C \arg f(z) = -2$$

FIGURE 6.20

As a specific illustration, consider

$$f(z) = z^3 \quad \text{and} \quad C: z = e^{it} \qquad (0 \le t \le 2\pi).$$

Notice that the image of this circle under $w = f(z)$ winds around the origin three times in the positive sense (see Fig. 6.21). Hence the net change in the argument of f is 6π, and Eq. (54) correctly predicts $N_0(f) - N_p(f) = 6\pi/2\pi = 3$.

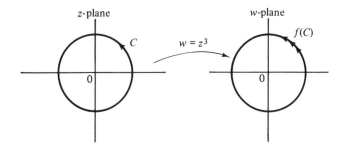

FIGURE 6.21

We can now prove an important theorem due to Rouché.

THEOREM 4 (Rouché's Theorem) *If f and g are each functions which are analytic inside and on a simple closed contour C and if the strict inequality*

(56)
$$|f(z) - g(z)| < |f(z)|$$

holds at each point on C, then f and g must have the same total number of zeros (counting multiplicity) inside C.

285   Observe that the inequality need only hold *on C*, not inside.

Proof First note that the strictness of the inequality (56) prevents f as well as g from being zero on C. Hence if we consider the function

$$F(z) \equiv \frac{g(z)}{f(z)},$$

then F will be analytic and nonzero on C. Furthermore, F is meromorphic inside C, and the difference in the number of its zeros and poles there satisfies

(57) $$N_0(F) - N_p(F) = N_0(g) - N_0(f),$$

where $N_0(g)$ and $N_0(f)$ are, respectively, the number of zeros of g and f inside C. [Think about this—if none of the zeros of g and f coincide, then Eq. (57) is obviously true. If some of the zeros are coincident, then multiplicities must be considered.]

Next we divide the inequality (56) by $|f(z)|$ to obtain $|1 - F(z)| < 1$ for z on C, showing that the image curve $F(C)$ must lie in the disk $|1 - w| < 1$ (see Fig. 6.22). But then $F(z)$ is restricted to lie in the right

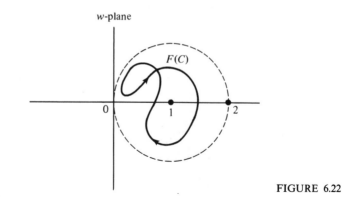

w-plane

F(C)

0   1   2

FIGURE 6.22

half-plane, and so the net change in its argument is zero. Hence by the Argument Principle

$$0 = \frac{1}{2\pi} \Delta_C \text{ arg } F(z) = N_0(F) - N_p(F).$$

This, together with Eq. (57), proves Rouché's Theorem. ∎

One typically uses Rouché's Theorem to deduce some information about the location of zeros of a given analytic function g. The technique involves producing an analytic function f whose zeros are known, and comparing it with g.

EXAMPLE 21 Prove that all five zeros of the polynomial

$$g(z) = z^5 + 3z + 1$$

lie in the disk $|z| < 2$.

SOLUTION Here we want to take C as the circle $|z| = 2$, and we need a comparison function $f(z)$ for which the inequality (56) of Rouché's Theorem is valid. An appropriate choice for $f(z)$ is

$$f(z) = z^5$$

because this function clearly has 5 zeros inside C and it yields a simple form for the difference $f - g$. In fact for $|z| = 2$ we have, by the triangle inequality,

$$|f(z) - g(z)| = |-3z - 1| \le 3|z| + 1 = 3 \cdot 2 + 1 = 7,$$

which is always strictly less than

$$|f(z)| = |z^5| = 2^5 = 32.$$

Hence g has the same number of zeros in $|z| < 2$ as does f, namely 5. ∎

EXAMPLE 22 Prove that the equation

$$z + 3 + 2e^z = 0$$

has precisely one root in the left half-plane.

SOLUTION Since Rouché's Theorem refers to domains bounded by contours, we cannot apply it directly to an unbounded set such as a half-plane. But by this time the reader will probably be able to anticipate our strategy; we choose C_ρ as in Fig. 6.23 and show that the function

$$g(z) = z + 3 + 2e^z$$

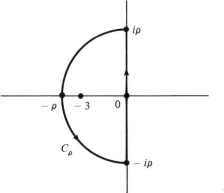

FIGURE 6.23

has exactly one zero inside C_ρ for every ρ sufficiently large. Let's take f to be

$$f(z) = z + 3,$$

which has exactly one zero in the left half-plane. Then for z on C_ρ we have

$$|f(z) - g(z)| = |-2e^z| = 2e^{\mathrm{Re}\,z} \le 2e^0 = 2,$$

while $f(z)$ is bounded below on C_ρ by

$$|f(z)| = |z + 3| \ge \begin{cases} 3, & \text{for } z = iy, \\ |z| - 3 = \rho - 3, & \text{for } |z| = \rho. \end{cases}$$

Thus when $\rho > 5$ we have

$$|f(z) - g(z)| < |f(z)| \qquad \text{(for all } z \text{ on } C_\rho\text{)}.$$

This implies that g also has precisely one (simple) zero inside C_ρ, and hence (letting $\rho \to \infty$) in the left half-plane. ∎

Rouché's Theorem can also be used to give an alternative proof of the Fundamental Theorem of Algebra.

EXAMPLE 23   Prove that every polynomial of degree n has n zeros.

SOLUTION   Consider any polynomial $g(z)$ of degree n,

$$g(z) = a_n z^n + a_{n-1} z^{n-1} + \cdots + a_1 z + a_0 \qquad (a_n \ne 0),$$

and for comparison take

$$f(z) = a_n z^n,$$

which has n zeros (at the origin). The difference

$$f(z) - g(z) = -(a_{n-1} z^{n-1} + \cdots + a_1 z + a_0)$$

is then a polynomial of degree at most $n - 1$ and hence does not grow as rapidly as the polynomial f of degree n. In more precise terms, on the circle C: $|z| = R$ we have

$$|f(z)| = |a_n| R^n,$$

and

$$|f(z) - g(z)| \le |a_{n-1}| R^{n-1} + \cdots + |a_1| R + |a_0|.$$

Hence if we choose R (> 1) so large that

$$\frac{|a_{n-1}|}{|a_n|} + \cdots + \frac{|a_1|}{|a_n|} + \frac{|a_0|}{|a_n|} < R,$$

then the inequality

$$|f(z) - g(z)| < |f(z)|$$

will be valid on C. This proves that g also has n zeros. ∎

Another interesting consequence of Rouché's Theorem is the *open mapping* property of analytic functions.

THEOREM 5 *If f is nonconstant and analytic in a domain D, then its range*

$$f(D) \equiv \{w \,|\, w = f(z) \text{ for some } z \text{ in } D\}$$

is an open set.

   Proof Let w_0 be any point in the range, and let z_0 be a preimage of w_0; i.e., $f(z_0) = w_0$. To show that $f(D)$ is open we must exhibit a neighborhood $\mathcal{N}$ of w_0 which is completely contained in $f(D)$. To this end observe that the function $F(z) \equiv f(z) - w_0$ is nonconstant and analytic in D, and has a zero at z_0. Since the zeros of such a function are isolated,† we can find a small circle C: $|z - z_0| = r$, with C and its interior contained in D, such that F is nonzero on C.
   Now let

$$m \equiv \min_{z \text{ on } C} |F(z)|,$$

and observe that by continuity m is positive. Our aim is to show that the neighborhood $\mathcal{N}$: $|w - w_0| < m$ is contained in $f(D)$. For any fixed w in $\mathcal{N}$ put

$$G(z) \equiv f(z) - w.$$

Then for all z on C we have

$$|F(z) - G(z)| = |[f(z) - w_0] - [f(z) - w]|$$
$$= |w - w_0| < m \leq |F(z)|$$

so that, by Rouché's Theorem, G has the same number of zeros inside C as does F. But since F has a zero at z_0, the function $G(z) = f(z) - w$ has at least one zero inside C. Therefore, $f(z) = w$ for some z inside C and so w lies in $f(D)$. Since w was an arbitrarily chosen point of $\mathcal{N}$, we conclude that $\mathcal{N}$ is completely contained in $f(D)$. ∎

---

†See Corollary 3 and Theorem 19 of Chapter 5.

Exercises 6.7

1. Which of the following functions are meromorphic in the whole plane?

   (a) $2z + z^3$ (b) $\text{Log } z$

   (c) $\dfrac{\sin z}{z^3 + 1}$ (d) $e^{1/z}$

   (e) $\tan z$ (f) $\dfrac{2i}{(z - 3)^2} + \cos z$

2. Evaluate

   $$\frac{1}{2\pi i} \oint_{|z| = 3} \frac{f'(z)}{f(z)} \, dz,$$

   where $f(z) = \dfrac{z^2(z - i)^3 e^z}{3(z + 2)^4(3z - 18)^5}$.

3. Let $P(z) = a_n z^n + a_{n-1} z^{n-1} + \cdots + a_1 z + a_0$, where $a_n \neq 0$. Explain why for each sufficiently large value of R

   $$\oint_{|z| = R} \frac{P'(z)}{P(z)} \, dz = 2n\pi i.$$

4. Let $f(z)$ be one-to-one and analytic on the closed disk $|z| \leq \rho$, and suppose that $f(z) \neq w_0$ for all z on the circle $|z| = \rho$. Explain why the value of the integral

   $$\frac{1}{2\pi i} \oint_{|z| = \rho} \frac{f'(z)}{f(z) - w_0} \, dz$$

   is either 0 or 1.

5. Prove that if $f(z)$ is analytic inside and on a simple closed contour C and is one-to-one on C, then $f(z)$ is one-to-one inside C. [HINT: Consider the image curve $f(C)$.]

6. Use Rouché's Theorem to show that the polynomial $z^6 + 4z^2 - 1$ has exactly two zeros in the disk $|z| < 1$.

7. Prove that the equation $z^3 + 9z + 27 = 0$ has no roots in the disk $|z| < 2$.

8. Find the number of roots of the equation $6z^4 + z^3 - 2z^2 + z - 1 = 0$ in the disk $|z| < 1$.

9. Prove that all the roots of the equation $z^6 - 5z^2 + 10 = 0$ lie in the annulus $1 < |z| < 2$.

10. Prove that the equation $z = 2 - e^{-z}$ has exactly one root in the right half-plane. Why must this root be real?

11. Prove that the polynomial $P(z) = z^4 + 2z^3 + 3z^2 + z + 2$ has exactly two zeros in the right half-plane. [HINT: Write $P(iy) = (y^2 - 2)(y^2 - 1) + iy(1 - 2y^2)$, and show that

$$\lim_{R \to \infty} \arg P(iy) \Big|_{-R}^{R} = 0.]$$

12. Suppose that $f(z)$ is analytic on $|z| \le 1$ and satisfies $|f(z)| < 1$ for $|z| = 1$. Prove that the equation $f(z) = z$ has exactly one root (counting multiplicity) in $|z| < 1$.

13. Give an example to show that the conclusion of Rouché's Theorem may be false if the strict inequality $|f(z) - g(z)| < |f(z)|$ is replaced by $|f(z) - g(z)| \le |f(z)|$ for z on C.

14. State and prove a generalization of Rouché's Theorem for meromorphic functions f and g which concludes that $N_0(f) - N_p(f) = N_0(g) - N_p(g)$.

15. Let $\lambda > 0$ be fixed, and let $f(z) = \tan z - \lambda z$. The zeros of f are important in certain problems related to heat flow. By completing each of the following steps, show that for n large $f(z)$ has exactly $2n + 1$ zeros inside the square Γ_n with vertices at $n\pi(1 \pm i)$, $n\pi(-1 \pm i)$.

    (a) Show that

    $$\tan(x + iy)$$
    $$= \left[\frac{\sin(2x)}{\cosh(2y) + \cos(2x)} \right] + i \left[\frac{\sinh(2y)}{\cosh(2y) + \cos(2x)} \right].$$

    (b) Prove that for all large integers n, the inequality $|\tan z| \le 2$ holds on the boundary of Γ_n. [HINT: Use the formula in part (a) for the horizontal segments.]

    (c) Show for all large integers n, the inequality $|f(z) + \lambda z| < \lambda |z|$ holds on the boundary of Γ_n.

    (d) Show that $f(z)$ has exactly $2n$ poles inside Γ_n, $n = 1, 2, \ldots$.

    (e) Conclude from the general form of Rouché's Theorem (Prob. 14) that $f(z)$ has $2n + 1$ zeros inside Γ_n for large integers n.

16. Let $f(z)$ be analytic in a domain D, and suppose that $f(z) - f(z_0)$ has a zero of order n at z_0 in D. Prove that for $\varepsilon > 0$ sufficiently small, there exists a $\delta > 0$ such that for all w in $|w - f(z_0)| < \delta$ the equation $f(z) - w$ has exactly n roots in $|z - z_0| < \varepsilon$.

17. Let $f_n(z)$, $n = 1, 2, \ldots$, be a sequence of functions analytic in the disk $D: |z| < R$ which converges uniformly to the analytic function $f(z)$ on each closed subset of D. Prove that if $f(z) \neq 0$ on $|z| = \delta$, $0 < \delta < R$, then for each n sufficiently large $f_n(z)$ has the same number of zeros in $|z| < \delta$ as does $f(z)$.

18. Let $P(z) = a_n z^n + \cdots + a_1 z + a_0$ and let

$$P^*(z) = z^n \overline{P(1/\bar{z})} = \bar{a}_0 z^n + \bar{a}_1 z^{n-1} + \cdots + \bar{a}_n.$$

Prove that if $|a_0 / a_n| > 1$, then $P(z)$ has the same number of zeros in $|z| < 1$ as does the polynomial $\bar{a}_0 P(z) - a_n P^*(z)$. [HINT: $|P(z)| = |P^*(z)|$ for $|z| = 1$.]

19. To establish the stability of feedback control systems one often has to ensure that a certain meromorphic function of the form $F(z) = 1 + P(z)$ has all its zeros in the left half-plane. The *Nyquist Stability Criterion* proceeds as follows: First consider the contour Γ_r shown in Fig. 6.24, and let m equal the net number of times that the image contour $P(\Gamma_r)$ encircles the point $w_0 = -1$ in a counterclockwise direction. Then let n equal the number of poles of $P(z)$ with positive real parts. Argue that if m equals n for all sufficiently large r, then all the zeros of $F(z)$ lie in the left half-plane (and thus the system is stable). [If for r large, $P(\Gamma_r)$ passes through the point $w_0 = -1$, then of course $F(z)$ has a zero on the imaginary axis; in such a case stability cannot be guaranteed.]

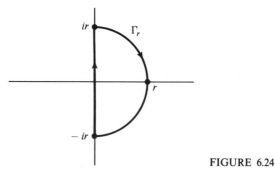

FIGURE 6.24

SUMMARY

A useful way to evaluate certain contour integrals is by means of residues. The residue of a function $f(z)$ at an isolated singularity z_0 is the coefficient a_{-1} of $1/(z - z_0)$ in the Laurent expansion for $f(z)$ about z_0. Simple formulas exist for computing the residues at the poles of $f(z)$. When all the singularities of $f(z)$ are isolated, its integral along a simple closed positively oriented contour is equal to $2\pi i$ times the sum of the residues at the singularities inside the contour.

Residue theory can be employed to evaluate certain integrals which arise in the calculus of functions of a real variable. For example, definite integrals over $[0, 2\pi]$ which involve $\sin \theta$ and $\cos \theta$ can be rewritten as contour integrals around the unit circle $C: |z| = 1$ after making the identification

$$\cos \theta = \frac{1}{2}\left(z + \frac{1}{z}\right), \qquad \sin \theta = \frac{1}{2i}\left(z - \frac{1}{z}\right) \qquad \text{(for } z = e^{i\theta}\text{)}.$$

Also certain improper integrals over infinite intervals, say over the real axis $(-\infty, +\infty)$, can be computed with the aid of expanding closed contours (such as semicircles or rectangles) which have a segment of the real axis as a component. For this approach to be successful the contours must be selected so that the sum of residues inside is easily computed, and so that the limiting values of the integral over the nonreal portions of the contours are known. Some specific results such as the Jordan Lemma are useful in this regard.

Sometimes these contours must be modified by indention to compensate for singularities on the original interval of integration. In the case when the complex version of an integrand is multiple-valued, it might be necessary to integrate along a branch cut. For this purpose the cut is regarded as having two distinct sides, with the integrand being defined differently on each side so as to preserve continuity.

When the only singularities of $f(z)$ are poles, the variation in the argument of $f(z)$ along a simple closed contour is related to the difference in the number of its zeros and poles inside the contour. A consequence of this fact is Rouché's Theorem, which provides a comparison technique for counting the number of zeros of an analytic function in a certain domain.

SUGGESTED READING

Further discussions on the calculus of residues can be found in the following texts:

[1]   CONWAY, J. B. *Functions of One Complex Variable*. Springer-Verlag New York, Inc., New York, 1973.

[2]   COPSON, E. T. *An Introduction to the Theory of Functions of a Complex Variable*. Oxford University Press, Inc., New York, 1962.

[3]   EVES, H. W. *Functions of a Complex Variable*, Vol. 2. Prindle, Weber & Schmidt, Inc., Boston, 1966.

[4]   DERRICK, W. R. *Introductory Complex Analysis and Applications*. Academic Press, Inc., New York, 1972.

7

Conformal Mapping

In this chapter we shift our point of view somewhat; rather than dealing with the algebraic properties of an analytic function $f(z)$, we are going to regard f as a *mapping* from its domain to its range, and consider its *geometric* properties. The ability to map one region onto another via an analytic function proves invaluable in applied mathematics, as we shall see in Sec. 7.1.

7.1 INVARIANCE OF LAPLACE'S EQUATION

One of the most valuable aspects of mappings generated by analytic functions is the persistence of Laplace's equation; roughly, this means that if $\phi(x, y)$ is harmonic in a certain domain D of the xy-plane (so that ϕ satisfies Laplace's equation

$$\frac{\partial^2 \phi}{\partial x^2} + \frac{\partial^2 \phi}{\partial y^2} = 0$$

in D), and if

(1) $$w = f(z)$$

is an analytic function mapping D onto a domain D' in the uv-plane, then ϕ is "carried over" by the mapping to a function which is harmonic in D'.

To express this fact precisely we must elaborate somewhat on the nature of the mapping. We assume that relation (1) provides a *one-to-one* correspondence between the points of D and those of D'.† Recall that this means $f(z_1) = f(z_2)$ only if $z_1 = z_2$. We also assume that the derivative df/dz is never zero in D; actually the latter is a consequence of the one-to-one assumption, but we won't prove it here.

Now since the mapping is one-to-one, it has an inverse; i.e., with each point of D' there can be associated a point of D, namely its preimage under f. This relationship, which is the inverse of $w = f(z)$, is suggestively written

(2) $$z = f^{-1}(w)$$

(recall, for instance, $\sin^{-1} w$). Figure 7.1 depicts both the relationships. Observe that f^{-1} is a single-valued function (since only one z is mapped to a particular w). The reader will probably not be surprised to learn that it is, in fact, an *analytic* function, and that its derivative is given by

(3) $$\frac{df^{-1}(w)}{dw} = \frac{1}{\dfrac{df(z)}{dz}} \qquad [\text{where } w = f(z)].$$

This equation can also be written in the form

(4) $$\frac{dz}{dw} = \frac{1}{\dfrac{dw}{dz}},$$

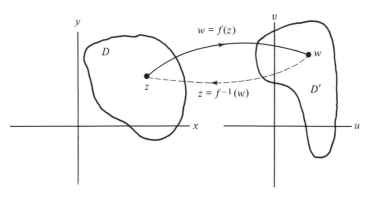

FIGURE 7.1

†Such a mapping is sometimes called *univalent* or *schlicht*.

keeping in mind the functional relationships. We have already proved a special case of Eq. (3) for the function $w = e^z$ and its inverse $z = \text{Log } w$, taking D to be the strip $|\text{Im } z| < \pi$ and D' to be the entire plane slit along the negative real axis (cf. Sec. 3.2). We shall invite the reader to prove Eq. (3) in general at the end of the next section.

It is sometimes helpful to indicate the mappings (1) and (2) in terms of real variables; thus Eq. (1) becomes

(5)
$$u = u(x, y),$$
$$v = v(x, y),$$

and its inverse, Eq. (2), becomes

(6)
$$x = x(u, v),$$
$$y = y(u, v).$$

Now we are prepared to demonstrate the property stated above about the persistence of Laplace's equation. Observe that if $\phi(x, y)$ is a function defined on D, then the domain D' "inherits" ϕ through the one-to-one mapping; that is, the function $\tilde{\phi}(u, v)$ defined by

$$\tilde{\phi}(u, v) \equiv \phi(x(u, v), y(u, v)),$$

or, equivalently,

$$\tilde{\phi}(w) \equiv \phi(f^{-1}(w)),$$

agrees with $\phi(x, y)$ at corresponding points. Our claim is that if ϕ is harmonic in D, then $\tilde{\phi}$ is harmonic in D'.

The proof is quite simple. Consider any point w_0 in D', say $w_0 = f(z_0)$. Now D is an open set, so there exists an open disk $\mathcal{N}$ centered at z_0 and lying entirely in D. Since $f^{-1}(w)$ is an analytic function, and thus continuous, there must be a sufficiently small neighborhood $\mathcal{N}'$ around w_0 whose image under f^{-1} lies entirely inside $\mathcal{N}$ (this is depicted in Fig. 7.2). To see that $\tilde{\phi}(u, v)$ is harmonic in the neighborhood $\mathcal{N}'$, recall that in Sec. 2.5 we proved that any harmonic function can be taken as the real part of an analytic function in "nice enough" domains—in particular, this is true for disks. Hence $\phi(x, y)$ is the real part of an analytic function $g(z)$ on $\mathcal{N}$. But then $\tilde{\phi}(u, v)$ is the real part of the composite function $g(f^{-1}(w))$ on $\mathcal{N}'$, and since the composition of analytic functions is again analytic, $\tilde{\phi}$ must be harmonic in the neighborhood $\mathcal{N}'$ of w_0. Consequently, as w_0 is an arbitrary point of D', the function $\tilde{\phi}$ must be harmonic everywhere on D'.

Another proof of this fact can be based upon the direct verification of Laplace's equation in the w-plane, using the Cauchy-Riemann equations (see Prob. 1).

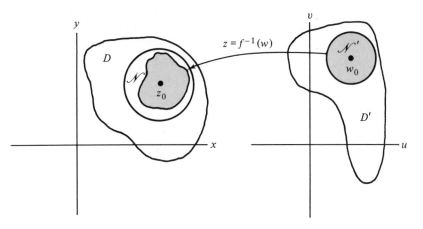

FIGURE 7.2

One can readily see why the above result is so useful in applications. Consider the *Dirichlet problem*, which requires us to find a function harmonic in a domain and taking specified values on its boundary [cf. (optional) Sec. 4.7]. Once we have solved this problem on a particular domain, we immediately have solutions on all the domains which we can map onto the original one via a one-to-one analytic function, as long as the boundary values correspond.

EXAMPLE 1   Find a function $\phi(x, y)$ harmonic inside the unit disk $|z| < 1$ and satisfying the boundary conditions

$$\phi(x, y) \to 0 \qquad \text{on the upper half-circle}$$

$$\phi(x, y) \to 1 \qquad \text{on the lower half-circle}$$

(see Fig. 7.3). (This function gives the electrostatic potential inside an infinitely long right circular cylinder cut in half lengthwise, with the two pieces maintained at different electrical potentials. Electrical insulation, of course, must be provided at the points of discontinuity $z = \pm 1$, and the potential is not specified there.)

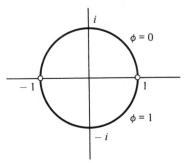

FIGURE 7.3

SOLUTION   This complicated-looking problem is solved by means of a mapping which carries the disk onto the upper half-plane. In Sec. 7.4 the reader will be instructed on how to construct such mappings, but for now we simply want to illustrate the power of the technique. So we state without proof that the function

(7)
$$w = f(z) = -i\,\frac{z-1}{z+1}$$

maps the unit circle onto the real axis and its interior onto the upper half-plane. (The exterior maps to the lower half-plane.) The correspondences are illustrated in Fig. 7.4. Observe that $|w|$ increases without bound as $z \to -1$, while $z = 1$ maps to $w = 0$.

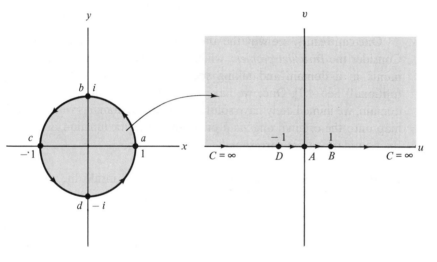

FIGURE 7.4

Moreover, the mapping (7) is one-to-one. In fact, the inverse is easily computed to be

(8)
$$z = \frac{i-w}{i+w}.$$

Notice that the upper half-circle, where the unknown function ϕ equals 0, is mapped to the positive real axis, while the lower half-circle (where $\phi = +1$) corresponds to the negative axis. Consequently, we can solve the given problem if we can solve the following simpler problem:

> Find a function $\psi(u, v)$ which is harmonic in the upper half-plane and approaches 0 on the positive real axis and $+1$ on the negative real axis.

The experience we have acquired by now tells us immediately that

$$\psi(u, v) = \frac{1}{\pi} \operatorname{Arg}(w).$$

Hence the solution to the original problem is derived from ψ by the mapping (7):

$$\phi(x, y) = \psi(u(x, y), v(x, y)) = \frac{1}{\pi} \operatorname{Arg}(f(z)) = \frac{1}{\pi} \operatorname{Arg}\left(-i \frac{z-1}{z+1}\right).$$

A little algebra results in the expression

$$\phi(x, y) = \frac{1}{\pi}\left(\frac{\pi}{2} - \tan^{-1} \frac{2y}{1-x^2-y^2}\right),\dagger$$

where the value of the arctangent is taken to be between $-\pi/2$ and $\pi/2$. Note that $\phi(x, 0) = \frac{1}{2}$, as we would expect from symmetry. ∎

With this example as motivation we devote the next few sections to a study of mappings given by analytic functions. The final two sections of the chapter will return us to applications, illustrating the power of this technique in handling many different situations. A table of some of the more useful mappings appears as Appendix II, for the reader's future convenience.

Exercises 7.1

1. Verify directly that if Eqs. (5) and (6) describe a one-to-one analytic mapping and $\phi(x, y)$ is harmonic, then

    $$\tilde{\phi}(u, v) \equiv \phi(x(u, v), y(u, v))$$

    satisfies Laplace's equation.

2. The exponential mapping $w = e^z$ carries the strip $0 < \operatorname{Im} z < 2\pi$ onto the w-plane cut along the positive real axis. What harmonic function $\tilde{\phi}(w)$ does the w-plane "inherit," via the mapping, from the harmonic function $\phi(z) = x + y$?

3. Find a function ϕ harmonic in the upper half-plane and taking the boundary values specified in Fig. 7.5. [HINT: Use a combination of $\arg(z - 2)$ and $\arg(z + 1)$.] Generalize this to the case of a function harmonic in the upper half-plane taking boundary values as indicated in Fig. 7.6.

---

†This solution can also be obtained from the Poisson Integral Formula described in (optional) Sec. 4.7.

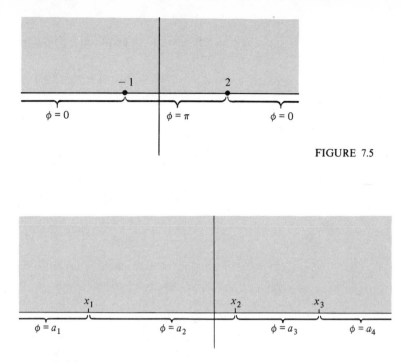

$\phi = 0$ $\phi = \pi$ $\phi = 0$

FIGURE 7.5

$\phi = a_1$ $\phi = a_2$ $\phi = a_3$ $\phi = a_4$

FIGURE 7.6

4. Consider the problem of finding a function ϕ which is harmonic in the upper half-plane and takes the values $\phi(x, 0) = x/(x^2 + 1)$ on the real axis. Observe that the obvious first guess,

$$\phi(z) = \text{Re} \frac{z}{z^2 + 1},$$

fails because $z/(z^2 + 1)$ is *not* analytic at $z = i$. However, the following strategy can be used.

(a) According to the text, the mappings (7) and (8) provide a correspondence between the upper half-plane and the unit disk. (Of course, one should interchange the roles of z and w in the formulas.) Thus the w-plane inherits from $\phi(z)$ a function $\tilde{\phi}(w)$ harmonic in the unit disk. Show that the values of $\tilde{\phi}(w)$ on the unit circle $w = e^{i\theta}$ must be given by

$$\tilde{\phi}(w) = \frac{\sin \theta}{2}.$$

(b) Argue that the harmonic function $\tilde{\phi}(w)$ must equal

$$\tilde{\phi}(w) = \tfrac{1}{2} \, \text{Im} \, w$$

throughout the unit disk.

(c) Use the mappings to carry $\tilde{\phi}(w)$ back to the z-plane, producing the function

$$\phi(z) = \frac{x}{x^2 + (y + 1)^2}$$

as a solution of the problem.

5. Use the strategy of Prob. 4 to find a function ϕ harmonic in the upper half-plane such that $\phi(x, 0) = 1/(x^2 + 1)$.

6. Suppose that the harmonic function $\phi(x, y)$ in the domain D is carried over to the harmonic function $\tilde{\phi}(u, v)$ in the domain D' via the one-to-one analytic mapping $w = f(z)$. Prove that if the normal derivative $d\phi/dn$ is zero on a curve Γ in D, then the normal derivative $d\tilde{\phi}/dn$ is zero on the image curve of Γ under f. [HINT: $d\phi/dn$ is the projection of the gradient $(\partial\phi/\partial x) + i(\partial\phi/\partial y)$ onto the normal, and the gradient is orthogonal to the level curves $\phi(x, y) = $ const.]

7. Let $\phi(x, y)$ satisfy the two-dimensional *Helmholtz equation*

$$\frac{\partial^2 \phi}{\partial x^2} + \frac{\partial^2 \phi}{\partial y^2} + k^2 \phi = 0$$

in the domain D. Show that if $z = g(w)$ is a one-to-one analytic mapping of the domain D' onto D, then the function $\tilde{\phi}(u, v) = \phi(g(w))$ satisfies

$$\frac{\partial^2 \tilde{\phi}}{\partial u^2} + \frac{\partial^2 \tilde{\phi}}{\partial v^2} + k^2 |g'(w)|^2 \tilde{\phi} = 0 \qquad \text{in } D'.$$

[NOTE: The Helmholtz equation arises in wave-guide analysis.]

7.2 GEOMETRIC CONSIDERATIONS

The geometric aspects of analytic mappings split rather naturally into two categories: *local* properties and *global* properties. Local properties need only hold in sufficiently small neighborhoods, while global properties hold throughout a domain. For example, consider the function e^z. It is one-to-one in any disk of diameter less than 2π, and hence it is locally one-to-one, but since $e^{z_1} = e^{z_2}$ when $z_1 - z_2 = 2\pi i$, the function is not globally one-to-one. On the other hand, sometimes local properties can be extended to global properties; in fact, this is the essence of *analytic continuation* [see (optional) Sec. 5.8].

Let us begin our study of local properties by considering "one-to-oneness." As the example e^z shows, a function may be locally one-to-one without being globally one-to-one. (Of course the opposite situation is impossible.) Furthermore, an analytic function may be locally one-to-one at some points but not at others. Indeed, consider

$$f(z) = z^2.$$

In any open set which contains the origin there will be distinct points z_1 and z_2 such that $z_2 = -z_1$, and hence (since $z_2^2 = z_1^2$) the function f will not be one-to-one. However, around any point other than the origin, we *can* find a neighborhood in which z^2 is one-to-one (any disk that excludes the origin will do; see Fig. 7.7). Thus $f(z) = z^2$ is locally one-to-one at every point other than the origin. An explanation of the exceptional nature of $z = 0$ in this example is provided by

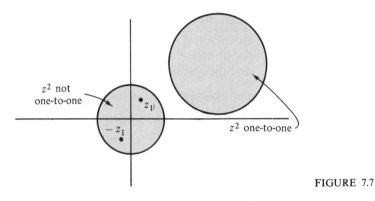

z^2 not one-to-one

z_1

$-z_1$

z^2 one-to-one

FIGURE 7.7

THEOREM 1 *If f is analytic at z_0 and $f'(z_0) \neq 0$, then there is some neighborhood $\mathcal{N}$ of z_0 such that f is one-to-one on $\mathcal{N}$.*

Proof If no such neighborhood existed, i.e., if inside *every* disk around z_0 one could find distinct points α and β such that $f(\alpha) = f(\beta)$, then by choosing a sequence of disks shrinking to z_0 we could construct an infinite number of pairs of complex numbers $\{\alpha_n, \beta_n\}_{n=1}^{\infty}$ such that for each n

(9) $$\alpha_n \neq \beta_n \quad \text{but} \quad f(\alpha_n) = f(\beta_n),$$

while the sequences $\{\alpha_n\}$ and $\{\beta_n\}$ each converged to z_0. See Fig. 7.8. But this contradicts the fact that $f'(z_0) \neq 0$, as we now show:

Let C be a circle centered at z_0 such that f is analytic inside and on C. Clearly each α_n and β_n will lie interior to C for n sufficiently large. Hence, for such n, we can use the Cauchy Integral Formula to write

$$\frac{f(\beta_n) - f(\alpha_n)}{\beta_n - \alpha_n} = \frac{1}{(\beta_n - \alpha_n)} \left[\frac{1}{2\pi i} \oint_C \frac{f(z)}{z - \beta_n} \, dz - \frac{1}{2\pi i} \oint_C \frac{f(z)}{z - \alpha_n} \, dz \right]$$

$$= \frac{1}{2\pi i} \oint_C \frac{f(z)}{(z - \beta_n)(z - \alpha_n)} \, dz.$$

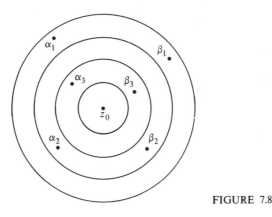

FIGURE 7.8

Now since each of the sequences $\{\alpha_n\}$ and $\{\beta_n\}$ converges to z_0, it is easy to see that the integrands

$$\frac{f(z)}{(z - \beta_n)(z - \alpha_n)}$$

converge uniformly (as $n \to \infty$) on C to $f(z)/(z - z_0)^2$. Hence the integrals themselves converge (cf. Theorem 8 in Chapter 5); i.e.,

$$\lim_{n \to \infty} \frac{f(\beta_n) - f(\alpha_n)}{\beta_n - \alpha_n} = \lim_{n \to \infty} \frac{1}{2\pi i} \oint_C \frac{f(z)}{(z - \beta_n)(z - \alpha_n)}\, dz$$

$$= \frac{1}{2\pi i} \oint_C \frac{f(z)}{(z - z_0)^2}\, dz.$$

But by (9) the left-hand side is zero, and by the generalized Cauchy formula the last integral equals $f'(z_0)$. Thus $f'(z_0) = 0$, contrary to the given hypothesis. ∎

Theorem 1 says that an analytic function is locally one-to-one at points where its derivative does not vanish. In advanced texts it is shown more generally that if z_0 is a zero of order m for f', then f is locally "$(m + 1)$-to-one" around (but excluding) z_0; in other words, each value of f is taken on $m + 1$ times. This is reinforced by the observation that $f(z) = z^2$ is two-to-one in any punctured neighborhood of the origin.

The next local property we shall discuss is *conformality*. Consider the following situation: $f(z)$ is analytic and one-to-one in a neighborhood of the point z_0, and γ_1 and γ_2 are two directed smooth curves (in this neighborhood) intersecting at z_0. Under the mapping f the image of these curves, γ_1' and γ_2', are also directed smooth curves, and they will intersect at $w_0 = f(z_0)$. At the point z_0 we construct vectors $\mathbf{v}_1$ and $\mathbf{v}_2$ tangent to γ_1 and γ_2, respectively, and pointing in the directions consistent with the orientations of the curves (see Fig. 7.9). Then the *angle from γ_1 to γ_2* is the

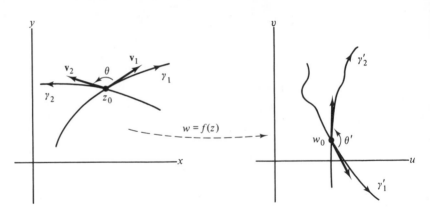

FIGURE 7.9

angle θ through which $\mathbf{v}_1$ must be rotated counterclockwise in order to lie along $\mathbf{v}_2$. The angle θ' from γ_1' to γ_2' is defined similarly.

Now the mapping f is said to be *conformal* at z_0 if these angles are preserved; i.e., $\theta = \theta'$ for every pair of directed smooth curves which intersect at z_0. For analytic mappings we have

THEOREM 2 *An analytic function f is conformal at every point z_0 for which $f'(z_0) \neq 0$.*

Proof By Theorem 1 we know that there is some open disk containing z_0 in which f is one-to-one. We will argue that every directed smooth curve through z_0 has its tangent (at z_0) turned through the same angle, under the mapping $w = f(z)$. Consequently, the angle between any two curves intersecting at z_0 will be preserved.

So let γ be any directed smooth curve through z_0 parametrized, say, by $z = z(t)$ with $z(t_0) = z_0$. The vector $z'(t_0)$ is then tangent to γ at z_0. Under the mapping the image, γ', of γ has the parametrization

$$w = w(t) = f(z(t))$$

with

$$w_0 \equiv f(z_0) = f(z(t_0)),$$

and the vector $w'(t_0)$ (if it is nonzero) is tangent to γ' at w_0. But the chain rule implies

(10) $$w'(t_0) = f'(z_0)z'(t_0),$$

so $w'(t_0)$ is nonzero since $f'(z_0) \neq 0$. Furthermore, we see from Eq. (10) that the angles which the tangent vectors $z'(t_0)$, $w'(t_0)$ make with the horizontal are related by

$$\arg w'(t_0) = \arg f'(z_0) + \arg z'(t_0).$$

Hence every curve through z_0 is rotated through the same angle $\arg f'(z_0)$ which, of course, is a constant independent of the particular curve. ∎

The condition that $f'(z_0)$ not vanish in Theorem 2 is crucial; the function $f(z) = z^2$, for which $f'(0) = 0$, does not preserve angles at the origin—it *doubles* them. However, it is common practice to call any mapping generated by a nonconstant analytic function a "conformal map," overlooking the violations occurring at the points where f' is zero. (Incidentally, these exceptional points will be reexamined in Sec. 7.5; they turn out to be quite important.)

Moving now to the global aspects of conformal mapping, we begin with a property which has both local and global ramifications: the *open mapping property*. A function is said to be an *open mapping* if the image of every open set in its domain is, itself, open; that is, the function maps open sets to open sets. For analytic functions we have

THEOREM 3 *Any nonconstant analytic function is an open mapping.*

A proof of this theorem was given in the (optional) Sec. 6.7. Note that the theorem prohibits, for instance, a nonconstant analytic mapping of a disk onto a portion of a line.

It is very useful in investigating conformal maps to exploit the concept of *connectivity*. For example, one can show that any nonconstant analytic function takes domains, i.e., open connected sets, to domains. Openness is preserved because of Theorem 3, and as for connectivity, we argue as follows. Let us say f maps the domain D onto the open set $\mathcal{O}$. To join $w_1 = f(z_1)$ to $w_2 = f(z_2)$ by a polygonal path in $\mathcal{O}$, first join z_1 to z_2 in D.

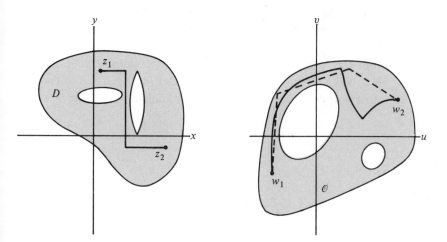

FIGURE 7.10

Then the image of this path is a path joining w_1 to w_2 in $\mathcal{O}$. Of course, the image path need not be polygonal, but any topologist worth his salt can prove that such a path, lying inside an *open* set, can be deformed into a polygonal path without leaving the set. Hence $\mathcal{O}$ is connected. (See Fig. 7.10.)

The remaining topic from the global theory of analytic functions which we shall consider here is the *Riemann Mapping Theorem*. Because it is primarily an *existence* theorem, its usefulness in applied mathematics is somewhat limited. However, it does elucidate one important fact, the *three-point normalization* property of conformal mappings.

THEOREM 4 (Riemann Mapping Theorem) *Let D be any simply connected domain in the plane other than the entire plane itself. Then there is a one-to-one analytic function which maps D onto the interior of the unit circle. Moreover, one can prescribe an arbitrary point of D and a direction through that point which are to be mapped to the origin and the direction of the positive real axis, respectively. Under such restrictions the mapping is unique.*

A direction through a point z_0 is specified, of course, by an angle ϕ as in Fig. 7.11. The Riemann Mapping Theorem allows us to specify z_0 and ϕ so that all curves through z_0 with tangent in the direction ϕ are mapped to curves through the origin with tangent along the positive real axis. The three-point normalization alluded to above refers to the fact that this yields three "degrees of freedom," or three choices to be made, in fixing the map: the real and the imaginary parts of the point which goes to 0, and the direction.

Again, we appeal to the references for a complete treatment of this theorem. (As an ominous note, we tantalize the reader by pointing out that the theorem makes no predictions about the boundary values of the function.)

From the Riemann Mapping Theorem we can conclude that any simply connected domain D_1 can be analytically mapped one-to-one onto any other simply connected domain D_2, assuming that neither D_1 nor D_2

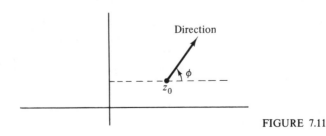

FIGURE 7.11

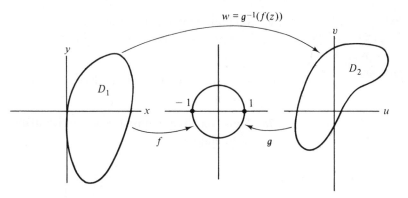

FIGURE 7.12

is the whole plane. Indeed, let f map D_1 onto the unit disk, and let g map D_2 onto this disk, in accordance with the theorem. Then $g^{-1}(f(z))$ maps D_1 onto D_2 and is one-to-one and analytic. See Fig. 7.12.

The latter part of this chapter will deal with constructing and applying *specific* conformal mappings.

Exercises 7.2

1. For each of the following functions, state the order m of the zero of the derivative f' at z_0 and show explicitly that the function is not one-to-one in any neighborhood of z_0.

   (a) $f(z) = z^2 + 2z + 1$, $z_0 = -1$

   (b) $f(z) = \cos z$, $z_0 = 0, \pm\pi, \pm 2\pi, \ldots$

   (c) $f(z) = e^{z^3}$, $z_0 = 0$

2. What happens to angles at the origin under the mapping $f(z) = z^\alpha$ for $\alpha > 1$? For $0 < \alpha < 1$?

3. Prove that if $w = f(z)$ is analytic at z_0 and $f'(z_0) \neq 0$, then $z = f^{-1}(w)$ is analytic at $w_0 = f(z_0)$, and

$$\frac{df^{-1}(w)}{dw} = \frac{1}{\dfrac{df(z)}{dz}}$$

   for $w = w_0$, $z = z_0$. [HINT: Theorem 1 guarantees that $f^{-1}(w)$ exists near w_0 and Theorem 3 implies that $f^{-1}(w)$ is continuous. Now generalize the proof in Sec. 3.2.]

4. Use the Open Mapping Theorem to prove the Maximum-Modulus Principle.

5. If $f(z)$ is analytic at z_0 and $f'(z_0) \neq 0$, show that the function $g(z) = \overline{f(z)}$ preserves the magnitude, but reverses the orientation, of angles at z_0.

6. Find all functions $f(z)$ analytic in D: $|z| < 1$ which assume only pure imaginary values in D.

7. Show that the mapping $w = z + 1/z$ maps circles $|z| = \rho$ ($\rho \neq 1$) onto ellipses

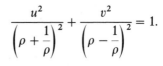

$$\frac{u^2}{\left(\rho + \dfrac{1}{\rho}\right)^2} + \frac{v^2}{\left(\rho - \dfrac{1}{\rho}\right)^2} = 1.$$

8. Let f be analytic at z_0 with $f'(z_0) \neq 0$. By considering the difference quotient, argue that "infinitesimal" lengths of segments drawn from z_0 are magnified by the factor $|f'(z_0)|$ under the mapping $w = f(z)$.

9. Let $w = f(z)$ be a one-to-one analytic mapping of the domain D onto the domain D', and let $A' = \text{area}(D')$. Using Prob. 8, argue the plausibility of the formula

$$A' = \iint_D |f'(z)|^2 \, dx \, dy.$$

10. Why is it impossible for D to be the whole plane in the Riemann Mapping Theorem? [HINT: Appeal to Liouville's theorem.]

11. Describe the image of each of the following domains under the mapping $w = e^z$.
    (a) the strip $0 < \text{Im } z < \pi$
    (b) the slanted strip between the two lines $y = x$ and $y = x + 2\pi$
    (c) the half-strip $\text{Re } z < 0$, $0 < \text{Im } z < \pi$
    (d) the half-strip $\text{Re } z > 0$, $0 < \text{Im } z < \pi$
    (e) the rectangle $1 < \text{Re } z < 2$, $0 < \text{Im } z < \pi$
    (f) the half-planes $\text{Re } z > 0$ and $\text{Re } z < 0$

12. Describe the image of each of the following domains under the mapping $w = \cos z = \cos x \cosh y - i \sin x \sinh y$.
    (a) the half-strip $0 < \text{Re } z < \pi$, $\text{Im } z < 0$
    (b) the half-strip $0 < \text{Re } z < \dfrac{\pi}{2}$, $\text{Im } z > 0$
    (c) the strip $0 < \text{Re } z < \pi$
    (d) the rectangle $0 < \text{Re } z < \pi$, $-1 < \text{Im } z < 1$

    [HINT: Consider the image of the boundary in each case.]

13. Let $P(z) = (z - \alpha)(z - \beta)$, and let L be any straight line through $(\alpha + \beta)/2$. Prove that P is one-to-one on each of the open half-planes determined by L.

14. Prove that if f has a simple pole at z_0, then there exists a punctured neighborhood of z_0 on which f is one-to-one.

15. A domain D is said to be *convex* if for any two points z_1, z_2 in D, the line segment joining z_1 and z_2 lies entirely in D. Prove the *Noshiro-Warschawski Theorem*: Let f be analytic in a convex domain D. If $\operatorname{Re} f'(z) > 0$ for all z in D, then f is one-to-one in D. [HINT: Write $f(z_2) - f(z_1)$ as an integral of f'.]

16. Let f be analytic at z_0 and suppose that $f'(z_0) = \lambda \neq 0$. Use the Noshiro-Warschawski Theorem (Prob. 15) to give another proof of the fact that there exists a neighborhood of z_0 on which f is one-to-one. [HINT: Choose a real number θ such that $e^{i\theta} f'(z_0)$ is real and positive. Then show that $e^{i\theta} f(z)$ is one-to-one in a neighborhood of z_0.]

17. (For students who have read Sec. 4.4a) Argue that a one-to-one analytic function will map simply connected domains to simply connected domains.

7.3 BILINEAR TRANSFORMATIONS

The problem of finding a one-to-one analytic function which maps one domain onto another can be quite perplexing, so it is worthwhile to investigate a few elementary mappings in order to compile some rules of thumb which we can draw upon. The basic properties of *bilinear transformations*, which we shall investigate in this section, constitute an essential portion of every analyst's bag of tricks.

First let's consider the simplest mapping of all, the *translation* defined by the function

$$(11) \qquad w = f(z) = z + c,$$

where c is a fixed complex number. Under this mapping every point is shifted by the vector corresponding to c. Its properties are quite apparent: The entire complex plane is mapped one-to-one onto itself, and every geometric object is mapped onto a congruent object. See Fig. 7.13.

Rotations are quite simple also. Observe that under the transformation

$$(12) \qquad w = f(z) = e^{i\phi} z,$$

with ϕ real, every point is rotated about the origin through the angle ϕ.

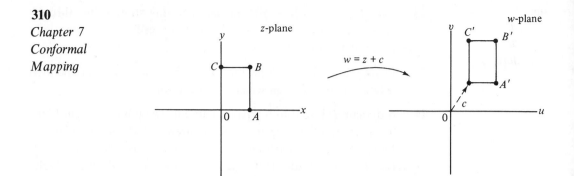

FIGURE 7.13

Such transformations are also one-to-one mappings of the complex plane onto itself and map geometric objects onto congruent objects. See Fig. 7.14.

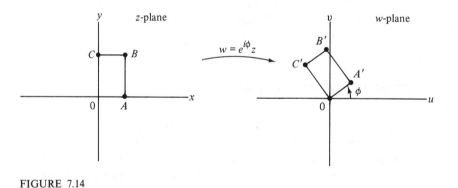

FIGURE 7.14

The mapping defined by

(13) $$w = f(z) = \rho z,$$

where ρ is a positive real constant, simply enlarges (or contracts) the distance of every point from the origin by the factor ρ; hence such a transformation is called a *magnification*. Observe that the distance between any two points is multiplied by this same constant, since

$$|w_1 - w_2| = |f(z_1) - f(z_2)| = |\rho z_1 - \rho z_2| = \rho |z_1 - z_2|.$$

Magnifications thus rescale distances and (since they are conformal) preserve angles; consequently any geometric object is mapped onto an object which is *similar* to the original. And again, the complex plane is mapped one-to-one onto itself. See Fig. 7.15.

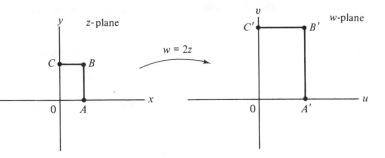

FIGURE 7.15

A *linear transformation*† is any mapping of the form

(14) $$w = f(z) = az + b,$$

where a and b are complex constants with $a \neq 0$. Such a transformation can be considered as the composition of a rotation, a magnification, and a translation (each of which is, of course, a special case of the linear transformation): Writing a in polar form as $a = \rho e^{i\phi}$, we express the linear transformation (14) by the sequence

$$w_1 = e^{i\phi} z,$$

$$w_2 = \rho w_1,$$

and, finally,

$$w = w_2 + b.$$

Hence the linear transformation is, once again, one-to-one in the complex plane, and the image of any object is geometrically similar to the original. See Fig. 7.16.

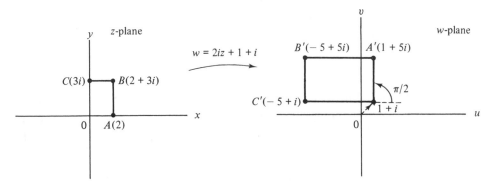

FIGURE 7.16

†This is, unfortunately, bad terminology, because in other branches of mathematics a "linear transformation" has the property $f(z_1 + z_2) = f(z_1) + f(z_2)$, while this is true of Eq. (14) only when $b = 0$. Worse yet, *bilinear* transformations are called linear transformations by some authors.

EXAMPLE 2 Find the linear transformation which rotates the entire complex plane through an angle θ about a given point z_0.

SOLUTION We know that the mapping

$$w_1 = e^{i\theta}z$$

rotates the plane through the angle θ about the origin. In particular, the point z_0 is mapped to the point $e^{i\theta}z_0$. If we now shift the whole plane so that the latter point is carried *back to* z_0, the net result will be a rotation of the plane about z_0. (Think about this; every straight line gets rotated through the angle θ, and z_0 is left fixed.) Thus the required answer is

$$w = w_1 + (z_0 - e^{i\theta}z_0) = e^{i\theta}z + (1 - e^{i\theta})z_0.$$

(The reader should observe the three degrees of freedom coming into play: θ, Re z_0, and Im z_0). ∎

EXAMPLE 3 Find a linear transformation which maps the circle C_1: $|z - 1| = 1$ onto the circle C_2: $|w - \frac{3}{2}i| = 2$.

SOLUTION Refer to Fig. 7.17. First we translate by -1 so that C_1 becomes a unit circle centered at the origin. Then we magnify by the

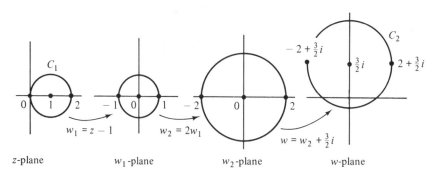

z-plane   *w_1-plane*   *w_2-plane*   *w-plane*

FIGURE 7.17

factor 2. Finally we translate $\frac{3}{2}$ units up the imaginary axis, bringing us to C_2. The mappings are

$$w_1 = z - 1,$$
$$w_2 = 2w_1 = 2z - 2,$$

and, last,

$$w = w_2 + \tfrac{3}{2}i = 2z - 2 + \tfrac{3}{2}i.$$

Moreover, any subsequent rotation about the point $\frac{3}{2}i$ can be permitted. ∎

Now we consider the *inversion* transformation defined by

(15)
$$w = f(z) = \frac{1}{z}.$$

It is easy to see that the inversion is a one-to-one mapping of the extended complex plane onto itself. Although it fails to preserve the shape of objects, this transformation does have an important geometrical property which will reveal itself if we look at what inversion does to lines and circles. For this purpose we regard ∞ as a member of every line.

(*i*) *Lines through the Origin.* The point $z = \rho e^{i\theta}$ is mapped to

$$w = \frac{1}{\rho e^{i\theta}} = \frac{1}{\rho} e^{-i\theta}.$$

Letting the real number ρ vary we see that the line making an angle θ with the real axis is mapped onto the line making the angle $-\theta$ with the real axis; the point at ∞ goes to the origin and vice versa. See Fig. 7.18.

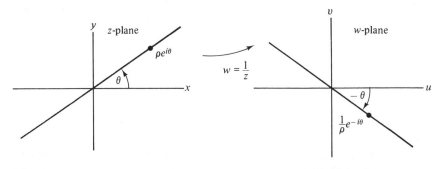

FIGURE 7.18

(*ii*) *Lines Not Passing through the Origin.* In the xy-plane, the equation of such a line is given by

(16) $Ax + By = C$, with $C \neq 0$.

Now since $w = 1/z$, we have

$$z = \frac{1}{w} = \frac{\overline{w}}{|w|^2} = \frac{u - iv}{u^2 + v^2}, \quad \text{for} \quad w = u + iv,$$

and so

(17) $$x = \frac{u}{u^2 + v^2}, \qquad y = \frac{-v}{u^2 + v^2}.$$

Making these substitutions into (16) we find

$$A\left(\frac{u}{u^2 + v^2}\right) + B\left(\frac{-v}{u^2 + v^2}\right) = C,$$

which can be written in the equivalent form

(18)
$$u^2 + v^2 - \frac{A}{C}u + \frac{B}{C}v = 0.$$

As can be seen by completing the squares, Eq. (18) describes a circle through the origin in the w-plane. Hence we conclude that the image points of the line (16) all lie on this circle. In fact, since both "ends" of the line (i.e., ∞) are mapped to $w = 0$, the image set is simple and closed and hence must be the *whole* circle having equation (18), as is indicated in Figs. 7.19 and 7.20.

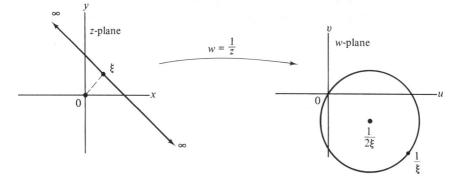

FIGURE 7.19   FIGURE 7.20

(iii) *Circles Passing through the Origin.* Because the inverse of the map $w = 1/z$ is itself an inversion, $z = 1/w$, this case is just the reverse of case (ii). In other words, from case (ii) the image of the circle

(19)
$$x^2 + y^2 - \frac{A}{C}x + \frac{B}{C}y = 0$$

under the map $w = 1/z$ is the line

$$Au + Bv = C.$$

This is all we need because any circle through $z = 0$ can be written in the form (19) for suitable choices of A, B, and C.

(iv) *Circles Not Passing through the Origin.* In the xy-plane, the equation of such a circle is of the form

$$x^2 + y^2 + Ax + By = C, \quad \text{with} \quad C \neq 0.$$

Using the relations (17), this equation becomes

$$\frac{u^2}{(u^2+v^2)^2} + \frac{v^2}{(u^2+v^2)^2} + \frac{Au}{u^2+v^2} - \frac{Bv}{u^2+v^2} = C,$$

which simplifies to

$$1 + Au - Bv = C(u^2+v^2).$$

As $C \neq 0$, we therefore obtain

$$u^2 + v^2 - \frac{A}{C}u + \frac{B}{C}v = \frac{1}{C}$$

for the equation satisfied by the image of the given circle. Such an equation describes a circle in the w-plane which does not pass through the origin and, as before, the image is the whole circle. See Fig. 7.21.

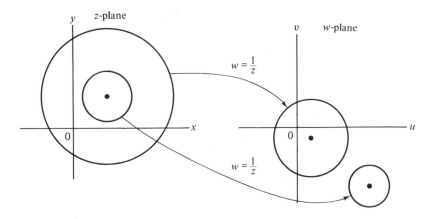

FIGURE 7.21

Summarizing, we see that the inversion mapping is one-to-one and carries the class of straight lines and circles into itself, a property shared with translations, rotations, and magnifications.

Now we are ready to define the bilinear transformations.

*DEFINITION 1 A **bilinear transformation** (sometimes known as a **linear fractional transformation** or **Möbius transformation**) is any function of the form*

(20)
$$w = f(z) = \frac{az+b}{cz+d}$$

with the restriction that $ad \neq bc$ (so that w is not a constant function).

Notice that since

$$f'(z) = \frac{ad - bc}{(cz + d)^2}$$

does not vanish, the bilinear transformation $f(z)$ is conformal at every point except its pole $z = -d/c$.

It is easy to see that the bilinear transformations include the previous elementary transformations of this section as special cases. More important is the fact that any bilinear transformation can be decomposed into a succession of these elementary transformations. If $c = 0$, we have the linear transformation which was treated earlier. For $c \neq 0$, the decomposition can be seen by writing

$$\frac{az + b}{cz + d} = \frac{\frac{a}{c}(cz + d) - \frac{ad}{c} + b}{cz + d} = \frac{a}{c} + \frac{b - \frac{ad}{c}}{cz + d},$$

which shows that the bilinear transformation can be expressed as a linear transformation (rotation + magnification + translation)

(21) $$w_1 = cz + d,$$

followed by an inversion

(22) $$w_2 = \frac{1}{w_1},$$

and then another linear transformation

(23) $$w = \left(b - \frac{ad}{c}\right)w_2 + \frac{a}{c}.$$

As a result of this decomposition and of our previous deliberations, we can summarize some properties of bilinear transformations.

THEOREM 5   *Let f be any bilinear transformation. Then*

(i) *f can be expressed as the composition of a finite sequence of translations, magnifications, rotations, and inversions.*

(ii) *f maps the extended complex plane one-to-one onto itself.*

(iii) *f maps the class of circles and lines to itself.*

(iv) *f is conformal at every point except its pole.*

One way of distinguishing the possibilities in property (*iii*) is as follows. If a line or circle passes through the pole ($z = -d/c$) of the bilinear transformation, it gets mapped to an unbounded figure. Hence its image is a straight line. A line or circle which avoids the pole, on the other hand, must be mapped to a circle.

EXAMPLE 4 Find the image of the *interior* of the circle $C: |z - 2| = 2$ under the bilinear transformation

$$w = f(z) = \frac{z}{2z - 8}.$$

SOLUTION First we find the image of the circle C. Since f has a pole at $z = 4$ and this point lies on C, the image has to be a straight line. To specify this line all we need is to determine two of its finite points. The points $z = 0$ and $z = 2 + 2i$ which lie on C have, as their images,

$$w = f(0) = 0 \quad \text{and} \quad w = f(2 + 2i) = \frac{2 + 2i}{2(2 + 2i) - 8} = -\frac{i}{2}.$$

Thus the image of C is the imaginary axis in the w-plane. From our discussion of connectivity in Sec. 7.2 we know that the interior of C is therefore mapped either onto the right half-plane Re $w > 0$ or onto the left half-plane Re $w < 0$. Since $z = 2$ lies inside C and

$$w = f(2) = \frac{2}{4 - 8} = -\frac{1}{2}$$

lies in the left half-plane, we conclude that the image of the interior of C is the left half-plane (see Fig. 7.22). ∎

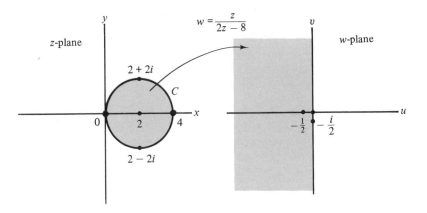

FIGURE 7.22

Now we shall present an example showing how to *construct* a conformal map of one region onto another.

EXAMPLE 5 Find a conformal map of the unit disk $|z| < 1$ onto the upper half-plane Im $w > 0$.

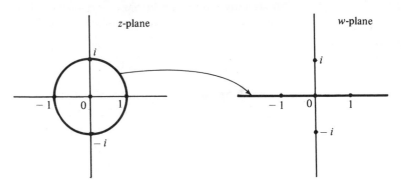

FIGURE 7.23

SOLUTION We are naturally led to look for a bilinear transformation which maps the circle $|z| = 1$ onto the real axis (Fig. 7.23). The transformation must therefore have a pole on the circle, according to our earlier remarks. Moreover, the origin $w = 0$ must also lie on the image of the circle. As a first step, let's look at

$$(24) \qquad w = f_1(z) = \frac{z - 1}{z + 1},$$

which maps -1 to ∞ and $+1$ to 0.

From the geometric properties of bilinear transformations we have learned, we can conclude that (24) maps $|z| = 1$ onto *some* straight line through the origin. To see *which* straight line, we plug in $z = i$ and find that the point

$$w = \frac{i - 1}{i + 1} = \frac{2i}{2} = i$$

also lies on the line. Hence the image of the circle under f_1 must be the imaginary axis. A rotation of $\pi/2$ is called for, and we conclude that

$$(25) \qquad w = f_2(z) = e^{(\pi/2)i} \cdot \frac{z - 1}{z + 1} = i\frac{z - 1}{z + 1}$$

maps the unit circle onto the real axis.

To see which half-plane is the image of the interior of the circle, we check the point $z = 0$. It is mapped by (25) to the point $w = -i$, in the *lower* half-plane. This is not what we want, but it can be corrected by a final rotation of π, yielding

$$(26) \qquad w = f(z) = -i \cdot \frac{z - 1}{z + 1}$$

as an answer to the problem. Of course, any subsequent translation to the right or left, or a magnification, can be permitted. Observe that (26) is precisely the mapping which was introduced in Example 1 to solve a problem in electrostatics, and we have thus verified the claims made there. ∎

Exercises 7.3

1. Find a linear transformation mapping the circle $|z| = 1$ onto the circle $|w - 5| = 3$ and taking the point $z = i$ to $w = 2$.

2. Discuss the image of the circle $|z - 2| = 1$ and its interior under the following transformations.

   (a) $w = z - 2i$ (b) $w = 3iz$

   (c) $w = \dfrac{z - 2}{z - 1}$ (d) $w = \dfrac{z - 4}{z - 3}$

   (e) $w = \dfrac{1}{z}$

3. Find the bilinear transformation which maps 0, 1, ∞ to the following respective points.

   (a) 0, i, ∞ (b) 0, 1, 2

   (c) $-i$, ∞, 1 (d) -1, ∞, 1

4. Find a bilinear transformation mapping the unit disk $|z| < 1$ onto the right half-plane and taking $z = -i$ to the origin.

5. Find a bilinear transformation mapping the lower half-plane to the disk $|z + 1| < 1$. [HINT: Do it in steps.]

6. A *fixed point* of a function $f(z)$ is a point z_0 satisfying $f(z_0) = z_0$. How many fixed points can a bilinear transformation have in the extended complex plane?

7. What is the image of the sector $-\pi/4 < \text{Arg } z < \pi/4$ under the mapping $w = z/(z - 1)$?

8. What is the image of the strip $0 < \text{Im } z < 1$ under the mapping $w = (z - i)/z$?

9. What is the image of the third quadrant under the mapping $w = (z + i)/(z - i)$?

10. Find a conformal map of the semidisk $|z| < 1$, $\text{Im } z > 0$, onto the upper half-plane. [HINT: Combine a bilinear transformation with

the mapping $w = z^2$. Make sure you cover the entire upper half-plane.]

11. Map the shaded region in Fig. 7.24 onto the upper half-plane. [HINT: Use a bilinear transformation to map the point 2 to ∞. Argue that the image region will be a *strip*. Then use the exponential map.]

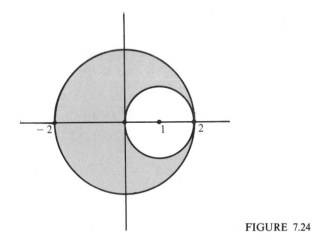

FIGURE 7.24

12. Find a bilinear transformation which takes the half-plane depicted in Fig. 7.25 onto the unit disk $|z| < 1$.

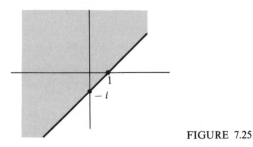

FIGURE 7.25

7.4 BILINEAR TRANSFORMATIONS, CONTINUED

We shall now explore some additional properties of bilinear transformations which enhance their usefulness as conformal mappings. These are the group properties, the cross-ratio formula, and the symmetry property.

Given any bilinear transformation

$$(27) \qquad w = f(z) = \frac{az + b}{cz + d} \qquad (ad \neq bc),$$

its inverse $f^{-1}(w)$ can be found by simply solving Eq. (27) for z in terms of w. This computation yields

$$z = f^{-1}(w) = \frac{dw - b}{-cw + a},$$

and we see that *the inverse of any bilinear transformation is again a bilinear transformation.* Furthermore, if we take the composition of two bilinear transformations, say

$$w = f_1(z) = \frac{a_1 z + b_1}{c_1 z + d_1} \quad \text{and} \quad \xi = f_2(w) = \frac{a_2 w + b_2}{c_2 w + d_2},$$

it can be readily shown that

$$\xi = f_2(f_1(z)) = \frac{(a_2 a_1 + b_2 c_1)z + (a_2 b_1 + b_2 d_1)}{(c_2 a_1 + d_2 c_1)z + (c_2 b_1 + d_2 d_1)}.$$

Hence *the composition of any two bilinear transformations is also a bilinear transformation.* [Rigorously speaking, we should make this claim only after verifying that $f_2(f_1)$ cannot reduce to a constant, but this is obviously true because the one-to-oneness of f_1 and f_2 implies that $f_2(f_1)$ is one-to-one.] Of course when we take $f_1^{-1}(f_1)$ we get the identity function $I(z) = z$, which is certainly a bilinear transformation. These facts are known as the *group properties* of the bilinear transformations.

We have already seen that bilinear transformations map the class of circles and lines to itself. Now we turn to the problem of finding a specific transformation which maps a *given* circle (or line) C_z in the z-plane to a *given* circle (or line) C_w in the w-plane. Recall from geometry that any three distinct noncollinear points uniquely determine a circle; if these points are collinear, then, of course, they uniquely determine a line (in particular this will be the case when one of them is ∞). Hence if we choose three points z_1, z_2, z_3 on C_z and three points w_1, w_2, w_3 on C_w and find a bilinear transformation f which satisfies

(28) $f(z_1) = w_1, \qquad f(z_2) = w_2, \qquad f(z_3) = w_3,$

then f must map C_z onto C_w.

It is not difficult to write down a bilinear transformation which satisfies Eqs. (28) in the case when $w_1 = 0$, $w_2 = 1$, and $w_3 = \infty$; this corresponds to the problem of mapping C_z onto the real axis and is reminiscent of the solution to Example 5 in the previous section. If all the points z_1, z_2, z_3 are finite, it is easy to see that

(29) $$T(z) = \frac{(z - z_1)(z_2 - z_3)}{(z - z_3)(z_2 - z_1)}$$

satisfies

(30) $\qquad T(z_1) = 0, \qquad T(z_2) = 1, \qquad T(z_3) = \infty,$

while if one of the z_i is ∞, the conditions (30) will be satisfied by

(31) $\qquad T(z) = \dfrac{z_2 - z_3}{z - z_3} \quad (z_1 = \infty), \qquad T(z) = \dfrac{z - z_1}{z - z_3} \quad (z_2 = \infty),$

$$\text{or} \qquad T(z) = \frac{z - z_1}{z_2 - z_1} \quad (z_3 = \infty).\dagger$$

We remark that the right-hand side of Eq. (29) [or its modified versions in Eqs. (31)] is called the *cross-ratio* of the four points z, z_1, z_2, z_3 and is sometimes abbreviated by writing (z, z_1, z_2, z_3); i.e.,

(32) $\qquad (z, z_1, z_2, z_3) \equiv \dfrac{(z - z_1)(z_2 - z_3)}{(z - z_3)(z_2 - z_1)}$

in the case of finite points. Notice that the order in which the points are listed is crucial in this notation. For example,

$$(z, 3, 0, i) = \frac{(z - 3)(0 - i)}{(z - i)(0 - 3)} = \frac{-iz + 3i}{-3z + 3i},$$

but

$$(z, i, 3, 0) = \frac{(z - i)(3 - 0)}{(z - 0)(3 - i)} = \frac{3z - 3i}{(3 - i)z}.$$

Now if we wish to solve the general problem of finding a bilinear transformation f which maps

$$z_1 \text{ to } w_1, \qquad z_2 \text{ to } w_2, \qquad z_3 \text{ to } w_3,$$

where w_1, w_2, w_3 are any three distinct points, we can proceed as follows: Let $T(z)$ be the bilinear transformation explicitly given above which takes

$$z_1 \text{ to } 0, \qquad z_2 \text{ to } 1, \qquad z_3 \text{ to } \infty,$$

and let $S(w)$ be the analogous transformation which maps

$$w_1 \text{ to } 0, \qquad w_2 \text{ to } 1, \qquad w_3 \text{ to } \infty.$$

Then the desired bilinear transformation f is given by the composition

(33) $\qquad w = f(z) = S^{-1}(T(z)),$

because

$$f(z_1) = S^{-1}(T(z_1)) = S^{-1}(0) = w_1,$$

$$f(z_2) = S^{-1}(T(z_2)) = S^{-1}(1) = w_2,$$

$$f(z_3) = S^{-1}(T(z_3)) = S^{-1}(\infty) = w_3.$$

---

†Note that the bilinear transformations in (31) can be obtained immediately from (29) by deleting the terms involving ∞.

Notice that Eq. (33) is equivalent to the equation

$$S(w) = T(z);$$

in other words, to map z_1, z_2, z_3 to the respective points w_1, w_2, w_3 we need merely equate the two cross-ratios

(34) $$(w, w_1, w_2, w_3) = (z, z_1, z_2, z_3)$$

and solve for w in terms of z.

EXAMPLE 6 Find a bilinear transformation which maps 0 to i, 1 to 2, and -1 to 4.

SOLUTION The appropriate cross-ratios are given by

$$(z, 0, 1, -1) = \frac{(z - 0)[1 - (-1)]}{[z - (-1)](1 - 0)} = \frac{2z}{z + 1}$$

and

$$(w, i, 2, 4) = \frac{(w - i)(2 - 4)}{(w - 4)(2 - i)} = \frac{-2(w - i)}{(w - 4)(2 - i)}.$$

Hence, solving the equation

$$\frac{-2(w - i)}{(w - 4)(2 - i)} = \frac{2z}{z + 1}$$

for w yields the desired transformation

$$w = \frac{(16 - 6i)z + 2i}{(6 - 2i)z + 2}. \quad \blacksquare$$

It is important to note that the circle or line Γ determined by the three points z_1, z_2, z_3 is also *oriented* by the order of these points. That is, Γ acquires the direction obtained by proceeding through the points z_1, z_2, z_3 in succession. [Notice that lines are regarded as "closed" at ∞ in the present context. Hence they, like circles, require a sequence of *three* points to determine a direction. See Fig. 7.26(d).] This orientation, in turn, uniquely specifies the "left region," the region which lies to the left of an observer traversing Γ (Fig. 7.26). Since bilinear transformations are conformal, it can be shown that a bilinear transformation which takes z_1, z_2, z_3 to the respective points w_1, w_2, w_3 must map the left region of the circle (or line) oriented by z_1, z_2, z_3 onto the left region of the circle (or line) oriented by w_1, w_2, w_3. To see this in the special case depicted in Fig. 7.27 imagine a short directed segment drawn from a point on the circle in the z-plane into the left region. The image of this segment is, by conformality, a curve from the image line into its left region. By connectivity, then, we conclude that the left region is mapped to the left region.

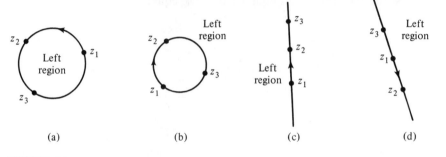

(a)　　　　　　(b)　　　　　(c)　　　　　(d)

FIGURE 7.26

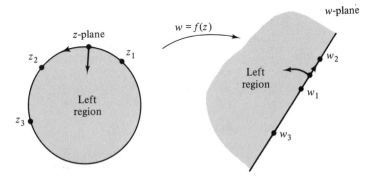

FIGURE 7.27

The other situations of mapping a circle to a circle, a line to a circle, and a line to a line can be treated in a similar manner.

Thus by judiciously selecting ordered points we can quickly write down an algebraic formula for a mapping of one "circular" region onto another.

EXAMPLE 7　Find a bilinear transformation which maps the region D_1: $|z| > 1$ onto the region D_2: Re $w < 0$.

SOLUTION　We shall take both D_1 and D_2 to be left regions. This is accomplished for D_1 by choosing any three points on the circle $|z| = 1$ which give it a negative orientation, say

$$z_1 = 1, \qquad z_2 = -i, \qquad z_3 = -1.$$

Similarly the three points

$$w_1 = 0, \qquad w_2 = i, \qquad w_3 = \infty$$

on the imaginary axis make D_2 its left region. Hence a solution to the problem is given by the transformation which takes

1 to 0,　　$-i$ to i,　　-1 to ∞.

This we obtain by setting

$$(w, 0, i, \infty) = (z, 1, -i, -1),$$

i.e.,

$$\frac{w - 0}{i - 0} = \frac{(z - 1)(-i + 1)}{(z + 1)(-i - 1)},$$

which yields

$$w = \frac{(z - 1)(1 + i)}{(z + 1)(-i - 1)} = \frac{1 - z}{1 + z}. \quad \blacksquare$$

Another important aspect of bilinear transformations is their symmetry-preserving property. First recall that two points z_1 and z_2 are symmetric with respect to a straight line $\mathscr{L}$ if $\mathscr{L}$ is the perpendicular bisector of the line segment joining z_1 and z_2 [see Fig. 7.28(a)]. From elementary geometry this is equivalent to saying that *every* line or circle through z_1 and z_2 intersects $\mathscr{L}$ orthogonally, i.e., at right angles. See Fig. 7.28(b). (Remember that a circle is orthogonal to $\mathscr{L}$ if and only if its center lies on $\mathscr{L}$.)

These considerations suggest the following definition of symmetry with respect to a circle C.

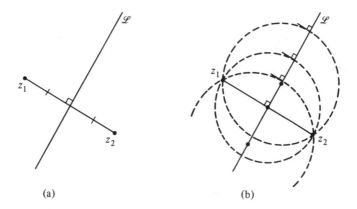

(a)   (b)

FIGURE 7.28

DEFINITION 2   *Two points z_1 and z_2 are said to be* **symmetric with respect to a circle** C *if every straight line or circle passing through z_1 and z_2 intersects C orthogonally* (Fig. 7.29).

In particular the center a of the circle C, and the point ∞, are symmetric with respect to C; there are no circles through these two points, and any line containing a (and necessarily ∞) is orthogonal to C, so the condition in Definition 2 holds.

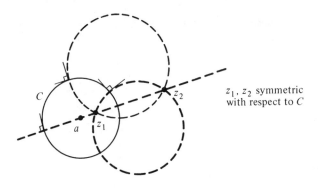

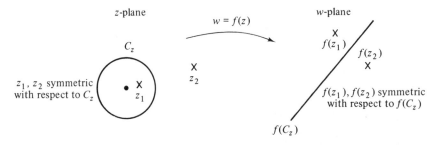

z_1, z_2 symmetric
with respect to C

FIGURE 7.29

Now we are in a position to state the symmetry-preserving property of bilinear transformations.

THEOREM 6 (Symmetry Principle) *Let C_z be a line or circle in the z-plane, and let $w = f(z)$ be any bilinear transformation. Then two points z_1 and z_2 are symmetric with respect to C_z if, and only if, their images $w_1 = f(z_1)$, $w_2 = f(z_2)$ are symmetric with respect to the image of C_z under f.*

The theorem is illustrated in Fig. 7.30 for the special case when C_z is a circle with image a straight line.

FIGURE 7.30

Proof This is easy; think about it. Two points are symmetric with respect to a circle (line) if every circle or line containing the points intersects the given circle (line) orthogonally. But bilinear transformations preserve the class of circles and lines, and they also preserve the orthogonality; hence they preserve the symmetry condition. ∎

Now given a circle C with center a and radius R, and given a point α, it would be convenient to have a formula for the point α^* symmetric to α with respect to C. For this purpose observe that the transformation

(35)
$$T(z) = (z, a - R, a + Ri, a + R)$$
$$= \frac{[z - (a - R)](Ri - R)}{[z - (a + R)](Ri + R)} = i\frac{z - (a - R)}{z - (a + R)}$$

maps three points of C, and hence all of C, onto the real axis. Thus by the Symmetry Principle α^* is symmetric to α with respect to C if and only if $T(\alpha^*)$ is symmetric to $T(\alpha)$ with respect to the real axis. But the latter condition is clearly equivalent to saying that $T(\alpha^*)$ and $T(\alpha)$ must be conjugate points, i.e.,

$$T(\alpha^*) = \overline{T(\alpha)},$$

or, using Eq. (35),

(36) $$i\frac{\alpha^* - (a - R)}{\alpha^* - (a + R)} = \overline{\left[i\frac{\alpha - (a - R)}{\alpha - (a + R)} \right]} = -i\frac{\bar{\alpha} - (\bar{a} - R)}{\bar{\alpha} - (\bar{a} + R)}.$$

Solving Eq. (36) for α^* yields the formula

(37) $$\alpha^* = \frac{R^2}{\bar{\alpha} - \bar{a}} + a.$$

(Notice that this also shows that the point symmetric to α with respect to C is *unique*.) From the representation (37) we see that

$$\arg(\alpha^* - a) = \arg\left(\frac{R^2}{\bar{\alpha} - \bar{a}} \right) = \arg\left[\frac{R^2(\alpha - a)}{|\alpha - a|^2} \right] = \arg(\alpha - a),$$

and

$$|\alpha^* - a| = \frac{R^2}{|\bar{\alpha} - \bar{a}|} = \frac{R^2}{|\alpha - a|},$$

implying that symmetric points α^* and α lie on the same ray from the center a and that the product of their distances from the center ($|\alpha^* - a| \cdot |\alpha - a|$) is equal to the radius squared. Figure 7.31 suggests a construction of symmetric points.

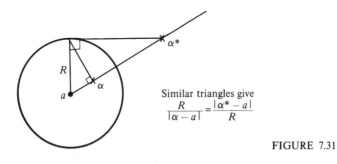

Similar triangles give
$$\frac{R}{|\alpha - a|} = \frac{|\alpha^* - a|}{R}$$

FIGURE 7.31

We remark that if C is regarded as the cross section of a mirrored sphere, then α^* corresponds to the reflection of α in the mirror.

EXAMPLE 8 Find all bilinear transformations that map $|z| < 1$ onto $|w| < 1$.

SOLUTION   Let $f(z)$ be any such bilinear transformation. Then f maps the circle C_z: $|z| = 1$ onto C_w: $|w| = 1$. Furthermore, there must be some point α, $|\alpha| < 1$, which is mapped to the origin, i.e., $f(\alpha) = 0$. According to formula (37) (with $a = 0$, $R = 1$) the point

$$\alpha^* = \frac{1^2}{\bar{\alpha} - 0} + 0 = \frac{1}{\bar{\alpha}}$$

is symmetric to α with respect to C_z. Hence $f(1/\bar{\alpha})$ must be symmetric to $f(\alpha) = 0$ with respect to C_w. But since the origin is the center of C_w, its symmetric point is ∞, i.e.,

$$f\left(\frac{1}{\bar{\alpha}}\right) = \infty.$$

Consequently, f has a zero at α and a pole at $1/\bar{\alpha}$, so f is of the form

$$f(z) = k \cdot \frac{z - \alpha}{z - \dfrac{1}{\bar{\alpha}}} = k\bar{\alpha}\frac{z - \alpha}{\bar{\alpha}z - 1}$$

for some constant k. Moreover, since $f(1)$ lies on C_w, we have

$$1 = |f(1)| = |k\bar{\alpha}| \cdot \left|\frac{1 - \alpha}{\bar{\alpha} - 1}\right| = |k\bar{\alpha}|.$$

Thus $k\bar{\alpha} = e^{i\theta}$ for some real θ, and we find

(38) $$f(z) = e^{i\theta} \cdot \frac{z - \alpha}{\bar{\alpha}z - 1} \qquad (|\alpha| < 1).$$

Conversely, the reader can easily show (Prob. 12) that any transformation of the form (38) maps $|z| < 1$ onto $|w| < 1$.   ∎

In fact it can be shown that the functions in Eq. (38) are the only one-to-one *analytic* mappings of the unit disk onto itself.

Exercises 7.4

1. Let $f_1(z) = (z + 2)/(z + 3)$, $f_2(z) = z/(z + 1)$. Find $f_1^{-1}(f_2(z))$.

2. Let f be a bilinear transformation such that $f(1) = \infty$ and f maps the imaginary axis onto the unit circle $|w| = 1$. What is the value of $f(-1)$?

3. Find the point symmetric to $4 - 3i$ with respect to each of the following circles.
   (a) $|z| = 1$ (b) $|z - 1| = 1$ (c) $|z - 1| = 2$

4. Given two nonintersecting circles C_1 and C_2, prove that there exist two distinct points z_1, z_2 which are symmetric with respect to both C_1 and C_2. Use this result to show that there is a bilinear transformation which maps C_1 and C_2 onto *concentric* circles.

5. Does there exist a bilinear transformation f which maps the real axis onto the unit circle $|w| = 1$ and satisfies $f(i) = 2$, $f(-i) = -\frac{1}{2}$?

6. Let $w = f(z)$ be the bilinear transformation mapping the points 0, λ, ∞ to $-i$, 1, i, respectively, where λ is real. For what values of λ is the upper half-plane mapped onto $|w| < 1$?

7. Using the cross-ratio notation, write down an equation defining a bilinear transformation which maps the half-plane below the line $y = 2x - 3$ onto the interior of the circle $|w - 4| = 2$. Repeat for the exterior of this circle.

8. Argue why the bilinear transformation defined by $(w, -i, 1, i) = (z, -i, i, 1)$ maps the unit circle onto itself but maps the interior onto the exterior. [HINT: Consider orientation.]

9. Let z_1, z_2, and z_3 be three distinct points which lie on a circle (or line) C. Prove that z and z^* are symmetric with respect to C if and only if $(z^*, z_1, z_2, z_3) = \overline{(z, z_1, z_2, z_3)}$.

10. Show that the distinct points w_1, w_2, w_3, and w_4 all lie on the same circle or line if and only if the cross-ratio (w_1, w_2, w_3, w_4) is real. [HINT: Consider the bilinear transformation defined by

$$(w, w_2, w_3, w_4) = (z, 0, 1, \infty)$$

and observe that $(z, 0, 1, \infty) \equiv z$.]

11. Repeat Example 8 using cross-ratios. [HINT: Think about the equation

$$(w, e^{i\theta_1}, e^{i\theta_2}, e^{i\theta_3}) = (z, -i, 1, i).]$$

12. Show that any transformation of the form (38) maps $|z| < 1$ onto $|w| < 1$.

13. Find a conformal map of the unit disk onto itself, taking the point $i/2$ to the origin.

14. Show that the bilinear transformation taking z_i to w_i ($i = 1, 2, 3$) can be expressed in determinant form as

$$\begin{vmatrix} 1 & z & w & zw \\ 1 & z_1 & w_1 & z_1 w_1 \\ 1 & z_2 & w_2 & z_2 w_2 \\ 1 & z_3 & w_3 & z_3 w_3 \end{vmatrix} = 0.$$

15. Show that every bilinear transformation which maps the upper half-plane onto the open unit disk must be of the form

$$f(z) = e^{i\theta} \frac{z - z_0}{z - \bar{z}_0}, \qquad \text{where } \text{Im}(z_0) > 0.$$

16. Find all bilinear transformations which map the upper half-plane onto itself.

17. Let z be fixed with $\text{Re } z \geq 0$, and let

$$T_0(w) = \frac{a_0}{z + a_0 + b_1 + w}, \qquad T_k(w) = \frac{a_k}{z + b_{k+1} + w}$$

$$(k = 1, 2, \ldots, n - 1)$$

be a sequence of bilinear transformations such that each a_k is real and positive and each b_k is pure imaginary or zero. Prove, by induction, that the composition

$$\xi = S(w) \equiv T_0(\cdots (T_{n-2}(T_{n-1}(w)))))$$

maps the half-plane $\text{Re } w > 0$ onto a region contained in the disk $|\xi - \frac{1}{2}| < \frac{1}{2}$.

18. Let $P(z) = z^n + c_1 z^{n-1} + c_2 z^{n-2} + \cdots + c_n$ be a polynomial of degree $n > 0$ with complex coefficients $c_k = p_k + iq_k$, $k = 1, 2, \ldots, n$. Set $Q(z) \equiv p_1 z^{n-1} + iq_2 z^{n-2} + p_3 z^{n-3} + iq_4 z^{n-4} + \cdots$. Prove *Wall's criterion* that if $Q(z)/P(z)$ can be written in the form

$$\frac{Q(z)}{P(z)} = \cfrac{a_0}{z + a_0 + b_1 + \cfrac{a_1}{z + b_2 + \cfrac{a_2}{z + b_3 + \ddots + \cfrac{a_{n-1}}{z + b_n}}}},$$

where each a_k is real and positive and each b_k is pure imaginary or zero, then all the zeros of $P(z)$ have negative real parts. [HINT: Write $Q(z)/P(z) = T_0(\cdots (T_{n-2}(T_{n-1}(0)))))$, where the transformations T_k are defined as in Prob. 17.]

19. Prove that $P(z) = z^3 + 3z^2 + 6z + 6$ has all its zeros in the left half-plane by applying the result of Prob. 18. [HINT: Use ordinary long division to obtain the representation for $Q(z)/P(z)$.]

7.5 THE SCHWARZ-CHRISTOFFEL TRANSFORMATION

We have seen that a function $f(z)$ is conformal at every point at which it is analytic and its derivative is nonzero. It is instructive to analyze what happens at certain points where these conditions are not met. For concreteness, let x_1 be a fixed point on the real axis and let $f(z)$ be a function whose *derivative* $f'(z)$ is given by $(z - x_1)^\alpha$ for some real α satisfying $-1 < \alpha < 1$. [To be precise, we shall take the argument of $z - x_1$ to lie between $-\pi/2$ and $3\pi/2$, introducing a branch cut vertically downward from x_1; see Fig. 7.32(a)]. We are going to use the equation

(39) $$f'(z) = (z - x_1)^\alpha$$

to determine certain features of the image of the real axis under the mapping f.

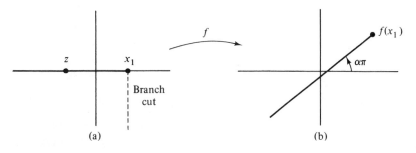

FIGURE 7.32

If z lies on the x-axis to the left of x_1, as in Fig. 7.32(a), then two observations follow from Eq. (39). Namely, f is conformal at z (since f' exists and is nonzero), and the argument of $f'(z)$ is constant for all such z:

$$\arg f'(z) = \arg(z - x_1)^\alpha = \alpha \arg(z - x_1) = \alpha\pi$$

(we shall ignore multiples of 2π in the derivation). Recalling the discussion of Sec. 7.2 we conclude that $f(z)$ maps the interval $(-\infty, x_1)$, considered as a curve whose tangents are all parallel to the real axis, onto another curve whose tangents all make an angle $\alpha\pi$ with the real axis. The latter curve must be, then, a portion of a straight line terminating at $f(x_1)$. See Fig. 7.32(b).

For z on the real axis to the right of x_1, we have

$$\arg f'(z) = \alpha \arg(z - x_1) = \alpha \cdot 0 = 0.$$

Hence, reasoning similarly, the interval (x_1, ∞) is mapped to a horizontal straight line, so that the whole picture looks like Fig. 7.33.

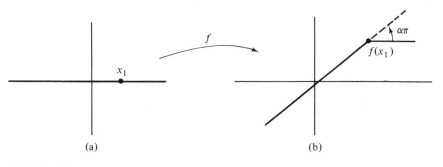

FIGURE 7.33

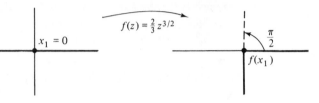

FIGURE 7.34

For the special case where $f(z) = \frac{2}{3}z^{3/2}$, which has $f'(z) = z^{1/2}$, the mapping (for the branch described earlier) is sketched in Fig. 7.34.

Now we start to generalize this model. If, instead of Eq. (39), we have

(40) $$f'(z) = A(z - x_1)^{\alpha}$$

for some complex constant A $(\neq 0)$, then

$$\arg f'(z) = \arg A + \alpha \arg(z - x_1),$$

and the mapping can be visualized by rotating Fig. 7.33(b) by an amount $\arg A$; see Fig. 7.35. In particular, the angle made by the image of the interval (x_1, ∞) is now $\arg A$, but the angle of the corner at $f(x_1)$ is unchanged.

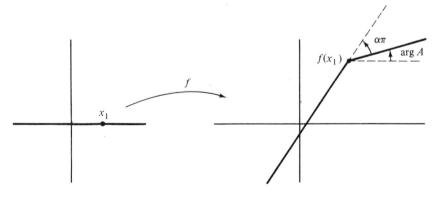

FIGURE 7.35

The next generalization is to consider a mapping given by a function f with

(41) $$f'(z) = A(z - x_1)^{\alpha_1}(z - x_2)^{\alpha_2} \cdots (z - x_n)^{\alpha_n};$$

here A $(\neq 0)$ is a complex constant, each α_i lies between -1 and $+1$, and the (real) x_i satisfy

$$x_1 < x_2 < \cdots < x_n.$$

(As before we take the argument of each $z - x_i$ to be between $-\pi/2$ and $3\pi/2$.) What does this mapping f do to the real axis?

From the equation

$$\arg f'(z) = \arg A + \alpha_1 \, \arg(z - x_1) + \alpha_2 \, \arg(z - x_2) + \cdots$$

$$+ \alpha_n \, \arg(z - x_n)$$

and the previous discussion we see that the images of the intervals $(-\infty, x_1), (x_1, x_2), \ldots, (x_n, \infty)$ are each portions of straight lines, making angles measured counterclockwise from the horizontal in accordance with the following prescription:

(42)

Interval	Angle of image
$(-\infty, x_1)$	$\arg A + \alpha_1 \pi + \alpha_2 \pi + \cdots + \alpha_n \pi$
(x_1, x_2)	$\arg A + \alpha_2 \pi + \cdots + \alpha_n \pi$
$\vdots$	$\vdots$
(x_{n-1}, x_n)	$\arg A + \alpha_n \pi$
(x_n, ∞)	$\arg A.$

Hence as z traverses the real axis from left to right $f(z)$ generates a polygonal path whose tangent at the point $f(x_i)$ has its angle decreased by the amount $\alpha_i \pi$; see Fig. 7.36.

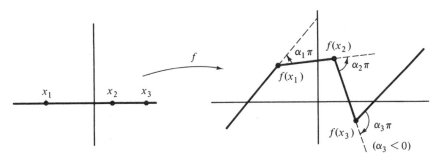

FIGURE 7.36

Now if the function $f(z)$ satisfies Eq. (41) it is, a priori, differentiable and hence analytic on the complex plane with the exception of the (downward) branch cuts from the points x_i. So for any z in the upper half-plane we can set

$$g(z) \equiv \int_\Gamma f'(\xi) \, d\xi,$$

where Γ is, for definiteness, the straight line segment from 0 to z, and conclude then that $f(z) = g(z) + B$ for some constant B. In particular, we can write

(43) $$f(z) = A \int_0^z (\xi - x_1)^{\alpha_1} (\xi - x_2)^{\alpha_2} \cdots (\xi - x_n)^{\alpha_n} \, d\xi + B.$$

Functions of the form (43) are known as *Schwarz-Christoffel transformations*. We have shown that such transformations map the real axis onto a polygonal line. This puts us in a good position to attack a specific and extremely important problem in conformal mapping, namely, the problem of mapping the upper half-plane one-to-one onto the interior of a given polygon. Observe that the *existence* of such a mapping is guaranteed by the Riemann Mapping Theorem stated in Sec. 7.2.

Suppose that the polygon P has vertices at the consecutive points w_1, $w_2, \ldots, w_n$ taken in counterclockwise order as in Fig. 7.37. We have thus given P a positive orientation.

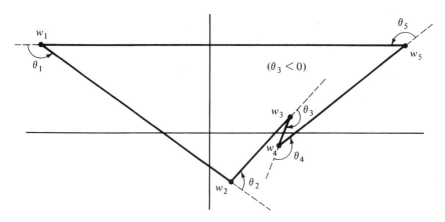

FIGURE 7.37

Now it is customary to call the *increase* in the angle of the tangent at the corner w_i the *exterior angle* at w_i; hence the exterior angle is measured (counterclockwise) from the dashed lines to the solid lines in Fig. 7.37. (This will dictate a change of sign when we relate these angles to the angles of Fig. 7.36, which measure the *decrease* in the angle of the tangent.) The exterior angles θ_i satisfy $-\pi < \theta_i < \pi$, with $\theta_i > 0$ when the corner is protruding and $\theta_i < 0$ when it is indented. Moreover, a traveler making a complete circuit around the perimeter of the polygon will be turned around one complete revolution, so that

(44) $$\theta_1 + \theta_2 + \cdots + \theta_n = 2\pi.$$

In attempting to map the upper half-plane onto the interior of P, let us pick real points

$$x_1 < x_2 < \cdots < x_{n-1}$$

(notice that we omit x_n) and consider the function

(45) $$g(z) = \int_0^z (\xi - x_1)^{-\theta_1/\pi} (\xi - x_2)^{-\theta_2/\pi} \cdots (\xi - x_{n-1})^{-\theta_{n-1}/\pi} \, d\xi.$$

Since $g(z)$ has the form (43) with $\alpha_i = -\theta_i/\pi$, we see from Fig. 7.36 that g maps the real axis onto a polygonal path whose *exterior angle* at $g(x_i)$ is $-\alpha_i\pi = \theta_i$ for $i = 1, 2, \ldots, n-1$ [that's why we put minus signs in (45)]. Since we are still experimenting, let's assume that g is one-to-one and that $\lim_{x\to+\infty} g(x)$ is finite and equals $\lim_{x\to-\infty} g(x)$. Then our mapping must look like Fig. 7.38; i.e., the initial and final segments intersect at $g(\infty)$, and the image of the real axis is a polygon P' with the same exterior angles as P, except possibly for the angle θ_∞, at $g(\infty)$. But then θ_∞ must equal θ_n, since they are both given by $2\pi - \sum_{i=1}^{n-1} \theta_i$.

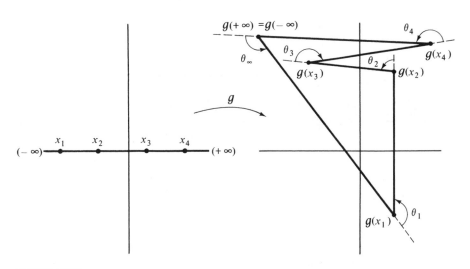

FIGURE 7.38

Now because P' has the same angles as P, by adjusting the lengths of the sides of P' we can make it *geometrically similar* to P. And it seems quite plausible that we could accomplish this by adjusting the points x_1, $x_2, \ldots, x_{n-1}$; after all, they determine where the corners of P' lie. Then, with the use of a rotation, a magnification, and a translation—in other words, a linear transformation—we could make these similar polygons coincide. Summarizing, we are led to speculate that with an appropriate choice of the constants we can construct a function

(46) $f(z) = Ag(z) + B$

$$= A \int_0^z (\xi - x_1)^{-\theta_1/\pi}(\xi - x_2)^{-\theta_2/\pi} \cdots (\xi - x_{n-1})^{-\theta_{n-1}/\pi}\, d\xi + B$$

i.e., a Schwarz-Christoffel transformation, which maps the real axis onto the perimeter of a given polygon P, with the correspondences

(47) $f(x_1) = w_1, \quad f(x_2) = w_2, \quad \ldots, \quad f(x_{n-1}) = w_{n-1}, \quad f(\infty) = w_n$.

Moreover, if our speculations are valid, we can use conformality and connectivity arguments to show that f maps the upper half-plane to the interior of P, as we wished, for observe that if γ is a segment as indicated in Fig. 7.39, conformality requires that its image, γ', have a tangent which initially points as shown, and connectivity completes the argument (assuming one-to-oneness). The whole story about Schwarz-Christoffel transformations is given in Theorem 7, whose proof can be found in the references.

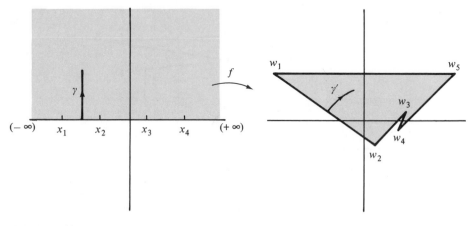

FIGURE 7.39

THEOREM 7 *Let P be a polygon having corners at $w_1, w_2, \ldots, w_n$ with corresponding exterior angles θ_i $(i = 1, 2, \ldots, n)$. Then there exists a function of the form (46) which is a one-to-one conformal map from the upper half-plane onto the interior of P. Furthermore, the correspondences (47) hold.*

Before we illustrate the technique, we must make two remarks. First, recall that in constructing the map we have three "degrees of freedom" at our disposal; thus we can specify three points on the real axis to be the preimages of three of the w_i. However, formula (46) already designates ∞ as the preimage of w_n, so we are free to choose only, say, x_1 and x_2, and the other x_i are then determined.

Second, in order to get a closed-form expression for the mapping we must be able to compute the integral in Eq. (46). A glance through a standard table of integrals shows that this is hopeless for $n > 4$, and not always possible even for smaller n. Numerical integration, however, is always possible.

EXAMPLE 9　Derive a Schwarz-Christoffel transformation mapping the upper half-plane onto the triangle in Fig. 7.40.

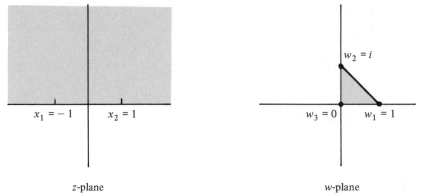

z-plane 　　　　　　　　　　　　w-plane

FIGURE 7.40

SOLUTION For the exterior angles we have $\theta_1 = \theta_2 = 3\pi/4$, $\theta_3 = \pi/2$. Hence, choosing $x_1 = -1$ and $x_2 = 1$ we have

$$f(z) = A \int_0^z (\xi + 1)^{-3/4} (\xi - 1)^{-3/4} \, d\xi + B$$

$$= A \int_0^z (\xi^2 - 1)^{-3/4} \, d\xi + B.$$

The integration must be performed numerically. To evaluate the constants we compute

$$f(x_1) = f(-1) = A \int_0^{-1} (\xi^2 - 1)^{-3/4} \, d\xi + B = A\eta + B,$$

where

$$\eta = \int_0^{-1} (\xi^2 - 1)^{-3/4} \, d\xi \approx -1.85(1 + i)$$

and

$$f(x_2) = f(1) = A \int_0^1 (\xi^2 - 1)^{-3/4} \, d\xi + B = -A\eta + B.$$

Setting these equal to w_1 and w_2, respectively, we find

$$A\eta + B = 1,$$

$$-A\eta + B = i.$$

Consequently,

$$A = \frac{1 - i}{2\eta}, \qquad B = \frac{1 + i}{2}. \qquad \blacksquare$$

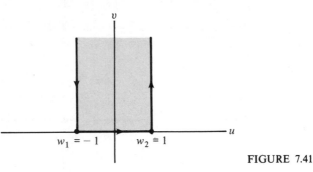

FIGURE 7.41

EXAMPLE 10 Determine a Schwarz-Christoffel transformation which maps the upper half-plane onto the semi-infinite strip $|\mathrm{Re}\ w| < 1$, $\mathrm{Im}\ w > 0$ (Fig. 7.41).

SOLUTION We return to the analysis surrounding Eq. (41) for mapping the real axis onto a polygonal path. To have the upper half-plane map onto the interior of the strip we choose the orientation indicated by the arrows in Fig. 7.41. Counterclockwise turns of $\pi/2$ radians at w_1 and w_2 can be accommodated by a mapping whose derivative is of the form

$$f'(z) = A(z - x_1)^{-1/2}(z - x_2)^{-1/2}.$$

Choosing $x_1 = -1$ and $x_2 = 1$ again, we compute

$$f(z) = A \int_0^z (\xi + 1)^{-1/2}(\xi - 1)^{-1/2}\,d\xi + B = \frac{A}{i} \int_0^z \frac{d\xi}{\sqrt{1 - \xi^2}} + B$$

$$= \frac{A}{i}\sin^{-1} z + B.$$

Setting $f(-1) = w_1 = -1$ and $f(1) = w_2 = 1$, we have

$$-iA\,\sin^{-1}(-1) + B = -1,$$

$$-iA\,\sin^{-1}(1) + B = 1,$$

which implies that $B = 0$ and $A = 2i/\pi$. Hence

$$f(z) = \frac{2}{\pi}\sin^{-1} z. \quad \blacksquare$$

EXAMPLE 11 Map the upper half-plane onto the domain consisting of the fourth quadrant plus the strip $0 < v < 1$.

SOLUTION The boundary of this domain consists of the line $v = 1$, the negative u-axis, and the negative v-axis. We shall regard this as the limiting form of the polygonal path indicated in Fig. 7.42, again choosing the orientation so that the specified domain lies to the left. A counterclockwise turn of π radians is called for at the corner "near $w = -\infty$,"

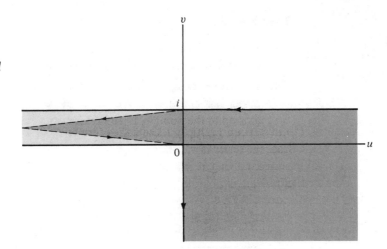

FIGURE 7.42

and a clockwise turn of $\pi/2$ radians occurs at $w = 0$. Selecting $x_1 = -1$ and $x_2 = 1$ as the preimages of these points we write, in accordance with Eq. (41),

$$f'(z) = A(z + 1)^{-1}(z - 1)^{1/2}.$$

Using integral tables, with some labor one arrives at

$$f(z) = Ai\left\{2\sqrt{1-z} + \sqrt{2} \log \frac{\sqrt{1-z} - \sqrt{2}}{\sqrt{1-z} + \sqrt{2}}\right\} + B.$$

The selection of branches is quite involved in this case, so we shall leave it to the industrious reader (Prob. 7) to verify that with the choice

$$\log \xi = \text{Log}|\xi| + i \arg \xi, \qquad -\frac{3}{2}\pi < \arg \xi \le \frac{\pi}{2},$$

$$\sqrt{\xi} = e^{(\log \xi)/2}, \qquad\qquad \log \xi \text{ as above,}$$

one finds that

$$f(z) = \frac{\sqrt{2}}{\pi}\sqrt{1-z} + \frac{1}{\pi}\log\frac{\sqrt{1-z} - \sqrt{2}}{\sqrt{1-z} + \sqrt{2}} + i$$

satisfies the required conditions

$$\text{Re } f(x) \to +\infty, \quad \text{Im } f(x) \to 1 \quad \text{as} \quad x \to -\infty,$$

$$\text{Re } f(x) \to -\infty, \quad \text{Im } f(x) \to 1 \quad \text{as} \quad x \to (-1)^-,$$

$$\text{Re } f(x) \to -\infty, \quad \text{Im } f(x) \to 0 \quad \text{as} \quad x \to (-1)^+,$$

$$f(1) = 0,$$

$$\text{Re } f(x) \to 0, \quad \text{Im } f(x) \to -\infty \quad \text{as} \quad x \to +\infty. \quad\blacksquare$$

Exercises 7.5

1. Use the Schwarz-Christoffel formula to derive the mapping $w = z^2$ of the first quadrant onto the upper half-plane.

2. Use the techniques in this section to find a conformal map of the upper half-plane onto the whole plane slit along the negative real axis up to the point -1. [HINT: Consider the slit as the limiting form of the wedge indicated in Fig. 7.43.]

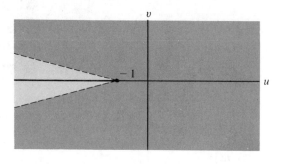

FIGURE 7.43

3. Map the upper half-plane onto the semi-infinite strip $u > 0$, $0 < v < 1$, indicated in Fig. 7.44.

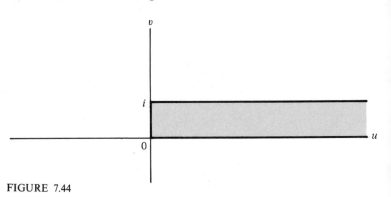

FIGURE 7.44

4. Map the upper half-plane onto the strip $0 < v < 1$, considered as the limiting form of Fig. 7.45.

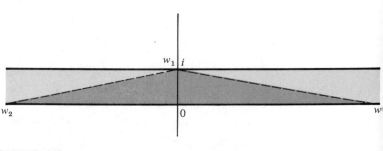

FIGURE 7.45

5. Map the upper half-plane onto the *exterior* of the semi-infinite strip in Fig. 7.41.

6. Map the upper half-plane onto the shaded region in Fig. 7.46.

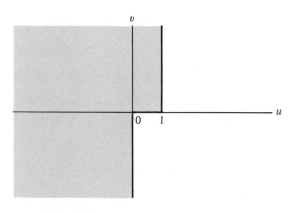

FIGURE 7.46

7. Verify that the choice of branches indicated in Example 11 yields the appropriate correspondences for $f(-\infty)$, $f(-1)$, $f(1)$, and $f(+\infty)$. [HINT: Argue that if z stays in the upper half-plane, $1 - z$ stays in the lower half-plane, $\sqrt{1 - z}$ stays in the fourth quadrant, and $(\sqrt{1 - z} - \sqrt{2})/(\sqrt{1 - z} + \sqrt{2})$ stays in the lower half-plane.]

8. Derive the expression

$$w = f(z) = \int_0^z \frac{d\xi}{\sqrt{(1 - \xi^2)(k^2 - \xi^2)}}$$

for a conformal map of the upper half-plane onto a rectangle, as indicated in Fig. 7.47. Show that the rectangular dimensions b and c

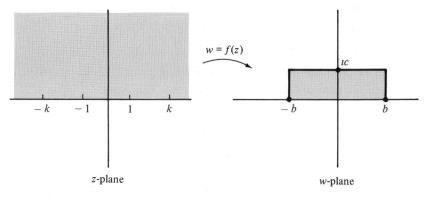

FIGURE 7.47

must be related, through k, by the equations

$$b = \frac{1}{k} \int_0^1 \frac{dx}{\sqrt{(1 - x^2)\left(1 - \dfrac{x^2}{k^2}\right)}},$$

$$c = \frac{1}{k} \int_1^k \frac{dx}{\sqrt{(x^2 - 1)\left(1 - \dfrac{x^2}{k^2}\right)}}.$$

(These are so-called "elliptic integrals"; see Ref. [2].)

9. Argue that a conformal mapping of the unit disk $|z| < 1$ onto the interior of a polygon with corners at $w_1, w_2, \ldots, w_n$ should have the form

$$w = A \int_0^z (\xi - z_1)^{-\theta_1/\pi} (\xi - z_2)^{-\theta_2/\pi} \cdots (\xi - z_n)^{-\theta_n/\pi} \, d\xi + B,$$

where the θ_i are the exterior angles and the points z_i on the unit circle are the preimages of the corresponding w_i.

10. Show that the transformation

$$w = \int_0^z \frac{d\xi}{(1 - \xi^2)^{2/3}}$$

maps the upper half-plane onto the interior of an equilateral triangle.

*11. What does the Schwarz Reflection Principle (Probs. 13 and 14 in Exercises 5.8) say about the image of the lower half-plane under the Schwarz-Christoffel transformations? (Consider, in turn, Example 10, then Example 9, and then the general case of Fig. 7.39.)

7.6 APPLICATIONS IN ELECTROSTATICS, HEAT FLOW, AND FLUID MECHANICS

The next two sections in this chapter are devoted to the solution of certain physical problems involving Laplace's equation

(48)
$$\frac{\partial^2 \phi}{\partial x^2} + \frac{\partial^2 \phi}{\partial y^2} = 0,$$

using conformal mapping techniques. We remind the reader that Eq. (48) governs two-dimensional steady-state phenomena in electrostatics, heat

343

*Section 7.6
Applications in
Electrostatics,
Heat Flow, and
Fluid Mechanics*

flow in a homogeneous medium, and fluid mechanics for an idealized fluid.

In electrostatics $\phi(x, y)$ is interpreted as the electric potential, or voltage, at the point (x, y), and its partial derivatives $\partial\phi/\partial x$ and $\partial\phi/\partial y$ are the components of the electric field intensity. Typically one specifies either the potential or the *normal* component of the intensity vector on the boundary of a domain and asks for the values of the potential inside (or outside) the domain. The curves defined by the equation $\phi = $ constant are called *equipotentials*.

In heat flow problems $\phi(x, y)$ is the temperature at the point (x, y); the curves $\phi = $ constant are then the *isotherms*. Usually one assumes that idealized heat sources or heat sinks are used to maintain fixed (specified) values of ϕ on certain parts of the boundary of a domain and that the rest of the boundary is thermally insulated. The latter condition is expressed mathematically by saying that the normal derivative of ϕ is zero; i.e., $\partial\phi/\partial n = 0$, where n is a coordinate measured perpendicular to the boundary. The problem, of course, is to find ϕ inside the domain.

The fluid mechanical interpretation of ϕ which we shall adopt is the following: The curves given by

$$\phi(x, y) = \text{constant}$$

are the paths which the fluid particles follow. In other words, they are *streamlines*. Thus in studying flow around a nonporous obstacle, the perimeter of that obstacle must constitute part of a streamline, and we specify $\phi = $ constant there. Sometimes $\phi(x, y)$ is known as the *stream function*.

A more detailed discussion of the physics of these phenomena, particularly the fluid mechanics interpretation, is given in Appendix I, but we hope that this rough sketch will aid the reader in visualizing the physical situations described herein.

As we indicated in Sec. 7.1, the basic strategy in solving these problems is to map the given domain conformally onto a simpler domain, to determine the harmonic function which satisfies the "transplanted" boundary conditions, and to carry this function back via the conformal map.

EXAMPLE 12   Find the function ϕ which is harmonic in the lens-shaped domain of Fig. 7.48(a) and takes the values 0 and 1 on the bounding circular arcs, as illustrated. Here ϕ can be interpreted as the steady-state temperature inside an infinitely long strip of material having this lens-shaped region as its cross section, with its sides maintained at the given temperatures.

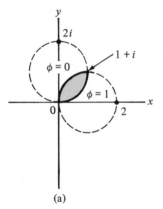

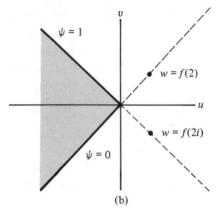

(a)   (b)

FIGURE 7.48

SOLUTION Because the domain is bounded by circular arcs, we are naturally inclined to see what we can do with bilinear transformations. If we choose the pole of the transformation to be at $z = 1 + i$, both circles will become straight lines—orthogonal lines, in fact, because conformality will preserve the right angle at $z = 0$. So let's consider

(49) $$w = f(z) = \frac{z}{z - (1 + i)},$$

which takes $z = 0$ to $w = 0$ and $z = 1 + i$ to $w = \infty$. To determine the image of the lens, we observe that since $z = 2$ goes to $w = 2/(1 - i) = 1 + i$ and $z = 2i$ goes to $w = 2i/(-1 + i) = 1 - i$, the lens is mapped onto the shaded region in Fig. 7.48(b), bounded by the rays Arg $w = 3\pi/4$ (the image of the arc where $\phi = 1$) and Arg $w = -3\pi/4$ (the image of the arc where $\phi = 0$). The corresponding harmonic function $\psi(w)$ in the w-plane is easily seen to be

$$\psi(w) = \frac{2}{\pi}\left(\frac{5\pi}{4} - \arg w\right),$$

taking the branch $0 < \arg w < 2\pi$. Carrying this back to the z-plane via Eq. (49), we find

$$\phi(x, y) = \frac{2}{\pi}\left(\frac{5\pi}{4} - \arg\frac{z}{z - (1 + i)}\right),$$

which can be expressed as

$$\phi(x, y) = \frac{2}{\pi}\left(\frac{\pi}{4} - \tan^{-1}\frac{x - y}{x(x - 1) + y(y - 1)}\right);$$

here $-\pi/2 < \tan^{-1}\theta < \pi/2$. ∎

EXAMPLE 13 Find the function ϕ which is harmonic in the shaded domain depicted in Fig. 7.49(a) and takes the value 0 on the inner circle

344

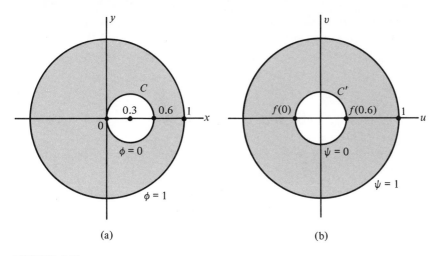

(a)   (b)

FIGURE 7.49

and 1 on the outer circle. One might interpret ϕ as the electrostatic potential inside a capacitor formed by two nested parallel (but noncentric) cylindrical conductors.

SOLUTION This problem would be trivial if the circles were concentric; recall that Log $|z|$ is harmonic (except at $z = 0$) and constant on circles centered at the origin, so we could construct a solution of the form a Log $|bz|$. Therefore it seems that we should try to map the given region onto an annulus as in Fig. 7.49(b). For this purpose we use one of the maps that take the unit disk onto itself, say

(50) $$w = f(z) = \frac{\alpha - z}{\bar{\alpha}z - 1} \qquad (|\alpha| < 1),$$

and we shall select the parameter α so that the inner circle C becomes a circle C' concentric with $|w| = 1$.

First observe that by taking α *real* we will have $f(z)$ real whenever z is real; i.e., f maps the x-axis onto the u-axis. By conformality, then, the image of the circle C is a circle which intersects the u-axis orthogonally. Hence its center lies on the u-axis. To ensure that it is concentric with $|w| = 1$ we simply enforce the condition $f(0) = -f(.6)$. Using Eq. (50), this requires

$$\frac{\alpha}{-1} = -\frac{\alpha - .6}{.6\alpha - 1},$$

or

$$.6\alpha^2 - 2\alpha + .6 = 0,$$

345

with solutions $\alpha_1 = \frac{1}{3}$ and $\alpha_2 = 3$. Since $|\alpha| < 1$, only α_1 is suitable and we have the map

$$w = f(z) = \frac{\frac{1}{3} - z}{\frac{z}{3} - 1} = \frac{1 - 3z}{z - 3}.$$

It takes the inner circle $|z - .3| = .3$ to the circle given by $|w| = |f(0)| = \frac{1}{3}$.

Now it is easy to see that the proper harmonic function, in the annulus, is

$$\psi(w) = \frac{\text{Log } |3w|}{\text{Log } 3},$$

so our problem is solved by

$$\phi(x, y) = \frac{\text{Log}\left(3 \left| \frac{1 - 3z}{z - 3} \right|\right)}{\text{Log } 3}$$

$$= \frac{1}{\text{Log } 3}\left\{\text{Log } 3 + \frac{1}{2}\text{Log}[(3x - 1)^2 + 9y^2]\right.$$

$$\left. - \frac{1}{2}\text{Log}[(x - 3)^2 + y^2]\right\}. \quad \blacksquare$$

EXAMPLE 14   Find a *nonconstant* function ϕ harmonic inside the infinite domain depicted in Fig. 7.50 and taking the value 0 on the indicated polygonal path. As the sketched lines indicate, the curves $\phi = $ constant will be streamlines for the flow of a deep river over a discontinuous stream bed.

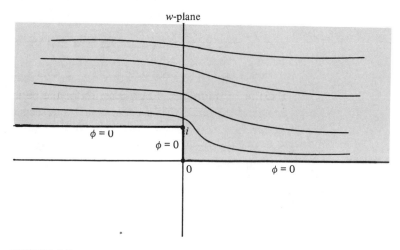

w-plane

$\phi = 0$

$\phi = 0$

i

0

$\phi = 0$

FIGURE 7.50

347

Section 7.6
Applications in
Electrostatics,
Heat Flow, and
Fluid Mechanics

SOLUTION   The Schwarz-Christoffel transformation is the tool for this geometry. Again we use the analysis surrounding Eq. (41) to map the real z axis onto the discontinuous "stream bed" in the w-plane. (Notice that a reversal of procedure occurs; the Schwarz-Christoffel transformation maps the *simple* region onto the *complicated* one.) To introduce the appropriate right-angle turns at, say, the images of $z = -1$ and $z = 1$, we must have

$$\frac{dw}{dz} = f'(z) = A(z + 1)^{+1/2}(z - 1)^{-1/2};$$

thus with the aid of an integral table we find

(51) $w = f(z) = A\{(z^2 - 1)^{1/2} + \log[z + (z^2 - 1)^{1/2}]\} + B.$

We take branches of these functions which are real and positive for large real z and which are analytic in the upper half-plane.† Then the correspondences $f(-1) = i, f(1) = 0$ require

$$A \log(-1) + B = Ai\pi + B = i,$$

$$A \log(1) + B = B = 0,$$

with solution $A = 1/\pi$ and $B = 0$. Hence the mapping from the z-plane to the flow region is

(52) $w = f(z) = \dfrac{1}{\pi}\{(z^2 - 1)^{1/2} + \log[z + (z^2 - 1)^{1/2}]\}.$

Now we must find a nonconstant harmonic function of z in the upper half-plane which vanishes on the real axis. The answer is obvious:

(53) $\psi(x, y) = y = \text{Im}(z).$

To complete the problem we must carry $\psi(x, y)$ back to the flow region in the w-plane. But since we have constructed the map from the simple domain to the complicated domain, we need the inverse of the function (52) to complete the problem. Rather than going through the details of solving Eq. (52) for z in terms of w, let's simply abbreviate the answer by stating that the harmonic function $\phi(u, v)$ is given by

$$\phi(u, v) = \text{Im}(f^{-1}(w)).$$

Actually, this is not a serious "cop-out." We have an explicit expression for the streamlines $\phi = $ constant simply by holding y constant and regarding x as a parameter in Eq. (52).   ∎

---

†We are omitting many important details here. Branching is often a very subtle business when Schwarz-Christoffel transformations are used. A painstaking analysis would reveal that here $(z^2 - 1)^{1/2}$ is positive for large positive z and negative for large negative z and that the log function is handled by the restriction $-\pi/2 < \arg \zeta < 3\pi/2$.

One of the classic problems in elementary physics involves the parallel-plate capacitor. Here we have two oppositely charged flat conducting sheets separated by a fixed distance, and we must determine the electrostatic potential in the region between them. In the simple case of infinite (square) plates, ϕ is proportional to y and the equipotentials are as in Fig. 7.51. This is a good approximation to the more realistic problem of two large plates separated by a relatively small distance; however, near the edges of the plates the potential behaves in a more complicated manner, and this can be computed by conformal mapping. Example 15 should thus be interpreted as finding the potential in the region around two *semi*-infinite conducting plates holding opposite charges.

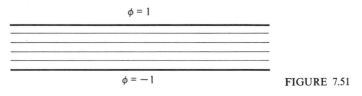

FIGURE 7.51

EXAMPLE 15   Find a function ϕ which is harmonic in the doubly slit plane and takes the values -1 and $+1$ on the two slits indicated in Fig. 7.52.

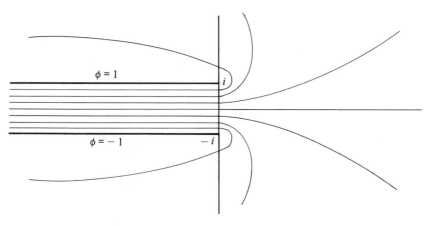

FIGURE 7.52

SOLUTION   We shall use the Schwarz-Christoffel transformation again, regarding the domain as the limiting form of the region sketched in Fig. 7.53, with the point w_0 going to $-\infty$. The limiting exterior angles are $-\pi$ at $w = i$, $+\pi$ at $w = w_0$, and $-\pi$ at $w = -i$. If the preimages of these points are $z = -1, 0$, and $+1$, respectively, we have [using Eq. (41) again]

$$\frac{dw}{dz} = A(z+1)z^{-1}(z-1) = A\left(z - \frac{1}{z}\right).$$

349

*Section 7.6
Applications in
Electrostatics,
Heat Flow, and
Fluid Mechanics*

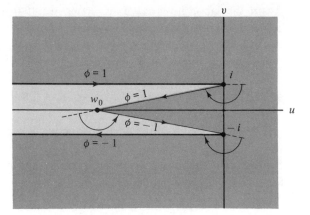

FIGURE 7.53

Consequently ,
$$w = f(z) = A\left(\frac{z^2}{2} - \log z\right) + B.$$

Enforcing $f(-1) = i$, $f(1) = -i$ and choosing the branch of $\log z$ which is positive for large positive z we get

$$A\left(\frac{1}{2} - i\pi\right) + B = i,$$

$$A\left(\frac{1}{2} - 0\right) + B = -i,$$

yielding $A = -2/\pi$, $B = 1/\pi - i$. Hence the mapping is

(54)
$$w = f(z) = -\frac{2}{\pi}\left(\frac{z^2}{2} - \log z\right) + \frac{1}{\pi} - i.$$

Notice that we have not checked the condition at $z = 0$. This is unnecessary because of the symmetry of the situation. At any rate, it is easy to see that $|w| \to \infty$ as $z \to 0$.

The transformed problem is depicted in Fig. 7.54. The obvious solution is

$$\psi(z) = \frac{2}{\pi}\,\text{Arg}\,z - 1.$$

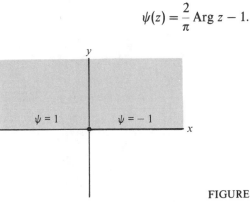

FIGURE 7.54

Again the labor involved in inverting Eq. (54) is prohibitive, but the curves $\psi = $ constant are given parametrically by writing Eq. (54) as

$$w = -\frac{2}{\pi}\left(\frac{r^2}{2}e^{2i\theta} - \text{Log } r - i\theta\right) + \frac{1}{\pi} - i$$

and holding θ constant while r varies from 0 to ∞. ∎

EXAMPLE 16   Find a nonconstant function ϕ which is harmonic in the slit upper half-plane of Fig. 7.55(a), taking the value $\phi = 0$ on the slit and the real axis. The lines $\phi = $ constant can be interpreted as the streamlines for fluid flow past a simple obstacle.

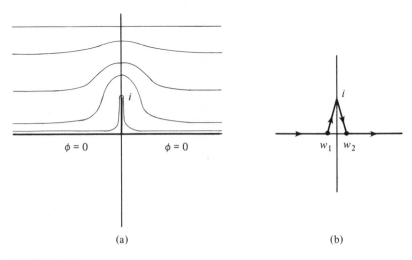

(a)   (b)

FIGURE 7.55

SOLUTION   Regarding the boundary as the limiting form of the polygonal path in Fig. 7.55(b), we construct a Schwarz-Christoffel transformation which maps -1 to w_1, 0 to i, and $+1$ to w_2, with limiting exterior angles $\pi/2$, $-\pi$, and $\pi/2$, respectively, as w_1 and w_2 approach 0. Hence

$$\frac{dw}{dz} = A(z + 1)^{-1/2}z(z - 1)^{-1/2} = \frac{Az}{(z^2 - 1)^{1/2}},$$

and

(55) $$w = f(z) = A(z^2 - 1)^{1/2} + B.$$

Taking the branch which is positive for large positive z, we make the correspondences

$$f(-1) = B = 0,$$

$$f(1) = B = 0,$$

$$f(0) = Ai + B = i.$$

351

Section 7.6
Applications in
Electrostatics,
Heat Flow, and
Fluid Mechanics

Hence $A = 1$, $B = 0$. (We are able to satisfy three conditions with only two constants in this case because of the symmetry of the region.)

In the z-plane the problem becomes, just as in Example 14, that of finding a nonconstant harmonic function vanishing on the real axis, so again we have $\psi(z) = \operatorname{Im} z = y$. In this case we can invert the map (55) to find

$$z = (w^2 + 1)^{1/2},$$

so that the required function is

$$\phi(u, v) = \operatorname{Im}(w^2 + 1)^{1/2},$$

taking the branch which is positive for large positive w. ∎

A final example which exploits the symmetry properties of bilinear transformations is

EXAMPLE 17   Find a function $\phi(x, y)$ which is harmonic in the portion of the upper half-plane exterior to the circle $C\colon |z - 5i| = 4$ and which takes the value $+1$ on the circle and 0 on the real axis (Fig. 7.56). The solution can be interpreted as the electric potential due to a charged conducting cylinder lying above a conducting plane.

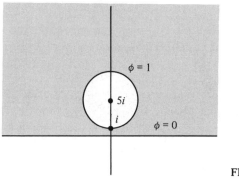

FIGURE 7.56

SOLUTION   We seek a bilinear transformation which maps the given region onto some simple domain where the solution is known. If we choose the pole of the transformation to lie on the real axis, the image region will be bounded by a line and a circle—just like what we started with. If the pole lies on the circle, the same thing happens. Any other location for the pole will yield a domain bounded by two circles; this, too, is difficult to work with *unless the circles are concentric*, as in the solution of Example 13.

How can we achieve this condition? If the pole is taken to be z_p, then the center of the real line's image-circle is the image of the point $\bar{z}_p$, which is symmetric to z_p with respect to the real line. Similarly, the center of the circle which is the image of C is the image of the point [recall formula (37)]

$$z_p^* = 5i + \frac{16}{(\bar{z}_p + 5i)}.$$

Hence the image-circles will be concentric if $\bar{z}_p = z_p^*$; i.e.,

$$\bar{z}_p = 5i + \frac{16}{(\bar{z}_p + 5i)},$$

which has solutions $z_p = \pm 3i$. If we take, say, $z_p = -3i$, then $\bar{z}_p = z_p^* = 3i$.

From the above, we conclude that the transformation

$$w = f(z) = \frac{z - 3i}{z + 3i}$$

maps the two conductors onto concentric circles *centered at the origin*. Therefore, the radius of the image of the real axis is

$$r_1 = |f(0)| = 1,$$

and the radius of the image of C is

$$r_2 = |f(i)| = \frac{1}{2}.$$

The function which is harmonic in the annulus $r_2 < |w| < r_1$ and takes the proper boundary values is

$$\psi(w) = \frac{\text{Log } |w|}{\text{Log } \frac{1}{2}} = \frac{-\text{Log } |w|}{\text{Log } 2},$$

so the solution to the problem is

$$\phi(x, y) = \frac{-1}{\text{Log } 2} \text{Log } \left| \frac{z - 3i}{z + 3i} \right|.$$

Note that the equipotentials are circles in the z-plane (Fig. 7.57). ∎

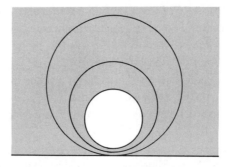

FIGURE 7.57

353

Section 7.6
Applications in
Electrostatics,
Heat Flow, and
Fluid Mechanics

Exercises 7.6

1. Find the electrostatic potential ϕ in the semidisk with the boundary values as shown in Fig. 7.58.

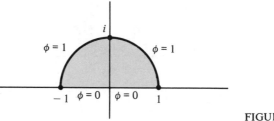

FIGURE 7.58

2. Find the temperature distribution in Fig. 7.59.

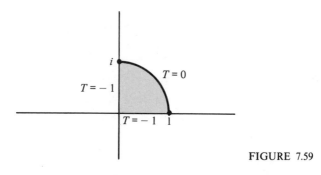

FIGURE 7.59

3. Find the electrostatic potential in the upper half-plane exterior to the unit circle under the conditions shown in Fig. 7.60.

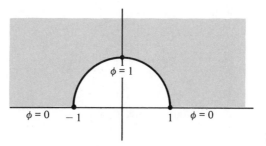

FIGURE 7.60

4. Find the electrostatic potential in the slit upper half-plane with the boundary values as depicted in Fig. 7.61.

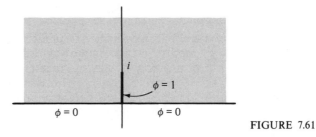

FIGURE 7.61

5. Find the temperature distribution in the unit disk with boundary values as shown in Fig. 7.62. [HINT: Map to the upper half-plane, then use Prob. 3 of Exercises 7.1.]

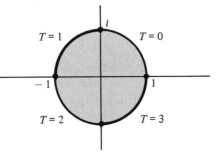

FIGURE 7.62

6. Find the electrostatic potential ϕ in the region between two conducting cylinders under the conditions shown in Fig. 7.63.

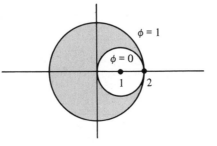

FIGURE 7.63

7. Find the temperature inside the infinite regions depicted in Fig. 7.64.

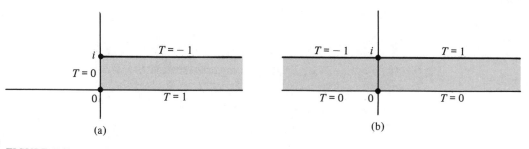

(a)

(b)

FIGURE 7.64

8. Find the temperature distribution in the crescent-shaped region given in Fig. 7.65.

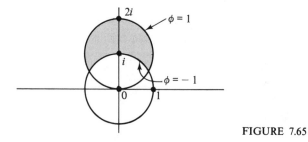

FIGURE 7.65

9. Find the potential between two nested nonconcentric conducting cylinders charged as shown in Fig. 7.66.

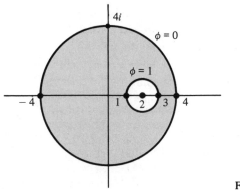

FIGURE 7.66

10. Find the potential in the region exterior to two conducting cylinders charged as shown in Fig. 7.67. [HINT: A bilinear transformation can be used to reduce this problem to the situation in Example 17.]

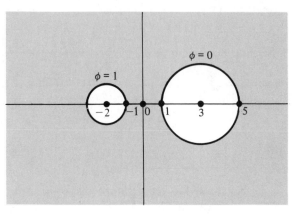

FIGURE 7.67

11. Rework Example 13 using the technique suggested by Prob. 4 in Exercises 7.4.

12. Find the streamlines for fluid flow in the region indicated in Prob. 6 in Exercises 7.5.

7.7 FURTHER PHYSICAL APPLICATIONS OF CONFORMAL MAPPING

The examples in the previous section were rather straightforward. We were assigned the task of solving a boundary value problem in an irregular region, and we constructed a mapping to a simpler region by techniques which we had learned earlier. In the present section we shall study some problems wherein the mappings are not found by straightforward methods but rather by techniques which may seem to arise from divine inspiration, dumb luck, or simply "experience."

The situation might be described as follows. When one is dealing with an area of mathematics which is so complicated that the direct solution of problems is not feasible, it is useful to try to gain some "feel" for the area by *postulating* a solution and then finding out what problem it solves. The mathematical community calls such procedures "inverse methods," as opposed to the "direct methods" illustrated in Sec. 7.6. As our first experiment with an inverse method, let us consider what situations can be analyzed with the mapping $w = \sin z$ and its inverse $z = \sin^{-1} w$. This transformation turns out to be quite a versatile tool, and we shall utilize it to solve four very different physical problems.

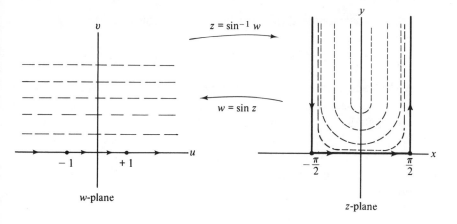

$z = \sin^{-1} w$

$w = \sin z$

-1 $+1$ u

w-plane

$-\frac{\pi}{2}$ $\frac{\pi}{2}$ x

z-plane

FIGURE 7.68

We saw in Sec. 7.5 that $z = \sin^{-1} w$ maps the upper half-plane onto the semi-infinite strip depicted in Fig. 7.68, with the real axis mapping to the sides and bottom of the strip. We can thus use the harmonic function $\psi_1(u, v) = v = \operatorname{Im} w$, which is zero on the real axis, to find the streamlines for flow in a blocked channel, as illustrated in the figure. They are given by the level curves

$$\phi_1(x, y) = \operatorname{Im}(\sin z) = \text{constant.}$$

On the other hand, the harmonic function $\phi_2(x, y) = y = \operatorname{Im} z$ in the z-plane is zero on the bottom of the strip, and the lines $\phi_2 = $ constant are perpendicular to the sides of the strip. Therefore their images, the curves

$$(56) \qquad \psi_2(u, v) = \operatorname{Im}(\sin^{-1} w) = \text{constant,}$$

intersect the u-axis orthogonally for $|u| > 1$, while $\psi_2 = 0$ on the segment $-1 \le u \le 1$; see Fig. 7.69. Hence Eq. (56) can be interpreted as the equipotentials around an infinitely long charged conducting strip of width 2.

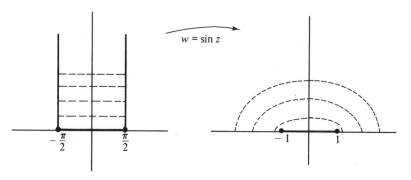

$w = \sin z$

$-\frac{\pi}{2}$ $\frac{\pi}{2}$

-1 1

FIGURE 7.69

The harmonic function which is conjugate to this $\psi_2(u, v)$ is, obviously,

$$(57) \qquad\qquad \eta(u, v) = \text{Re}(\sin^{-1} w)$$

(to be accurate, we should say that η is *one* of the harmonic conjugates to ψ_2; recall Sec. 2.5). The level curves of $\eta(u, v)$ intersect those of $\psi_2(u, v)$ orthogonally, so they look like the dashed curves in Fig. 7.70. In particular, $\eta(u, v) = -\pi/2$ on the u-axis to the left of -1, and $\eta = \pi/2$ on the u-axis to the right of $+1$. Thus (57) can be interpreted as the potential due to two oppositely charged semi-infinite conducting plates lying side by side.

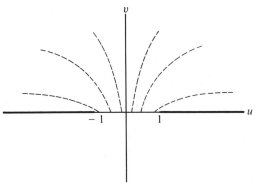

FIGURE 7.70

Finally, it is easy to verify that the mapping $w = \sin z$ as depicted in Fig. 7.68 is symmetric about the y-axis, so that we can restrict it to determine a one-to-one map of the strip in Fig. 7.71 onto the first quadrant. Consider the harmonic function $\phi_3(x, y) = 2x/\pi = (2 \, \text{Re} \, z)/\pi$. It is zero on the y-axis and $+1$ on the line $x = \pi/2$, and its level curves intersect the x-axis orthogonally. Hence the "inherited" function

$$\psi_3(u, v) = \frac{2}{\pi} \text{Re}(\sin^{-1} w)$$

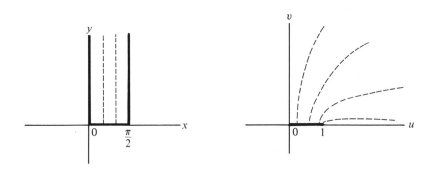

FIGURE 7.71

359

Section 7.7
Further Physical
Applications of
Conformal
Mapping

is zero on the v-axis and $+1$ on the u-axis for $u > 1$, while the curves $\psi_3 =$ constant intersect the segment $0 < u < 1$ orthogonally. The latter condition implies that the normal derivative of ψ_3 is zero on the segment. As a result, we can interpret ψ_3 as the steady-state temperature in the first quadrant when the v-axis is held at 0 degrees, the u-axis is held at 1 degree for $u > 1$, and the portion $0 < u < 1$ of the u-axis is thermally insulated.

Continuing in this vein, observe that we can solve an interesting variety of problems with the function $f(z) = z^\alpha$. For $\alpha > 0$, we have seen that a suitable branch of f maps the wedge $0 < \arg z < \pi/\alpha$ onto the upper half-plane, and thus Im z^α is the stream function for fluid flow in a wedge. But of course *any* of the streamlines Im $z^\alpha =$ constant could be considered as the perimeter of an obstacle placed in the flow. For example, if $\alpha = 2$,

$$\text{Im } z^2 = 2xy,$$

and we have a stream function for flow inside, say, the rectilinear hyperbola $xy = 1$ (cf. Fig. 2.5). For $\alpha = -1$, the level curves

$$\text{Im } z^{-1} = \frac{-y}{x^2 + y^2} = \text{constant}$$

are all circles through the origin, as shown in Fig. 7.72. One can visualize these curves as equipotentials for a (two-dimensional) dipole at $z = 0$. (The fluid analogy is a bit strained.)

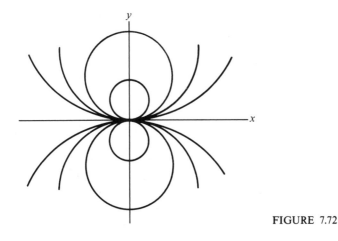

FIGURE 7.72

The level curves for the harmonic function $\text{Log}\,|z|$ have an interesting interpretation. Recall that they are concentric circles (Fig. 7.73). Considered as streamlines, they represent the flow due to a *vortex*, or whirlpool, at the origin. On the other hand, if they are interpreted as isotherms

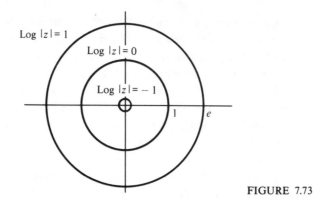

FIGURE 7.73

or equipotentials, we infer the existence of a heat source or point charge† at $x = y = 0$. As we shall illustrate below, it is often instructive to study the effect of superimposing vortices or sources on a given pattern.

A very important application of the inverse method, which we shall describe presently, springs from considerations involving the following simple example.

EXAMPLE 18   Find the streamlines for flow around a cylindrical obstacle as depicted in Fig. 7.74.

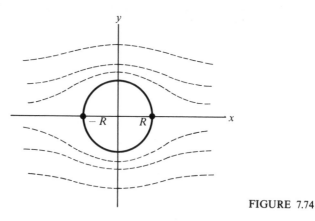

FIGURE 7.74

SOLUTION   We assume for the moment that the flow is symmetric with respect to the x-axis. (We shall see below that other interpretations are possible.) Then we only have to deal with $y \geq 0$.

To map the flow region onto the upper half-w-plane we need a (non-constant) analytic function $f(z)$ which is real on the x-axis (at least for $|x| > R$) and on the circle $|z| = R$. The first condition is satisfied by a large class of functions, e.g., rational functions with real coefficients. To

†Remember that a point charge in two dimensions corresponds to a *line* charge in three dimensions.

361

Section 7.7
Further Physical
Applications of
Conformal
Mapping

handle the second condition, we observe that since the circle can be described by $z\bar{z} = R^2$, we have $\bar{z} = R^2/z$ there; hence the rational function $z + R^2/z$ is equal to $z + \bar{z} = 2 \operatorname{Re} z$ on the circle. Consequently we are led to the mapping

$$(58) \qquad\qquad w = f(z) = z + \frac{R^2}{z}.$$

It is easily verified that (58) produces a one-to-one map of the flow region onto the upper half-plane. Thus the appropriate stream function corresponds to $\psi(u, v) = v = \operatorname{Im} w$, yielding the streamlines

$$\phi(x, y) = \operatorname{Im}\left(z + \frac{R^2}{z}\right) = \text{constant.} \quad \blacksquare$$

If we drop the symmetry assumption, we require only that the circle itself be a streamline. Hence we can add any constant multiple of $\operatorname{Log}|z|$ (since it has $|z| = R$ as a streamline) to the stream function and obtain "circulating" flow patterns as in Fig. 7.75.

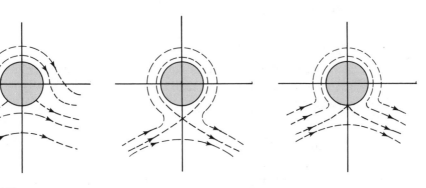

FIGURE 7.75

In 1908 the mathematician N. Joukowski had the inspiration to see what the mapping (58) would do, not to the circle $|z| = R$, but to an off-center circle such as C in Fig. 7.76(a). The result was an *airfoil*, as indicated in Fig. 7.76(b). By starting with different circles C we can generate a variety of these so-called Joukowski airfoils. Furthermore, since we have already found a wide class of flows around cylinders (Figs. 7.74 and 7.75), we can use the "Joukowski transformation" (58) to carry them over and compute flows around these airfoils! For instance, we can adjust the "point of attachment" P in Fig. 7.77 by modifying the constant multiple of the Log term in the "original" stream function. And we can shape the airfoil to meet certain specifications by

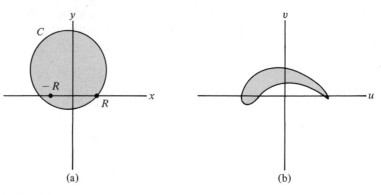

FIGURE 7.76

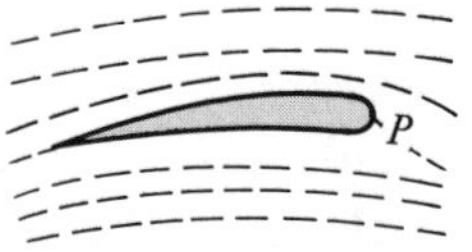

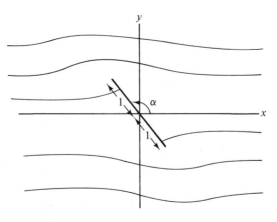

FIGURE 7.77

trying different choices for C and introducing modifications into the mapping; e.g.,

$$w = f(z) = z + \frac{R^2}{z} + \frac{a}{z^2} + \frac{b}{z^3} + \cdots.$$

Needless to say, this technique has been extremely useful in aircraft design, and we direct the interested reader to the specialized literature for further study.

Exercises 7.7

1. Consider the problem of fluid flow around a straight obstacle inclined at an angle α, as in Fig. 7.78. The stream function ψ must be

FIGURE 7.78

363

Section 7.7
Further Physical
Applications of
Conformal
Mapping

constant on the obstacle, and the streamlines ($\psi =$ constant) must tend to horizontal straight lines at large distances from the origin. Show that

$$\psi(x, y) = \text{Im}\left[e^{-i\alpha}z\left(\cos \alpha + i \sin \alpha \sqrt{1 - \frac{e^{2i\alpha}}{z^2}}\right)\right]$$

satisfies these conditions.

2. Analyze the temperature distribution in the plate depicted in Fig. 7.79. [HINT: Use the solution to Prob. 6 in Exercises 7.5 and the sine function.]

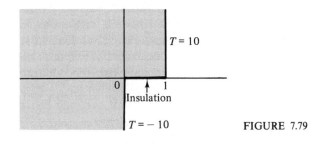

$T = 10$

0 1
Insulation

$T = -10$ FIGURE 7.79

3. Find the temperature distribution in the first quadrant under the boundary conditions indicated in Fig. 7.80.

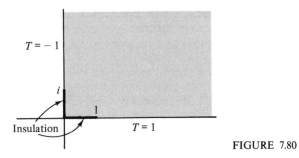

$T = -1$

i

1

Insulation $T = 1$

FIGURE 7.80

4. Analyze the temperature distribution in the slab $0 < y < 1$ under the conditions shown in Fig. 7.81.

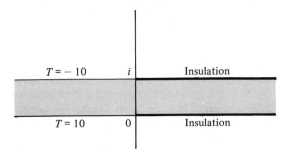

$T = -10$ i Insulation

$T = 10$ 0 Insulation

FIGURE 7.81

5. Another feasible approach to Example 18 is to map the shaded region in Fig. 7.82 to the upper half-plane as follows: First use a bilinear transformation to map R to ∞ and $-R$ to 0. Argue that the shaded region then maps onto a 90-degree wedge, which can be rotated if necessary to coincide with the first quadrant. Squaring then maps onto the upper half-plane, and taking the imaginary part

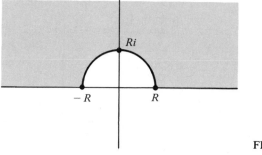

FIGURE 7.82

of the whole transformation should solve the problem of finding the stream function. Show that the implementation of this scheme leads to the mapping

$$w = -\left(\frac{z+R}{z-R}\right)^2$$

and the stream function

$$\psi(z) = \operatorname{Im} w = \frac{4yR(x^2 + y^2 - R^2)}{[(x-R)^2 + y^2]^2}.$$

Although this function is harmonic in the shaded region and zero on the boundary, we reject it on the physical basis that the flow we seek must have nearly horizontal streamlines for large y. That is, the curves $\psi(x, y) = $ constant must approximate $y = $ constant far away from the obstacle. This is obviously not the case for the above solution. (This problem illustrates an additional complication which we have ignored in our elementary treatment, namely, consideration of boundary conditions "at infinity" for unbounded domains.)

6. Show that the mapping (58) takes two concentric circles, $|z| = R$ and $|z| = R' > R$, onto a line segment and an ellipse as shown in Fig. 7.83. Use this to find the electrostatic potential between a conducting elliptic cylinder surrounding a conducting strip.

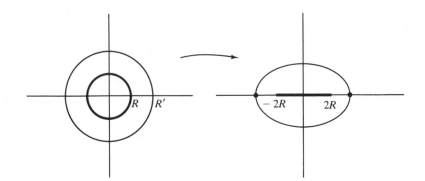

FIGURE 7.83

7. Using the mapping (58), find the streamlines for the flow indicated in Fig. 7.84.

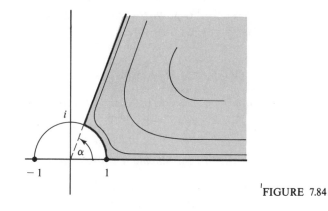

FIGURE 7.84

SUMMARY

One of the most important aspects of mappings generated by analytic functions is the persistence of solutions of Laplace's equation. That is, if $w = f(z)$ is a one-to-one analytic function mapping one domain onto another [which implies that $z = f^{-1}(w)$ is analytic also] and if $\phi(z)$ is harmonic in the first domain, then $\psi(w) \equiv \phi(f^{-1}(w))$ is harmonic in the second domain. This leads to a useful technique for solving boundary value problems for Laplace's equation; one performs a preliminary mapping to a domain such as a quadrant, half-plane, or annulus, where the corresponding problem is easy to solve, and then one carries the solution function back via the mapping.

An analytic mapping is conformal, i.e., it preserves angles, at all points where the derivative is nonzero. Using this property and connectivity considerations one can often determine the mapped image of a domain from the image of its boundary.

One important category of conformal mappings is the bilinear transformations. They are functions of the form $(az + b)/(cz + d)$, with $ad \neq bc$, and they can be expressed as compositions of translations, rotations, magnifications, and inversions. Because they preserve the class of straight lines and circles and their associated symmetries, they are usually the method of choice for solving problems in domains bounded by such figures.

On the other hand, Schwarz-Christoffel transformations map half-planes to polygons, and thus they are the appropriate tool for such geometries. These mappings are computed by determining the conditions imposed on the derivative f' at the corners of the polygon.

With these devices one can solve many two-dimensional problems in electrostatics, heat flow, and fluid dynamics. Sometimes insight into the nature of more complicated situations is provided by experimenting with inverse methods, where one first postulates solutions and then analyzes what problems they solve.

SUGGESTED READING

Most of the references listed previously treat conformal mapping, but the following texts are particularly useful:

Riemann Mapping Theorem and Geometric Considerations

[1]   GOLUZIN, G. M. *Geometric Theory of Functions of a Complex Variable.* American Mathematical Society, 1969.

[2]   NEHARI, Z. *Conformal Mapping.* McGraw-Hill Book Company, New York, 1966.

Schwarz-Christoffel Transformation

Reference 2 above.

[3]   LEVINSON, N., and REDHEFFER, R. *Complex Variables.* Holden-Day, Inc., San Francisco, 1970.

Compendium of Conformal Maps

[4]   CHURCHILL, R. V. *Complex Variables and Applications*, 2nd ed. McGraw-Hill Book Company, New York, 1960.

[5]   KOBER, H. *Dictionary of Conformal Representations*, 2nd ed. Dover Publications, Inc., New York, 1957.

[6] COURANT, R. *Dirichlet's Principle, Conformal Mapping, and Minimal Surfaces.* John Wiley & Sons, Inc. (Interscience Division), New York, 1950.

[7] DETTMAN, J. W. *Applied Complex Variables.* The Macmillan Company, New York, 1965.

[8] ENGLAND, A. H. *Complex Variable Methods in Elasticity.* John Wiley & Sons, Inc. (Interscience Division), New York, 1971.

[9] KYRALA, E. *Applied Functions of a Complex Variable.* John Wiley & Sons, Inc., New York, 1972.

[10] MARSDEN, J. E. *Basic Complex Analysis.* W. H. Freeman and Company, Publishers, San Francisco, 1973.

[11] MILNE-THOMPSON, L. M. *Theoretical Hydrodynamics.* The Macmillan Company, New York, 1960.

[12] ROTHE, R., OLLENDERF, F., and POHLHAUSEN, K. *Theory of Functions.* Dover Publications, Inc., New York, 1961.

*8

Fourier Series
and
Integral Transforms

In Sec. 3.4 we gave some indication of why, when analyzing linear time-invariant systems, it is particularly advantageous to deal with sinusoidal functions as inputs. Briefly, the virtues of employing an input of the form $Ae^{i\omega t}$ are

(*i*) Compactness of notation—a real expression such as $\alpha \cdot \cos(\omega t + \phi) + \beta \sin(\omega t + \psi)$ can be represented simply by $\text{Re}(Ae^{i\omega t})$.

(*ii*) The fact that differentiation amounts to multiplication by $i\omega$—thus, in a sense, replacing calculus by algebra.

(*iii*) The fact that the steady-state response of the system to this input will have the same form, a complex constant times $e^{i\omega t}$.

For these reasons it would be very helpful if a general input function $F(t)$ could be expressed as a sum of these sinusoids. One could then determine the output by finding the response to each sinusoidal component (which is an easier problem) and then adding these responses together (recall that superposition of solutions is permissible in a linear system). The *Fourier* theory is devoted to this problem.

8.1 FOURIER SERIES

As indicated in the introduction, the main goal of this chapter is to establish the possibility of expressing a (possibly complex-valued) function of a real variable, $F(t)$, as a sum of sinusoidal functions of the form $e^{i\omega t}$. The present section is devoted to the special case when $F(t)$ is periodic with real period L; i.e., $F(t) = F(t + L)$ for all t.

Naturally we are inclined to seek a decomposition of F into sinusoids with the same period; i.e., only those values of ω should occur such that $e^{i\omega(t+L)} = e^{i\omega t}$. This implies that $e^{i\omega L} = 1$, so that ω must be one of the numbers

$$\omega_n = \frac{2\pi n}{L} \qquad (n = 0,\ \pm 1,\ \pm 2,\ \ldots).$$

To be specific, we assume that $L = 2\pi$ (one can always rescale to achieve this condition). Our problem is thus to find (complex) numbers c_n such that

$$(1) \qquad\qquad F(t) = \sum_{n=-\infty}^{\infty} c_n e^{int}.$$

Suppose, for the moment, that the series in Eq. (1) converges *uniformly* to $F(t)$ for $-\pi \le t \le \pi$ (and hence for all t). For any fixed integer m we can multiply by e^{-imt} to obtain

$$(2) \qquad\qquad F(t)e^{-imt} = \sum_{n=-\infty}^{\infty} c_n e^{i(n-m)t},$$

again converging uniformly, from which it follows that $F(t)e^{-imt}$ is a continuous function and that termwise integration of the series is valid (recall Theorem 8 of Chapter 5). Integrating Eq. (2) over the interval $[-\pi, \pi]$ yields

$$(3) \qquad \int_{-\pi}^{\pi} F(t)e^{-imt}\, dt = \sum_{n=-\infty}^{\infty} c_n \int_{-\pi}^{\pi} e^{i(n-m)t}\, dt;$$

however,

$$\int_{-\pi}^{\pi} e^{i(n-m)t}\, dt = \begin{cases} \dfrac{e^{i(n-m)t}}{i(n-m)}\bigg|_{-\pi}^{\pi} = 0 & \text{if } n \ne m, \\[4mm] t\,\bigg|_{-\pi}^{\pi} = 2\pi & \text{if } n = m. \end{cases}$$

Hence only the term $2\pi c_m$ survives on the right-hand side of Eq. (3). As a result we have the following formula for the coefficient c_m:

$$(4) \qquad c_m = \frac{1}{2\pi} \int_{-\pi}^{\pi} F(t)e^{-imt}\, dt,$$

valid whenever the series in Eq. (1) is uniformly convergent.

Whether or not the series is convergent, we use the following terminology:

DEFINITION 1 *If $F(t)$ has period 2π and is integrable over $[-\pi, \pi]$, the (formal) series $\sum_{n=-\infty}^{\infty} c_n e^{int}$ with coefficients given by Eq. (4) is called the* **Fourier series** *for F; the numbers c_n are called the* **Fourier coefficients** *of F.*

[More generally, if $F(t)$ has period L the Fourier series looks like $\sum_{n=-\infty}^{\infty} c_n e^{in2\pi t/L}$, and the Fourier coefficients become

$$c_n = \frac{1}{L}\int_{-L/2}^{L/2} F(t)e^{-in2\pi t/L}\, dt.]$$

What we have shown is that under the assumption that $F(t)$ has a representation of the form (1) which is known to be uniformly convergent, then the series in question must be the Fourier series. Now we must investigate this assumption and try to determine *under what conditions the Fourier series will converge to F.*

A partial answer to this question can be derived from analytic function theory. Consider a function $f(z)$ analytic in some annulus, such as D in Fig. 8.1, which contains the unit circle. Then, of course, f can be represented by a Laurent series:

$$(5) \qquad f(z) = \sum_{n=-\infty}^{\infty} a_n z^n \qquad (z \text{ in } D).$$

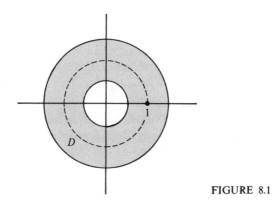

FIGURE 8.1

We shall be particularly concerned with the values of f on the unit circle where the series converges uniformly. Parametrizing this circle by $z = e^{it}$, $-\pi \leq t \leq \pi$, we introduce the notation $F(t) = f(e^{it})$ and rewrite Eq. (5) as

$$(6) \qquad\qquad F(t) = \sum_{n=-\infty}^{\infty} a_n e^{int}.$$

Observe our good fortune; the function $F(t)$ has period 2π, and Eq. (6) is a decomposition of F into a series of sinusoids, converging uniformly! Thus Eq. (6) must be the Fourier series for $F(t)$.

In fact, we can even present an independent derivation of formula (4) for the Fourier coefficients in this case. According to Theorem 14 of Chapter 5 the coefficients in Eq. (5) are given by

$$a_n = \frac{1}{2\pi i} \oint_{|z|=1} \frac{f(\xi)}{\xi^{n+1}} \, d\xi,$$

and inserting the parametrization we find

$$a_n = \frac{1}{2\pi i} \int_{-\pi}^{\pi} f(e^{it}) e^{-it(n+1)} i e^{it} \, dt$$

$$= \frac{1}{2\pi} \int_{-\pi}^{\pi} F(t) e^{-int} \, dt,$$

in agreement with Eq. (4).

Of course, we have not proved a great deal; we have only shown that *the Fourier series of a function F converges uniformly to F in those cases when the values of F(t) coincide with the values of an analytic function f(z) for z = e^{it}$. Furthermore, this technique of finding a Fourier series by way of a Laurent expansion is usually of more theoretical than practical value.

EXAMPLE 1   Find the Fourier series for the periodic function

$$F(t) = e^{2 \cos t}$$

using the above technique.

SOLUTION   First we must find an analytic function $f(z)$ which matches the values of $F(t)$ for z on the unit circle. This is easy; since

$$\cos t = \frac{e^{it} + e^{-it}}{2},$$

we see that

$$F(t) = e^{(z + 1/z)} = f(z)$$

when $z = e^{it}$. Hence the Fourier series for F can be obtained from the Laurent series for f. We have

$$e^{(z + 1/z)} = e^z e^{1/z}$$

$$= \left(\sum_{m=0}^{\infty} \frac{z^m}{m!} \right) \left(\sum_{l=0}^{\infty} \frac{z^{-l}}{l!} \right),$$

and we can multiply these series termwise. (Termwise multiplication of Laurent series is valid, although we have not proved it in this text.) The term involving z^n in the result comes from the sum of products of terms $z^m/m!$ times $z^{-l}/l!$ with $m - l = n$; collecting these we have

$$e^{(z + 1/z)} = \sum_{n=-\infty}^{\infty} z^n \left(\sum_{m=n}^{\infty} \frac{1}{m!} \cdot \frac{1}{(m-n)!} \right).$$

Hence the Fourier series for $F(t)$ is

$$F(t) = \sum_{n=-\infty}^{\infty} c_n e^{int}$$

with

$$c_n = \sum_{m=n}^{\infty} \frac{1}{m!\,(m-n)!}. \qquad \blacksquare$$

EXAMPLE 2   Show that when $F(t) = f(e^{it})$ with f analytic, termwise differentiation of the Fourier series for F is valid.

SOLUTION   We know that the Laurent series (5) can be differentiated termwise:

(7) $$\frac{df(z)}{dz} = \sum_{n=-\infty}^{\infty} n a_n z^{n-1}.$$

For $z = e^{it}$, the chain rule yields

$$\frac{df}{dt} = \frac{df}{dz} \frac{dz}{dt} = \frac{df}{dz} i e^{it}.$$

Inserting Eq. (7) for df/dz and identifying $f(e^{it})$ as $F(t)$, we find

$$\frac{d}{dt} f(e^{it}) = \frac{dF(t)}{dt} = \sum_{n=-\infty}^{\infty} n a_n e^{i(n-1)t} i e^{it},$$

or

$$\frac{dF(t)}{dt} = \sum_{n=-\infty}^{\infty} i n a_n e^{int},$$

which agrees with termwise differentiation of Eq. (6). $\blacksquare$

As another illustration of the fertility of this approach, we shall present a heuristic derivation of *Poisson's formula* for harmonic functions on the unit disk. Since the validity of the formula has been stated in (optional) Sec. 4.7, we shall proceed formally and not worry about the rigorous justification of each detail.

We are given a continuous real-valued function $U(\theta)$ having period 2π, and we want to find a function $u(z)$ which is harmonic for $|z| < 1$ and which approaches the value $U(\theta)$ as $z \to e^{i\theta}$; in other words, we want to solve the Dirichlet problem for the unit disk (see Sec. 4.7). First we assume that $U(\theta)$ has a Fourier expansion

$$U(\theta) = \sum_{n=-\infty}^{\infty} c_n e^{in\theta} = \sum_{n=-\infty}^{\infty} \left[\frac{1}{2\pi} \int_{-\pi}^{\pi} U(\phi)e^{-in\phi} \, d\phi \right] e^{in\theta},$$

where we have inserted the coefficient formula (4). If we combine the terms for n and $-n$, we derive (observe that $n = 0$ is exceptional)

$$U(\theta) = \frac{1}{2\pi} \int_{-\pi}^{\pi} U(\phi) \, d\phi + \sum_{n=1}^{\infty} \frac{1}{2\pi} \int_{-\pi}^{\pi} U(\phi)(e^{in(\theta-\phi)} + e^{-in(\theta-\phi)}) \, d\phi$$

$$= \frac{1}{2\pi} \int_{-\pi}^{\pi} U(\phi) \, d\phi + 2 \sum_{n=1}^{\infty} \frac{1}{2\pi} \int_{-\pi}^{\pi} U(\phi)\cos n(\theta - \phi) \, d\phi.$$

Now we use a device known to mathematicians as "Abel-Poisson summation" to sum the series. First we artificially introduce the variable r to obtain a function $g(r, \theta)$:

$$(8) \quad g(r, \theta) \equiv \frac{1}{2\pi} \int_{-\pi}^{\pi} U(\phi) \, d\phi + \frac{2}{2\pi} \sum_{n=1}^{\infty} \int_{-\pi}^{\pi} U(\phi)r^n \cos n(\theta - \phi) \, d\phi.$$

This yields three dividends; first, observe that the series

$$(9) \qquad\qquad 1 + 2 \sum_{n=1}^{\infty} r^n \cos n(\theta - \phi)$$

converges uniformly in ϕ, if $0 \le r < 1$. Hence it can be multiplied by $U(\phi)$ and integrated termwise. But this results in 2π times the right-hand side of Eq. (8). Thus we can rewrite Eq. (8) as

$$(10) \qquad g(r, \theta) = \frac{1}{2\pi} \int_{-\pi}^{\pi} U(\phi) \left\{ 1 + 2 \sum_{n=1}^{\infty} r^n \cos n(\theta - \phi) \right\} d\phi.$$

Second, observe that the series (9) is, in fact, the real part of the series

$$(11) \qquad\qquad 1 + 2 \sum_{n=1}^{\infty} r^n e^{in\theta}e^{-in\phi} = 1 + 2 \sum_{n=1}^{\infty} z^n e^{-in\phi},$$

a power series in $z = re^{i\theta}$. Since the latter series converges for $|z| < 1$, it defines an analytic function inside the unit disk, and consequently its real part, (9), is harmonic! As a result, $g(r, \theta)$ is the real part of an analytic function [since $U(\phi)$ is real], and hence g is a harmonic function of $z = re^{i\theta}$ for $r < 1$. The formal substitution $r = 1$ in Eq. (10) yields the Fourier series for $U(\theta)$, so we are led to postulate that Eq. (10) solves the Dirichlet problem; i.e., $u(z) = u(re^{i\theta}) = g(r, \theta)$ is a function which is harmonic for $|z| < 1$ and approaches $U(\theta)$ as $|z| \to 1$.

Finally, the third dividend of our labors follows from the equality

$$(12) \qquad 1 + 2 \sum_{n=1}^{\infty} r^n \cos n(\theta - \phi) = \frac{1 - r^2}{1 - 2r \cos(\theta - \phi) + r^2},$$

which we invite the reader to prove as Prob. 2. Using this in Eq. (10), we arrive at the *Poisson formula*

$$u(re^{i\theta}) = \frac{1 - r^2}{2\pi} \int_{-\pi}^{\pi} \frac{U(\phi)}{1 - 2r \cos(\theta - \phi) + r^2} \, d\phi,$$

expressing a harmonic function inside the unit disk in terms of its "boundary values." As we indicated earlier, we refer the reader to Sec. 4.7 for a more precise statement of the validity of Poisson's formula.

At this point our achievements can be summarized as follows: Subject to some fairly restrictive analyticity assumptions, the equation

$$(13) \qquad F(t) = \sum_{n=-\infty}^{\infty} c_n e^{int}$$

is valid when

$$(14) \qquad c_n = \frac{1}{2\pi} \int_{-\pi}^{\pi} F(t) e^{-int} \, dt \qquad \text{(for all } n\text{)}.$$

Now notice that there is nothing in Eqs. (13) or (14) which would indicate the necessity of any analytic properties of F. Indeed, the coefficients (14) can be evaluated for any integrable F. So, we speculate, why should the validity of Eq. (13) hinge on analyticity? Shouldn't we expect that the Fourier series converges under weaker conditions? The answer is yes, but the proofs of the more general convergence theorems lie outside analytic function theory. We shall simply quote some of these results without proof.

The first theorem is more or less in line with our speculations. It postulates only the integrability of $|F|^2$, but it pays the price in that a much weaker type of convergence occurs.

THEOREM 1 *If the integral $\int_{-\pi}^{\pi} |F(t)|^2 \, dt$ exists, then the Fourier series defined by Eqs. (13) and (14) exists and converges to F in the mean square sense; i.e.,*

$$\lim_{N \to \infty} \int_{-\pi}^{\pi} \left| F(t) - \sum_{n=-N}^{N} c_n e^{int} \right|^2 dt = 0.$$

EXAMPLE 3   Prove *Parseval's identity* for the Fourier coefficients:

(15)
$$\int_{-\pi}^{\pi} |F(t)|^2 \, dt = \lim_{N \to \infty} 2\pi \sum_{n=-N}^{N} |c_n|^2,$$

if $|F|^2$ is integrable over $[-\pi, \pi]$.

SOLUTION   We have

$$\int_{-\pi}^{\pi} \left| F(t) - \sum_{n=-N}^{N} c_n e^{int} \right|^2 dt$$

$$= \int_{-\pi}^{\pi} \left[F(t) - \sum_{n=-N}^{N} c_n e^{int} \right]\left[\overline{F(t)} - \sum_{n=-N}^{N} \bar{c}_n e^{-int} \right] dt.$$

Since the conjugate of $F(t) \cdot \sum_{n=-N}^{N} \bar{c}_n e^{-int}$ is $\overline{F(t)} \cdot \sum_{n=-N}^{N} c_n e^{int}$, the right-hand side becomes

(16)
$$\int_{-\pi}^{\pi} |F(t)|^2 \, dt - 2 \operatorname{Re} \sum_{n=-N}^{N} \bar{c}_n \int_{-\pi}^{\pi} F(t) e^{-int} \, dt$$

$$+ \int_{-\pi}^{\pi} \left(\sum_{n=-N}^{N} c_n e^{int} \right)\left(\sum_{n=-N}^{N} \bar{c}_n e^{-int} \right) dt.$$

Recognizing the expression for the Fourier coefficient [Eq. (14)] in the above, we can write the second term as $-2(2\pi) \sum_{n=-N}^{N} |c_n|^2$. The third term can be expanded, but we must change one of the summation indices to avoid confusion; this term then becomes

$$\sum_{n=-N}^{N} c_n \sum_{m=-N}^{N} \bar{c}_m \int_{-\pi}^{\pi} e^{i(n-m)t} \, dt.$$

Recalling our previous evaluation of this integral, we see that this reduces to $2\pi \sum_{n=-N}^{N} c_n \bar{c}_n$. Thus we have shown

$$\int_{-\pi}^{\pi} \left| F(t) - \sum_{n=-N}^{N} c_n e^{int} \right|^2 dt = \int_{-\pi}^{\pi} |F(t)|^2 \, dt - 2\pi \sum_{n=-N}^{N} |c_n|^2.$$

According to Theorem 1, the left-hand side approaches zero as $N \to \infty$. Hence

$$\int_{-\pi}^{\pi} |F(t)|^2 \, dt - \lim_{N \to \infty} 2\pi \sum_{n=-N}^{N} |c_n|^2 = 0,$$

and Eq. (15) results.   ∎

The next Fourier convergence theorem is valuable in electrical engineering applications, where switching circuits may produce (theoretically) discontinuous input functions, such as the periodic step function illustrated in Fig. 8.2.

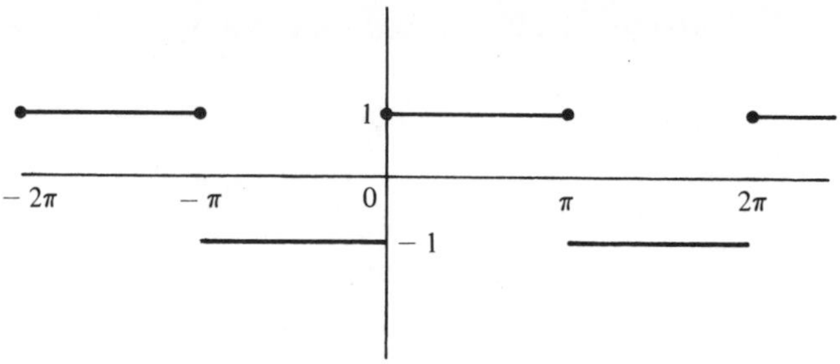

FIGURE 8.2

We restrict ourselves to periodic functions $F(t)$ with a finite number of discontinuities in any period. Specifically, we assume that $F(t)$ has period 2π and that there is a finite subdivision of the interval $[-\pi, \pi]$ given by

$$-\pi = \tau_0 < \tau_1 < \tau_2 < \cdots < \tau_{n-1} < \tau_n = \pi$$

such that

(i) $F(t)$ is continuously differentiable on each open subinterval (τ_j, τ_{j+1}) for $j = 0, 1, 2, \ldots, n - 1$,

(ii) As t approaches any subdivision point τ_j from the left, $F(t)$ and $F'(t)$ approach limiting values denoted by $F(\tau_j -)$ and $F'(\tau_j -)$, respectively, and

(iii) As t approaches any τ_j from the right, $F(t)$ and $F'(t)$ again approach limiting values denoted $F(\tau_j +)$ and $F'(\tau_j +)$, respectively.

Such a function is said to be *piecewise smooth*.

Of course, if F is continuous at τ_j, then $F(\tau_j -) = F(\tau_j +) = F(\tau_j)$. For the step function in Fig. 8.2, $F(0-) = F(\pi+) = -1$, and $F(0+) = F(\pi-) = +1 = F(0) = F(\pi)$. Moreover, $F'(0-) = F'(0+) = 0$, but $F'(0)$ does not exist.

THEOREM 2 Suppose that $F(t)$ is periodic and piecewise smooth. Then the Fourier series for F converges to $F(t)$ at all points t where F is continuous, and converges to $\frac{1}{2}[F(\tau_j +) + F(\tau_j -)]$ at the subdivision points.

EXAMPLE 4   Compute the Fourier series for the step function in Fig. 8.2, and state its convergence properties.

SOLUTION   The Fourier coefficients are given by

$$c_n = \frac{1}{2\pi} \int_{-\pi}^{\pi} F(t)e^{-int} \, dt = \frac{1}{2\pi} \int_{-\pi}^{0} (-1)e^{-int} \, dt + \frac{1}{2\pi} \int_{0}^{\pi} (1)e^{-int} \, dt$$

$$= \begin{cases} 0 & \text{if } n = 0, \\ \dfrac{i\{(-1)^n - 1\}}{\pi n} & \text{otherwise.} \end{cases}$$

Hence the Fourier series is

(17) $$\frac{i}{\pi} \sum_{\substack{n=-\infty \\ n \neq 0}}^{\infty} \left[\frac{(-1)^n - 1}{n} \right] e^{int}.$$

According to Theorem 2, it converges to $+1$ for $0 < t < \pi$, to -1 for $-\pi < t < 0$, and to the average, 0, for $t = 0$ and $t = \pi$. ∎

When Fourier analysis (or *frequency analysis*, as it is sometimes called) is used to solve linear systems governed by differential equations, the question naturally arises as to whether or not a Fourier series can legitimately be differentiated termwise. (Obviously, the result of Example 2 is much too restrictive.) The following argument seems to cover a great many cases of interest to engineers: Suppose that $F(t)$ has a convergent Fourier series expansion

(18) $$F(t) = \sum_{n=-\infty}^{\infty} c_n e^{int},$$

and suppose furthermore that the termwise-differentiated series

(19) $$\sum_{n=-\infty}^{\infty} inc_n e^{int}$$

can be shown (by, say, the M-test) to be uniformly convergent on $[-\pi, \pi]$. Under such circumstances we know the "derived series" (19) can be legitimately *integrated* from $-\pi$ to t, termwise. But the result of this integration is the original series (18), up to a constant. Hence the sum function of (19) must be the *derivative* of $F(t)$, and we have proved

THEOREM 3 *Suppose that the Fourier expansion (18) is valid and that the derived series (19) converges uniformly on $[-\pi, \pi]$. Then*

$$\sum_{n=-\infty}^{\infty} inc_n e^{int} = \frac{d}{dt} \sum_{n=-\infty}^{\infty} c_n e^{int}.$$

EXAMPLE 5   Find the Fourier series for the periodic function

$$F(t) = \left| \sin \frac{t}{2} \right|^5,$$

and state the convergence properties for the derived series.

SOLUTION   [Observe that $F(t)$ has period 2π.] The Fourier coefficients are given by

$$c_n = \frac{1}{2\pi} \int_{-\pi}^{\pi} \left| \sin \frac{t}{2} \right|^5 e^{-int} dt.$$

These integrals can be evaluated by standard techniques after applying the identity

$$\sin^5 \theta = \tfrac{5}{8} \sin \theta - \tfrac{5}{16} \sin 3\theta + \tfrac{1}{16} \sin 5\theta.$$

With some labor one finds that the Fourier series for $F(t)$ is given by

$$(20) \qquad \sum_{n=-\infty}^{\infty} \frac{240/\pi}{225 - 1036n^2 + 560n^4 - 64n^6} e^{int},$$

and it converges to $F(t)$ according to Theorem 2. Differentiating termwise we derive

$$(21) \qquad \sum_{n=-\infty}^{\infty} \frac{in(240/\pi)}{225 - 1036n^2 + 560n^4 - 64n^6} e^{int}.$$

This series converges uniformly, as can be seen by comparing its increasing and decreasing parts with the (convergent) series $\sum_{n=1}^{\infty} 2 \cdot 240/\pi 64n^5$ [the factor 2 ensures that the terms of this series dominate those of (21) for large n]. Hence (21) represents $F'(t)$. Moreover, termwise differentiation of (21) can be justified by comparing the result with $\sum_{n=1}^{\infty} 2 \cdot 240/\pi 64n^4$; hence

$$F''(t) = \sum_{n=-\infty}^{\infty} \frac{-n^2(240/\pi)}{225 - 1036n^2 + 560n^4 - 64n^6} e^{int}.$$

Clearly two more termwise differentiations are justified, leading to Fourier series for $F^{(3)}(t)$ and $F^{(4)}(t)$. In fact, the student should verify that the original function $F(t)$ is continuously differentiable exactly four times! (The fifth derivative jumps from $-\tfrac{15}{4}$ to $+\tfrac{15}{4}$ as t increases through 0, $\pm 2\pi, \pm 4\pi$, etc.) ∎

The final example in this section illustrates how Fourier series are used in practice to solve linear problems.

EXAMPLE 6   Find a function $f(t)$ which satisfies the differential equation

$$(22) \qquad \frac{d^2 f(t)}{dt^2} + 2 \frac{df(t)}{dt} + 2f(t) = F(t),$$

where $F(t)$ is the periodic "sawtooth" function prescribed by

$$F(t) = \begin{cases} -1 - \dfrac{2t}{\pi}, & -\pi \le t \le 0, \\[2mm] -1 + \dfrac{2t}{\pi}, & 0 \le t \le \pi \end{cases}$$

(see Fig. 8.3).

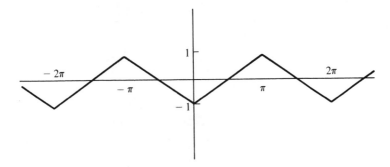

FIGURE 8.3

SOLUTION First we show how a solution can be found to the simpler equation

(23)
$$\frac{d^2g(t)}{dt^2} + 2\frac{dg(t)}{dt} + 2g(t) = e^{i\omega t},$$

where the "forcing function" on the right-hand side has been replaced by a simple sinusoid. The considerations outlined in the introduction to this chapter indicate that Eq. (23) has a solution of the form $g(t) = Ae^{i\omega t}$. To find A, we plug this expression into Eq. (23) and obtain

$$-\omega^2 Ae^{i\omega t} + 2i\omega Ae^{i\omega t} + 2Ae^{i\omega t} = e^{i\omega t},$$

or (dividing by $e^{i\omega t}$)

$$(-\omega^2 + 2i\omega + 2)A = 1.$$

Solving for A, we deduce that

(24)
$$g(t) = \frac{e^{i\omega t}}{-\omega^2 + 2i\omega + 2}$$

solves Eq. (23).

Next we expand the given function $F(t)$ into a Fourier series. Using formula (14) for the coefficients, we find

$$c_n = \frac{1}{2\pi} \int_{-\pi}^{\pi} F(t)e^{-int}\, dt$$

$$= \frac{1}{2\pi} \int_{-\pi}^{0} \left(-1 - \frac{2t}{\pi}\right)e^{-int}\, dt + \frac{1}{2\pi} \int_{0}^{\pi} \left(-1 + \frac{2t}{\pi}\right)e^{-int}\, dt$$

$$= \begin{cases} 0 & \text{if } n = 0, \\ \dfrac{2}{\pi^2 n^2}\{(-1)^n - 1\} & \text{if } n \neq 0. \end{cases}$$

Hence

(25)
$$F(t) = \sum_{\substack{n=-\infty \\ (n \neq 0)}}^{\infty} \frac{2\{(-1)^n - 1\}}{\pi^2 n^2} e^{int},$$

which is valid by Theorem 2. Now we argue as follows: We have Eq. (25) expressing $F(t)$ as a linear combination, albeit infinite, of sinusoids, and we have Eq. (24) expressing a solution for a single sinusoid. By linearity, then, we are led to postulate that the same linear combination of these solutions ought to solve the given equation; i.e.,

(26)
$$f(t) = \sum_{\substack{n=-\infty \\ (n \neq 0)}}^{\infty} \frac{2\{(-1)^n - 1\}}{\pi^2 n^2} \frac{e^{int}}{-n^2 + 2in + 2}$$

should be valid. To complete the argument, we first make the observation that Eq. (26) certainly solves Eq. (22) termwise. Furthermore, by comparing the series in Eq. (26) with the convergent series $\sum_{n=-\infty, n \neq 0}^{\infty} (8/\pi^2 n^4)$, we conclude that the former converges and that it can be legitimately differentiated termwise twice. Since Eq. (22) involves no derivatives higher than the second, we are done.

[The reader should observe that termwise differentiation of the saw-tooth series (25) yields $2/\pi$ times the step function series (17), which is consistent with the fact that the derivative of the sawtooth is $2/\pi$ times the step function, except at the "break points" 0, $\pm\pi$, $\pm 2\pi$, ... This phenomenon, which is not predicted by Theorem 3, reflects the fact that there are more powerful convergence results for Fourier series. Some of these can be found in the references.] ∎

In concluding this section, we note that a Fourier series is sometimes called a *trigonometric series*, because any series of the form

$$\sum_{n=-\infty}^{\infty} c_n e^{int}$$

can be written in the alternative form

$$\sum_{n=0}^{\infty} \alpha_n \cos nt + \sum_{n=1}^{\infty} \beta_n \sin nt.$$

Indeed, in Prob. 1 we ask the reader to verify that

$$\alpha_0 = c_0$$

and that for $n > 0$

$$\alpha_n = c_n + c_{-n},$$
$$\beta_n = i(c_n - c_{-n}),$$

or, equivalently,

$$c_n = \frac{\alpha_n - i\beta_n}{2},$$

$$c_{-n} = \frac{\alpha_n + i\beta_n}{2}.$$

Exercises 8.1

1. Rewrite the series

    $$\sum_{-\infty}^{\infty} c_n e^{int}$$

    as a trigonometric series of the form

    $$\sum_{n=0}^{\infty} \alpha_n \cos nt + \sum_{n=1}^{\infty} \beta_n \sin nt,$$

    deriving the relations

    $$\alpha_0 = c_0,$$

    $$\alpha_n = c_n + c_{-n} \qquad (n \geq 1),$$

    $$\beta_n = i(c_n - c_{-n}) \qquad (n \geq 1).$$

    What are the conditions on the coefficients c_n such that the sum of the series is a real function?

2. Prove Eq. (12). [HINT: Use Eq. (11).]

3. Verify the Fourier representations of the indicated functions and state the convergence properties on the interval $[-\pi, \pi]$.

    (a) $\displaystyle\sum_{\substack{n=-\infty \\ n \text{ even}}}^{\infty} \frac{-2}{\pi(n^2 - 1)} e^{int} = |\sin t|$

    (b) $\displaystyle\sum_{n=-\infty}^{\infty} \frac{(-1)^n \sinh \pi}{(1 - in)\pi} e^{int} = e^t$

    (c) $\displaystyle\frac{1}{\pi} + \frac{1}{2}\sin t - \frac{2}{\pi}\sum_{n=1}^{\infty} \frac{\cos 2nt}{4n^2 - 1} = \begin{cases} 0, & -\pi \leq t \leq 0 \\ \sin t, & 0 \leq t \leq \pi \end{cases}$

    (d) $\displaystyle\frac{\pi}{4} + \sum_{n=1}^{\infty} \frac{(-1)^n - 1}{\pi n^2}\cos nt - \sum_{n=1}^{\infty} \frac{(-1)^n}{n}\sin nt$

    $$= \begin{cases} 0, & -\pi < t \leq 0 \\ t, & 0 \leq t < \pi \end{cases}$$

4. Compute the Fourier series for the following functions.

(a) $F(t) = \sin^3 t$

(b) $F(t) = \left| \cos^3 \dfrac{t}{3} \right|$

(c) $F(t) = t^2 \qquad (-\pi < t < \pi)$

(d) $F(t) = t|t| \qquad (-\pi < t < \pi)$

5. Which of the series in Prob. 3 can be differentiated termwise?

6. (a) If $F(t)$ is defined only for $0 \le t \le \pi$, show that by defining $F(-t) \equiv -F(t), 0 < t \le \pi$, and constructing the Fourier series for this function over the interval $[-\pi, \pi]$, one arrives at a *Fourier sine series*

$$\sum_{n=1}^{\infty} \beta_n \sin nt$$

for F, with coefficients given by

$$\beta_n = \frac{2}{\pi} \int_0^{\pi} F(t) \sin nt \, dt.$$

State conditions for the Fourier sine series to converge to $F(t)$ for $0 \le t \le \pi$.

(b) As in part (a), show that the definition $F(-t) \equiv F(t), 0 < t \le \pi$, produces a *Fourier cosine series*

$$\sum_{n=0}^{\infty} \alpha_n \cos nt$$

with coefficients

$$\alpha_n = \frac{2}{\pi} \int_0^{\pi} F(t) \cos nt \, dt.$$

State the conditions for convergence on $[0, \pi]$.

7. Show that the Fourier sine and cosine series for the function $F(t) = t, 0 \le t \le \pi$, are given by

$$\sum_{n=1}^{\infty} \frac{2(-1)^{n+1}}{n} \sin nt$$

and

$$\frac{\pi}{2} - \frac{4}{\pi} \sum_{\substack{n=1 \\ n \text{ odd}}}^{\infty} \frac{\cos nt}{n^2},$$

respectively.

8. Find the Fourier representation for the periodic solutions of the following equations.

(a) $\dfrac{d^2f}{dt^2} + 3f = \sin^4 t$

(b) $\dfrac{d^2f}{dt^2} + \dfrac{df}{dt} + f = t^2$, $-\pi \le t \le \pi$, continued with period 2π

(c) $\dfrac{d^2f}{dt^2} + 4\dfrac{df}{dt} + 2f =$ (the step function in Fig. 8.2)

9. Suppose that we wish to solve the Dirichlet problem for the unit disk and that the boundary values of the desired harmonic function for $z = e^{i\theta}$ are represented by the series

$$U(\theta) = \sum_{n=0}^{\infty} \alpha_n \cos n\theta + \sum_{n=1}^{\infty} \beta_n \sin n\theta.$$

Argue that the solution to the problem is given by

$$u(re^{i\theta}) = \sum_{n=0}^{\infty} \alpha_n r^n \cos n\theta + \sum_{n=1}^{\infty} \beta_n r^n \sin n\theta.$$

10. As an illustration of the power of Fourier methods in solving partial differential equations, consider the *non*static problem of heat flow along a uniform rod of length π, whose ends are maintained at zero degrees temperature. The temperature T is now a function of position x along the rod $(0 \le x \le \pi)$ and time t. If the initial $(t = 0)$ temperature distribution is specified to be $f(x)$, the equations which T must satisfy are

$$\frac{\partial T(x, t)}{\partial t} = \frac{\partial^2 T(x, t)}{\partial x^2}$$

$$T(0, t) = T(\pi, t) = 0$$

$$T(x, 0) = f(x)$$

for $0 < x < \pi$, $t > 0$. Assuming the validity of termwise differentiation and Fourier expansions, show that

$$T(x, t) = \sum_{n=1}^{\infty} a_n \sin nx \, e^{-n^2 t}$$

solves the equations, where a_n is defined by

$$a_n = \frac{2}{\pi} \int_0^{\pi} f(\xi) \sin n\xi \, d\xi.$$

[HINT: You will need the Fourier sine series, Prob. 6.] What is the limiting value of $T(x, t)$ as $t \to \infty$? Interpret this.

11. Another illustration of the power of Fourier methods is provided by the vibrating string problem. A taut string fastened at $x = 0$ and $x = \pi$ is initially distorted into the shape $u = f(x)$, where u is the displacement of the string at the point x, and then the string is released. The equations governing the displacement $u(x, t)$ of the string are

$$\frac{\partial^2 u(x, t)}{\partial x^2} = \frac{\partial^2 u(x, t)}{\partial t^2}$$

$$u(0, t) = u(\pi, t) = 0$$

$$u(x, 0) = f(x)$$

$$\frac{\partial u(x, 0)}{\partial t} = 0$$

for $0 < x < \pi$, $t > 0$. Again assuming the validity of termwise differentiation and Fourier expansions, show that

$$u(x, t) = \sum_{n=1}^{\infty} b_n \sin nx \cos nt$$

solves the equations, where b_n is defined by

$$b_n = \frac{2}{\pi} \int_0^\pi f(\xi) \sin n\xi \, d\xi.$$

[HINT: Use the Fourier sine series again.] How would you modify this representation if the "initial conditions" were interchanged to read

$$u(x, 0) = 0$$

$$\frac{\partial u(x, 0)}{\partial t} = f(x)?$$

Combine these formulas to satisfy the more general set of initial conditions

$$u(x, 0) = f_1(x)$$

$$\frac{\partial u(x, 0)}{\partial t} = f_2(x).$$

8.2 THE FOURIER TRANSFORM

We move on to the next stage in our program of decomposing arbitrary functions into sinusoids. We have seen how a periodic function can be expressed as a Fourier series, so now we seek a similar representation for nonperiodic functions.

To begin with, let's assume we are given a nonperiodic function $F(t)$, $-\infty < t < \infty$, which is, say, continuously differentiable. Then if we pick an interval of the form $(-L/2, L/2)$ we can represent $F(t)$ by a Fourier series *for t in this interval*:

$$(27) \qquad F(t) = \sum_{n=-\infty}^{\infty} c_n e^{in2\pi t/L}, \qquad \frac{-L}{2} < t < \frac{L}{2},$$

with coefficients given by

$$(28) \qquad c_n = \frac{1}{L} \int_{-L/2}^{L/2} F(t) e^{-in2\pi t/L}\, dt \qquad (n = 0, \pm 1, \pm 2, \ldots).$$

Actually the series in Eq. (27) defines a *periodic* function $F_L(t)$, $-\infty < t < \infty$, which coincides with $F(t)$ on $(-L/2, L/2)$; see Fig. 8.4. [Notice that $F_L(t)$ may be discontinuous even though $F(t)$ is smooth.]

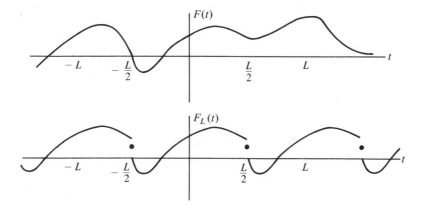

FIGURE 8.4

Thus we have a sinusoidal representation of $F(t)$ over an interval of length L. If we now let $L \to \infty$, it seems reasonable to conjecture that this might evolve into a sinusoidal representation of $F(t)$ valid for *all t*. Let's explore this possibility.

We are going to rewrite these equations in what will seem at first like a rather bizarre form, but it will aid in interpreting them as $L \to \infty$. We define g_n to be $c_n L/2\pi$, and introduce the factor $[(n+1) - n] \equiv 1$ into the series in Eq. (27). Then we have

$$(29) \qquad F_L(t) = \sum_{n=-\infty}^{\infty} g_n e^{in2\pi t/L} \frac{[(n+1) - n]2\pi}{L}$$

and

$$(30) \qquad g_n = \frac{1}{2\pi} \int_{-L/2}^{L/2} F(t) e^{-in2\pi t/L}\, dt.$$

Now write $\omega_n = n2\pi/L$, producing

(31)
$$F_L(t) = \sum_{n=-\infty}^{\infty} G_L(\omega_n) e^{i\omega_n t} (\omega_{n+1} - \omega_n),$$

where the function $G_L(\omega)$ can be defined for *any* real ω by

(32)
$$G_L(\omega) = \frac{1}{2\pi} \int_{-L/2}^{L/2} F(t) e^{-i\omega t} \, dt.$$

As L goes to infinity, $G_L(\omega)$ evolves rather naturally into a function $G(\omega)$ which is known as the *Fourier transform* of F:

(33)
$$G(\omega) = \frac{1}{2\pi} \int_{-\infty}^{\infty} F(t) e^{-i\omega t} \, dt.$$

Moreover, since $\Delta\omega_n \equiv \omega_{n+1} - \omega_n$ goes to zero as $L \to \infty$ and since ω_n ranges from $-\infty$ to $+\infty$, Eq. (31) begins to look very much like a Riemann sum for the integral

$$\int_{-\infty}^{\infty} G(\omega) e^{i\omega t} \, d\omega.$$

Thus we are led to propose the equality

(34)
$$F(t) = \int_{-\infty}^{\infty} G(\omega) e^{i\omega t} \, d\omega$$

for nonperiodic F, when G is defined by Eq. (33). Equation (34) is called the *Fourier inversion formula*.

Equations (33) and (34) are the essence of Fourier transform theory. As is suggested by this discussion, it is often profitable to indulge one's whimsy and think of the integral in Eq. (34) as a generalized "sum" of sinusoids, summed over a *continuum* of frequencies ω. Equation (33) then gives the "coefficients," $G(\omega)$, in the sum.

EXAMPLE 7   Find the Fourier transform and verify the inversion formula for the function

$$F(t) = \frac{1}{t^2 + 4}.$$

SOLUTION   Observe that

$$F(t) = \frac{1}{t^2 + 4} = \frac{1}{(t - 2i)(t + 2i)}$$

is analytic except for simple poles at $t = \pm 2i$. We shall use residue theory to evaluate the Fourier transform, interpreting the integral as a principal value:

$$G(\omega) = \frac{1}{2\pi} \text{p.v.} \int_{-\infty}^{\infty} \frac{e^{-i\omega t}}{t^2 + 4} \, dt.$$

If $\omega \geq 0$, we close the contour with expanding semicircles in the lower half-plane; by the techniques of Chapter 6 we find

$$G(\omega) = \frac{1}{2\pi}(-2\pi i)\text{Res}\left(\frac{e^{-i\omega t}}{t^2 + 4}; -2i\right)$$

$$= -i \cdot \lim_{t \to -2i} \frac{e^{-i\omega t}}{t - 2i} = \frac{e^{-2\omega}}{4} \qquad (\omega \geq 0).$$

Similarly, for $\omega < 0$ we close in the upper half-plane and find

$$G(\omega) = \frac{1}{2\pi}(2\pi i)\text{Res}\left(\frac{e^{-i\omega t}}{t^2 + 4}; 2i\right) = \frac{e^{2\omega}}{4} \qquad (\omega < 0).$$

In short,

$$G(\omega) = \frac{e^{-2|\omega|}}{4}.$$

To verify the Fourier inversion formula we compute

$$\int_{-\infty}^{\infty} G(\omega)e^{i\omega t}\,d\omega = \int_{-\infty}^{\infty} \frac{e^{-2|\omega|}}{4} \cdot e^{i\omega t}\,d\omega.$$

By symmetry, the imaginary part vanishes, and this integral equals

$$\text{Re}\int_{-\infty}^{\infty} \frac{e^{-2|\omega|}}{4} e^{i\omega t}\,d\omega = 2 \cdot \text{Re}\int_{0}^{\infty} \frac{e^{-2\omega}}{4} e^{i\omega t}\,d\omega = \frac{1}{2}\,\text{Re}\,\frac{e^{(-2+it)\omega}}{-2 + it}\bigg|_{\omega = 0}^{\infty}$$

$$= \frac{1}{t^2 + 4}.$$

Hence

(35) $$\frac{1}{t^2 + 4} = \int_{-\infty}^{\infty} \frac{e^{-2|\omega|}}{4} \cdot e^{i\omega t}\,d\omega. \qquad \blacksquare$$

As in the case of Fourier series, a wealth of theorems have been discovered stating conditions under which the Fourier integral representations (33) and (34) are valid. A very useful one for applications deals with piecewise smooth functions $F(t)$ like those in Theorem 2 of the previous section; that is, on every bounded interval $F(t)$ is continuously differentiable for all but the finite number of values $t = \tau_1, \tau_2, \ldots, \tau_n$, and at each τ_j the "one-sided limits" of $F(t)$ and $F'(t)$ exist.

THEOREM 4 *Suppose that $F(t)$ is piecewise smooth on every bounded interval and that $\int_{-\infty}^{\infty} |F(t)|\,dt$ exists. Then the Fourier transform, $G(\omega)$, of F exists and*

$$\int_{-\infty}^{\infty} G(\omega)e^{i\omega t}\,d\omega = \begin{cases} F(t) & \text{where } F \text{ is continuous,} \\[2mm] \dfrac{F(t+) + F(t-)}{2} & \text{otherwise.} \end{cases}$$

EXAMPLE 8   Find the Fourier transform of the function

$$F(t) = \begin{cases} 1, & -\pi \le t \le \pi, \\ 0, & \text{otherwise} \end{cases}$$

(Fig. 8.5), and express the inversion formula.

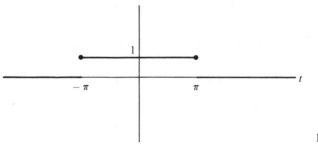

FIGURE 8.5

SOLUTION   We have

$$G(\omega) = \frac{1}{2\pi} \int_{-\pi}^{\pi} (1)e^{-i\omega t}\, dt = \frac{\sin \omega\pi}{\omega\pi}.$$

Hence Theorem 4 tells us that

(36) $$\int_{-\infty}^{\infty} \frac{\sin \omega\pi}{\omega\pi} e^{i\omega t}\, d\omega = \begin{cases} 1, & |t| < \pi, \\ 0, & |t| > \pi, \\ \dfrac{1}{2}, & t = \pm\pi. \end{cases}$$ ∎

EXAMPLE 9   Find the Fourier transform of the function

$$F(t) = \begin{cases} \sin t, & |t| \le 6\pi, \\ 0, & \text{otherwise} \end{cases}$$

(Fig. 8.6), and express the inversion formula. (Physicists call this function a "finite wave train.")

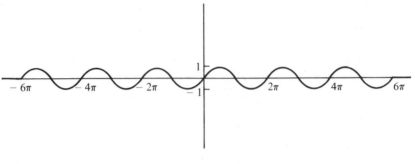

FIGURE 8.6

SOLUTION   We have

$$G(\omega) = \frac{1}{2\pi} \int_{-6\pi}^{6\pi} (\sin t) e^{-i\omega t} \, dt$$

$$= \frac{i \sin 6\pi\omega}{\pi(1 - \omega^2)}.$$

Since $F(t)$ is continuous everywhere,

(37) $$\int_{-\infty}^{\infty} \frac{i \sin 6\pi\omega}{\pi(1 - \omega^2)} e^{i\omega t} \, d\omega = F(t). \quad \blacksquare$$

The Fourier transform equations are used just like Fourier series in solving linear systems; as an illustration, consider

EXAMPLE 10   Find a function which satisfies the differential equation

(38) $$\frac{d^2 f(t)}{dt^2} + 2 \frac{df(t)}{dt} + 2 f(t) = \begin{cases} \sin t, & |t| \le 6\pi, \\ 0, & \text{otherwise.} \end{cases}$$

SOLUTION   In Example 6 we learned that a solution to $f'' + 2f' + 2f = e^{i\omega t}$ is

$$\frac{e^{i\omega t}}{-\omega^2 + 2i\omega + 2}.$$

Now the right-hand side of Eq. (38) is the function $F(t)$ in the previous example, and Eq. (37) can be interpreted as expressing F as a "superposition" of sinusoids of the form $e^{i\omega t}$. Hence we propose that the corresponding superposition of solutions

(39) $$f(t) = \int_{-\infty}^{\infty} \frac{i \sin 6\pi\omega}{\pi(1 - \omega^2)} \left(\frac{e^{i\omega t}}{-\omega^2 + 2i\omega + 2} \right) d\omega$$

solves the given equation. As before, we should establish that this expression converges and can be differentiated twice under the integral sign, but, instead, we invite the student to verify that residue theory yields the expression (see Prob. 3)

$$f(t) = \begin{cases} 0, & t \le -6\pi, \\ \frac{2}{5}(e^{-6\pi-t} - 1)\cos t + \frac{1}{5}(e^{-6\pi-t} + 1)\sin t, & -6\pi \le t \le 6\pi, \\ -(e^{6\pi} - e^{-6\pi})e^{-t}(\frac{2}{5}\cos t + \frac{1}{5}\sin t), & t > 6\pi, \end{cases}$$

which, as direct computation shows, solves the differential equation (38). $\blacksquare$

As an amusing exercise in the manipulation of contour integrals we now present an informal derivation of an identity which bears close resemblence to a Fourier expansion theorem, if we are lenient in interpreting relations (33) and (34) between the function and its transform.

EXAMPLE 11 Suppose that the function $F(t)$ is analytic and bounded by a constant M in an open strip $|\operatorname{Im} t| < \delta$, and define $G_L(\omega)$ as in Eq. (32). Argue that, as $L \to \infty$,

$$(40) \qquad \text{p.v.} \int_{-\infty}^{\infty} G_L(\omega) e^{i\omega t}\, d\omega \to F(t)$$

for each real t.

SOLUTION Notice, first of all, that if $F(t)$ has a Fourier transform, it will be given by the limit of $G_L(\omega)$ as $L \to \infty$. Hence (40) looks very much like a Fourier inversion formula. In fact, we shall argue that the members of (40) are equal whenever $L > 2|t|$.

For this purpose we write

$$(41) \qquad \text{p.v.} \int_{-\infty}^{\infty} G_L(\omega) e^{i\omega t}\, d\omega$$

$$= \lim_{r \to \infty} \frac{1}{2\pi} \int_{-r}^{r} \left[\int_{-L/2}^{L/2} F(\tau) e^{-i\omega\tau}\, d\tau \right] e^{i\omega t}\, d\omega \equiv \lim_{r \to \infty} I_r\,.$$

We state without proof that the order of integration can legitimately be reversed under these circumstances, producing

$$I_r = \frac{1}{2\pi} \int_{-r}^{r} \left[\int_{-L/2}^{L/2} F(\tau) e^{-i\omega\tau}\, d\tau \right] e^{i\omega t}\, d\omega$$

$$= \frac{1}{2\pi} \int_{-L/2}^{L/2} F(\tau) \left[\int_{-r}^{r} e^{i\omega(t-\tau)}\, d\omega \right] d\tau$$

$$= \frac{1}{2\pi} \int_{-L/2}^{L/2} \frac{F(\tau)}{i(t-\tau)} \left(e^{ir(t-\tau)} - e^{-ir(t-\tau)} \right) d\tau.$$

We write this as

$$(42) \qquad I_r = \frac{1}{2\pi i} \int_{-L/2}^{L/2} \frac{F(\tau)}{t-\tau} \left(e^{ir(t-\tau)} - 1 \right) d\tau$$

$$+ \frac{1}{2\pi i} \int_{-L/2}^{L/2} \frac{F(\tau)}{t-\tau} \left(1 - e^{-ir(t-\tau)} \right) d\tau,$$

for reasons which will become apparent shortly. Observe that each integrand is analytic in τ as long as we stay in the strip $|\operatorname{Im} \tau| < \delta$, because the singularity at $\tau = t$ is removable. Hence the integrals are independent

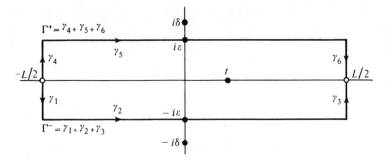

FIGURE 8.7

of path. We choose to evaluate the first integral in Eq. (42) along the contour Γ^- in Fig. 8.7, and the second along Γ^+:

$$I_r = \frac{1}{2\pi i} \int_{\Gamma^-} \frac{F(\tau)}{t - \tau} (e^{ir(t-\tau)} - 1) \, d\tau + \frac{1}{2\pi i} \int_{\Gamma^+} \frac{F(\tau)}{t - \tau} (1 - e^{-ir(t-\tau)}) \, d\tau.$$

If we rewrite this as

$$I_r = \frac{1}{2\pi i} \int_{\Gamma^-} \frac{F(\tau)}{\tau - t} \, d\tau - \frac{1}{2\pi i} \int_{\Gamma^+} \frac{F(\tau)}{\tau - t} \, d\tau + \frac{1}{2\pi i} \int_{\Gamma^-} \frac{F(\tau)}{t - \tau} e^{ir(t-\tau)} \, d\tau$$

$$- \frac{1}{2\pi i} \int_{\Gamma^+} \frac{F(\tau)}{t - \tau} e^{-ir(t-\tau)} \, d\tau,$$

we recognize that the first two integrals combine to give the integral of $F(\tau)/2\pi i(\tau - t)$ around the simple closed positively oriented contour $(\Gamma^-, -\Gamma^+)$, and according to Cauchy's integral formula, this is precisely $F(t)$. Thus (40) will follow if we show that the last two integrals go to zero as $r \to \infty$. This is most encouraging, since $\exp[ir(t - \tau)] \to 0$ for τ in the lower half-plane and $\exp[-ir(t - \tau)] \to 0$ for τ in the upper half-plane.

To fill in the details, consider first the part of the integration along γ_1: $\tau = -L/2 - iy$, $0 \le y \le \varepsilon$ (see Fig. 8.7). We have

$$\left| \frac{1}{2\pi i} \int_{\gamma_1} \frac{F(\tau)}{t - \tau} e^{ir(t-\tau)} \, d\tau \right| \le \frac{1}{2\pi} \int_0^\varepsilon \frac{M}{L/2 - |t|} \left| e^{irt} e^{irL/2} e^{-ry} \right| \, dy$$

$$= \frac{1}{\pi} \frac{M}{L - 2|t|} \int_0^\varepsilon e^{-ry} \, dy$$

$$= \frac{1}{\pi} \frac{M}{L - 2|t|} \frac{1 - e^{-r\varepsilon}}{r},$$

where M is the bound for $|F|$. This certainly approaches zero as $r \to \infty$. For the integration along γ_2: $\tau = x - i\varepsilon$, $-L/2 \le x \le L/2$ (Fig. 8.7 again),

$$\left| \frac{1}{2\pi i} \int_{\gamma_2} \frac{F(\tau)}{t - \tau} e^{ir(t - \tau)} \, d\tau \right| \le \frac{1}{2\pi} \int_{-L/2}^{L/2} \frac{M}{\varepsilon} \left| e^{ir(t - x + i\varepsilon)} \right| \, dx$$

$$= \frac{1}{2\pi} \frac{Me^{-r\varepsilon}}{\varepsilon} L,$$

again vanishing as $r \to \infty$. The integrals over γ_3, γ_4, and γ_6 in Fig. 8.7 are similar to that over γ_1, and the integral over γ_5 is handled like that over γ_2. ∎

Exercises 8.2

1. Find the Fourier transform of the following functions and express the inversion formula.

   (a) $F(t) = e^{-|t|}$

   (b) $F(t) = e^{-t^2}$ [HINT: Complete the square and use the fact that $\int_{-\infty}^{\infty} e^{-x^2} \, dx = \sqrt{\pi}$.]

   (c) $F(t) = te^{-t^2}$ [HINT: Integrate by parts and use part (b).]

   (d) $F(t) = \dfrac{\sin t}{t}$ [HINT: Use Eq. (36).]

   (e) $F(t) = \dfrac{\sin \pi t}{1 - t^2}$ [HINT: Rescale Example 9.]

2. Find the Fourier transform and verify the inversion formula (by residue theory or some other method) for the following functions.

   (a) $F(t) = \dfrac{1}{t^4 + 1}$

   (b) $F(t) = \dfrac{t}{t^4 + 1}$

   (c) $F(t) = e^{-t^2}$ [HINT: See Prob. 1(b).]

3. Use residue theory to evaluate the integral in Eq. (39).

4. Find the Fourier integral representations for solutions to the following differential equations.

   (a) $\dfrac{d^2 f}{dt^2} + \dfrac{df}{dt} + f = e^{-t^2}$

(b) $\dfrac{d^2f}{dt^2} + 4\dfrac{df}{dt} + f = \begin{cases} 0, & t < 0 \\ e^{-t}, & t \geq 0 \end{cases}$

(c) $\dfrac{d^2f}{dt^2} + 2\dfrac{df}{dt} + 3f = \begin{cases} 1, & |t| < 1 \\ 0, & \text{otherwise} \end{cases}$

5. If $f(x)$ is a given function whose Fourier transform is $G(\omega)$, show that the function

$$T(x, t) = \int_{-\infty}^{\infty} G(\omega)e^{i\omega x}e^{-\omega^2 t}\, d\omega$$

solves the partial differential equation

$$\frac{\partial T(x, t)}{\partial t} = \frac{\partial^2 T(x, t)}{\partial x^2}$$

and the initial condition

$$T(x, 0) = f(x)$$

for $t > 0$ and $-\infty < x < \infty$. [These equations describe the flow of heat in an infinite rod heated initially to the temperature $T = f(x)$. Recall Prob. 10 in Exercises 8.1.] Assume the validity of the Fourier representations and differentiation under the integral sign.

Insert the expression for the transform $G(\omega)$, interchange the order of integration, and use the hint accompanying Prob. 1(b) to derive the formula

$$T(x, t) = \frac{1}{2\sqrt{\pi t}} \int_{-\infty}^{\infty} f(\xi)e^{-(x-\xi)^2/4t}\, dt.$$

6. If $f(x)$ is a given function whose Fourier transform is $G(\omega)$, show that the function

$$u(x, t) = \int_{-\infty}^{\infty} G(\omega)e^{i\omega x} \cos \omega t\, d\omega$$

solves the partial differential equation

$$\frac{\partial^2 u(x, t)}{\partial x^2} = \frac{\partial^2 u(x, t)}{\partial t^2}$$

and the initial conditions

$$u(x, 0) = f(x)$$

$$\frac{\partial u}{\partial t}(x, 0) = 0$$

for $t > 0$ and $-\infty < x < \infty$. [These equations govern the motion of an infinite taut string initially displaced to the configuration

$u = f(x)$ and released at $t = 0$. Recall Prob. 11 in Exercises 8.1.] Assume the validity of the Fourier representations and differentiation under the integral sign.

How would you modify this representation if the initial conditions were

$$u(x, 0) = 0$$

$$\frac{\partial u(x, 0)}{\partial t} = f(x)?$$

Combine these results to solve the general set of initial conditions

$$u(x, 0) = f_1(x)$$

$$\frac{\partial u}{\partial t}(x, 0) = f_2(x).$$

8.3 THE LAPLACE TRANSFORM

In the two previous sections we were motivated by the desire to solve linear systems by means of frequency analysis. The strategy we were employing can be stated as follows: If a linear system is forced by a sinusoidal input function, $e^{i\omega t}$, then we expect that there ought to be a solution which is a sinusoid having the same frequency.

Now this is probably not the *only* solution; e.g., consider the problem of finding a function $f(t)$ which satisfies the differential equation

(43)
$$\frac{d^2 f(t)}{dt^2} + 2\frac{df(t)}{dt} + f(t) = e^{i2t}.$$

It has a solution of the form Ae^{i2t} with $A = 1/(4i - 3)$. But if $g(t)$ is a solution of the so-called "associated homogeneous equation"

$$\frac{d^2 g(t)}{dt^2} + 2\frac{dg(t)}{dt} + g(t) = 0,$$

then the function g may be added to a solution f of Eq. (43) to produce another solution of Eq. (43). For example, the function

(44)
$$\frac{1}{4i - 3} e^{i2t} + 7e^{-t}$$

also solves Eq. (43), since $7e^{-t}$ is a "homogeneous solution." The reader should verify the solution (44) by direct computation to see exactly what's going on.

Now for most *physical* systems, these homogeneous solutions are transient in nature; i.e., they die out as time increases [like e^{-t} in (44)]. This is evidenced by the fact that most physical systems, if not forced, eventually come to rest (due to dissipative phenomena such as resistance, damping, radiation loss, etc.). Such systems are called *asymptotically stable*. In these cases, we argue that the analysis of the preceding sections provides the *unique* solutions for the types of problems formulated there, because for both the periodic functions and the functions integrable over the whole real line the inputs have been driving the system "since $t = -\infty$" and hence the transients must have died out by any (finite) time.

Now it is time to become more flexible and to develop some mathematical machinery which will handle the transients. That is, we must take into account two considerations: The input is "turned on" at $t = 0$ and has *not* been driving the system for all time, and the system starts in some "initial configuration" at $t = 0$ which is probably *not* in agreement with the steady-state solution. The *Laplace transform*, as we shall see, handles both these effects.

Let us begin by dealing with the input function. We have $F(t)$ defined for all $t \geq 0$. For this discussion it is convenient to set $F(t) = 0$ for all negative t (so that F is defined on the whole real line) and then consider the Fourier transform of F:

$$G(\omega) = \frac{1}{2\pi} \int_{-\infty}^{\infty} F(t)e^{-i\omega t} \, dt.$$

In our case,

(45)
$$G(\omega) = \frac{1}{2\pi} \int_{0}^{\infty} F(t)e^{-i\omega t} \, dt.$$

Now if F is sufficiently well behaved near infinity (we shall not be precise here), one can show that Eq. (45) defines a function of ω which is analytic in the lower half-plane Im $\omega < 0$. Indeed, the derivative is given, as expected, by

$$\frac{dG(\omega)}{d\omega} = \frac{-i}{2\pi} \int_{0}^{\infty} tF(t)e^{-i\omega t} \, dt;$$

the *lower* half-plane is appropriate because

$$\left| e^{-i\omega t} \right| = e^{(\text{Im } \omega)t}$$

is bounded there. If we let ω be pure imaginary, say $\omega = -is$ with s nonnegative, we create

(46)
$$g(s) \equiv 2\pi G(-is),$$

a function which is called the *Laplace transform* of $F(t)$:

(47)
$$g(s) = \int_0^\infty F(t)e^{-st}\, dt.$$

It is often useful to indicate the relation between $g(s)$ and $F(t)$ by employing the notation

$$g(s) = \mathcal{L}\{F\}(s).$$

As an example, consider $F(t) = e^{-t}$; its Laplace transform is

$$g(s) = \mathcal{L}\{e^{-t}\}(s) = \int_0^\infty e^{-t}e^{-st}\, dt = -\frac{e^{-(s+1)t}}{s+1}\Big|_0^\infty = \frac{1}{s+1}.$$

We remark that the integral in Eq. (47) may converge even if F does not approach zero as $t \to \infty$, provided that s is sufficiently large. Indeed, for the function $F(t) = e^{7t}$, we have

$$\int_0^b e^{7t}e^{-st}\, dt = \frac{e^{(7-s)b} - 1}{7 - s},$$

and if $s > 7$, this approaches the limit $(s - 7)^{-1}$ as $b \to \infty$. In fact, whenever there exist two positive numbers M and α such that

$$|F(t)| \le Me^{\alpha t}, \qquad \text{for all } t \ge 0,$$

one can show that the integral in Eq. (47) converges for any *complex s* satisfying $\mathrm{Re}(s) > \alpha$. Accordingly, we shall say that Eq. (47) defines the Laplace transform $\mathcal{L}\{F\}(s)$ for any (complex) value of s for which the integral converges.

As a simple extension of the above computation shows, the Laplace transform of the function e^{at} is $1/(s - a)$ for $\mathrm{Re}(s) > \mathrm{Re}(a)$. By interpreting this statement with various choices of the constant a we are able to derive the first eight entries in the accompanying Laplace transform table. Entry (*ix*) is obtained by integration by parts, and it leads immediately to entries (*x*), (*xi*), and (*xii*).

The derivation of (*xiii*) proceeds as follows:

$$\mathcal{L}\{F(t)e^{-at}\}(s) = \int_0^\infty F(t)e^{-at}e^{-st}\, dt = \int_0^\infty F(t)e^{-(s+a)t}\, dt = \mathcal{L}\{F\}(s + a).$$

Entry (*xiv*) says, of course, that the Laplace transform is linear.

By looking at the transform of the derivative $F'(t)$, we can see how the Laplace transform takes initial configurations into account; we have

$$\mathcal{L}\{F'\}(s) = \int_0^\infty e^{-st}F'(t)\, dt.$$

(i) $\mathscr{L}\{e^{at}\} = \dfrac{1}{s-a}$ $[\operatorname{Re}(s) > \operatorname{Re}(a)]$

(ii) $\mathscr{L}\{1\} = \mathscr{L}\{e^{0t}\} = \dfrac{1}{s}$ $[\operatorname{Re}(s) > 0]$

(iii) $\mathscr{L}\{\cos \omega t\} = \operatorname{Re} \mathscr{L}\{e^{i\omega t}\} = \dfrac{s}{s^2 + \omega^2}$ $[\omega \text{ real}, \operatorname{Re}(s) > 0]$

(iv) $\mathscr{L}\{\sin \omega t\} = \operatorname{Im} \mathscr{L}\{e^{i\omega t}\} = \dfrac{\omega}{s^2 + \omega^2}$ $[\omega \text{ real}, \operatorname{Re}(s) > 0]$

(v) $\mathscr{L}\{\cosh \omega t\} = \mathscr{L}\{\cos i\omega t\} = \dfrac{s}{s^2 - \omega^2}$ $[\omega \text{ real}, \operatorname{Re}(s) > |\omega|]$

(vi) $\mathscr{L}\{\sinh \omega t\} = \mathscr{L}\{-i \sin i\omega t\} = \dfrac{\omega}{s^2 - \omega^2}$ $[\omega \text{ real}, \operatorname{Re}(s) > |\omega|]$

(vii) $\mathscr{L}\{e^{-\lambda t} \cos \omega t\} = \operatorname{Re} \mathscr{L}\{e^{-(\lambda + i\omega)t}\} = \dfrac{s + \lambda}{(s + \lambda)^2 + \omega^2}$

$$[\omega, \lambda \text{ real}, \operatorname{Re}(s) > -\lambda]$$

(viii) $\mathscr{L}\{e^{-\lambda t} \sin \omega t\} = \operatorname{Im} \mathscr{L}\{e^{-(\lambda + i\omega)t}\} = \dfrac{\omega}{(s + \lambda)^2 + \omega^2}$

$$[\omega, \lambda \text{ real}, \operatorname{Re}(s) > -\lambda]$$

(ix) $\mathscr{L}\{t^n e^{at}\} = \dfrac{n!}{(s - a)^{n+1}}$ $[\operatorname{Re}(s) > \operatorname{Re}(a)]$

(x) $\mathscr{L}\{t^n\} = \dfrac{n!}{s^{n+1}}$ $[\operatorname{Re}(s) > 0]$

(xi) $\mathscr{L}\{t \cos \omega t\} = \operatorname{Re} \mathscr{L}\{te^{i\omega t}\} = \dfrac{s^2 - \omega^2}{(s^2 + \omega^2)^2}$ $[\omega \text{ real}, \operatorname{Re}(s) > 0]$

(xii) $\mathscr{L}\{t \sin \omega t\} = \operatorname{Im} \mathscr{L}\{te^{i\omega t}\} = \dfrac{2s\omega}{(s^2 + \omega^2)^2}$ $[\omega \text{ real}, \operatorname{Re}(s) > 0]$

(xiii) $\mathscr{L}\{F(t)e^{-at}\}(s) = \mathscr{L}\{F\}(s + a)$

(xiv) $\mathscr{L}\{aF(t) + bH(t)\} = a\mathscr{L}\{F(t)\} + b\mathscr{L}\{H(t)\}$

Now if F' is sufficiently well behaved so that integration by parts is permitted, this becomes

$$\mathscr{L}\{F'\}(s) = -\int_0^\infty (-s)e^{-st}F(t)\, dt + e^{-st}F(t)\Big|_0^\infty,$$

and assuming that $e^{-st}F(t) \to 0$ as $t \to \infty$, we find

(48) $$\mathscr{L}\{F'\}(s) = s\mathscr{L}\{F\}(s) - F(0).$$

Iterating this equation results in

$$(49) \qquad \mathscr{L}\{F''\}(s) = s\mathscr{L}\{F'\}(s) - F'(0)$$

$$= s^2\mathscr{L}\{F\}(s) - sF(0) - F'(0),$$

and, in general,

$$(50) \qquad \mathscr{L}\{F^{(k)}\}(s) = s^k\mathscr{L}\{F\}(s) - s^{k-1}F(0) - s^{k-2}F'(0) - \cdots - F^{(k-1)}(0).$$

Sufficient conditions for the validity of these equations are given in

THEOREM 5 *Suppose that the function $F(t)$ and its first $n - 1$ derivatives are continuous for $t \geq 0$ and that $F^{(n)}(t)$ is piecewise smooth on every finite interval $[0, b]$. Also, suppose that there are positive constants M, α such that for $k = 0, 1, \ldots, n - 1$*

$$|F^{(k)}(t)| \leq Me^{\alpha t} \qquad (t \geq 0).$$

Then the Laplace transforms of F, F', F'', $\ldots$, $F^{(n)}$ exist for $\mathrm{Re}(s) > \alpha$, and Eq. (50) is valid for $k = 1, 2, \ldots, n$.

The reader is invited to prove this theorem in Prob. 3.

To illustrate how the Laplace transform is used in solving the so-called "initial-value problems," we consider

EXAMPLE 12 Find the function $f(t)$ which satisfies

$$(51) \qquad \frac{d^2 f(t)}{dt^2} + 2\frac{df(t)}{dt} + f(t) = \sin t$$

for $t \geq 0$ and which at $t = 0$ has the properties

$$(52) \qquad f(0) = 1, \qquad f'(0) = 0.$$

SOLUTION We begin by taking the transform of the Eq. (51). Thanks to the linearity property *(xiv)* we have

$$\mathscr{L}\{f''(t)\} + 2\mathscr{L}\{f'(t)\} + \mathscr{L}\{f(t)\} = \mathscr{L}\{\sin t\}.$$

Using Eq. (50) and the initial conditions (52) we find

$$\mathscr{L}\{f'(t)\} = s\mathscr{L}\{f(t)\} - 1,$$

$$\mathscr{L}\{f''(t)\} = s^2\mathscr{L}\{f(t)\} - s \cdot 1 - 0.$$

Thus our equation is transformed to

$$(s^2 + 2s + 1)\mathscr{L}\{f(t)\} - s - 2 = \mathscr{L}\{\sin t\},$$

or, using entry *(iv)* of the table,

$$\mathscr{L}\{f(t)\} = \frac{s + 2}{s^2 + 2s + 1} + \frac{1}{(s^2 + 2s + 1)(s^2 + 1)}.$$

Writing the first term on the right as

$$\frac{s+2}{s^2+2s+1} = \frac{s+1}{(s+1)^2} + \frac{1}{(s+1)^2} = \frac{1}{s+1} + \frac{1}{(s+1)^2},$$

we find that, according to entries (*i*) and (*ix*), it is the Laplace transform of the function

$$e^{-t} + te^{-t}.$$

To analyze the second term, we use partial fractions to express

$$\frac{1}{(s^2+2s+1)(s^2+1)} = \frac{1}{2}\frac{1}{s+1} + \frac{1}{2}\frac{1}{(s+1)^2} - \frac{1}{2}\frac{s}{s^2+1},$$

which is the Laplace transform of

$$\frac{1}{2}e^{-t} + \frac{1}{2}te^{-t} - \frac{1}{2}\cos t$$

[see entries (*i*), (*ix*), and (*iii*)].

Hence the solution is

$$f(t) = e^{-t} + te^{-t} + \frac{1}{2}e^{-t} + \frac{1}{2}te^{-t} - \frac{1}{2}\cos t$$

$$= \frac{3}{2}e^{-t} + \frac{3}{2}te^{-t} - \frac{1}{2}\cos t,$$

which can be directly verified. ∎

In the example we found it fairly easy to solve for the Laplace transform of the solution; to find the solution itself we had to invert the transform. Often, as illustrated, this can be done by referring to a table of Laplace transforms. However, since this transform was derived from the Fourier transform, which has an inversion formula, we suspect that a formula also exists for the inverse Laplace transform. To see this, recall that by Theorem 4 the Fourier inversion formula for a continuously differentiable, integrable F is

$$F(t) = \int_{-\infty}^{\infty} G(\omega)e^{i\omega t}\, d\omega.$$

Recall, also, that the Laplace transform $\mathcal{L}\{F\}$ was expressed in terms of the Fourier transform by formula (46), or, equivalently,

$$G(\omega) = \frac{1}{2\pi}\mathcal{L}\{F\}(i\omega).$$

Hence we have immediately

$$F(t) = \frac{1}{2\pi}\int_{-\infty}^{\infty}\mathcal{L}\{F\}(i\omega)e^{i\omega t}\, d\omega$$

for such functions. This formula is often written (substituting $-is$ for ω) as

$$(53) \qquad F(t) = \frac{1}{2\pi i} \int_{-i\infty}^{i\infty} \mathcal{L}\{F\}(s)e^{st}\,ds,$$

with the obvious interpretation of these imaginary limits of integration.

We would like to generalize formula (53) to cover *nonintegrable* functions whose Laplace transforms are defined only for Re(s) sufficiently large. This is easy to achieve if we can find a positive number a sufficiently large so that $F(t)e^{-at}$ is integrable. Then we write the inversion formula (53) for the function $F(t)e^{-at}$:

$$(54) \qquad F(t)e^{-at} = \frac{1}{2\pi i} \int_{-i\infty}^{i\infty} \mathcal{L}\{F(t)e^{-at}\}(s)e^{st}\,ds.$$

Inserting entry (*xiii*) of the table in Eq. (54) we multiply by e^{at} to derive

$$(55) \qquad F(t) = \frac{1}{2\pi i} \int_{-i\infty}^{i\infty} \mathcal{L}\{F\}(s+a)e^{(s+a)t}\,ds,$$

which we interpret as

$$F(t) = \frac{1}{2\pi i} \int_{a-i\infty}^{a+i\infty} \mathcal{L}\{F\}(s)e^{st}\,ds.$$

A rigorous analysis produces the following generalization:

THEOREM 6 *Suppose that $F(t)$ is piecewise smooth on every finite interval $[0, b]$ and that $|F(t)|$ is bounded by $Me^{\alpha t}$ for $t \geq 0$. Then $\mathcal{L}\{F\}(s)$ exists for Re(s) > α, and if $a > \alpha$, then for all $t > 0$*

$$\frac{F(t+) + F(t-)}{2} = \frac{1}{2\pi i} \int_{a-i\infty}^{a+i\infty} \mathcal{L}\{F\}(s)e^{st}\,ds.$$

EXAMPLE 13   Find the piecewise smooth function whose Laplace transform is

$$\frac{1}{s^4 - 1}.$$

SOLUTION   It is possible to employ partial fractions and the transform table to solve this problem, but we shall illustrate the use of the inversion formula. Observe that this function is certainly analytic for Re(s) > 1. To get the inverse transform, let us evaluate the integral

$$I = \int_{2-i\infty}^{2+i\infty} \frac{1}{s^4 - 1} e^{st}\,ds \qquad (t > 0).$$

This can be done by residue theory. I is the limit, as $\rho \to \infty$, of the contour integral

$$I_\rho = \int_{\gamma_\rho} \frac{e^{zt}}{z^4 - 1} \, dz,$$

where γ_ρ is the vertical segment from $2 - i\rho$ to $2 + i\rho$. We can close the contour with the half-circle C_ρ: $z = 2 + \rho e^{i\theta}$, $\pi/2 \le \theta \le 3\pi/2$; see Fig. 8.8.

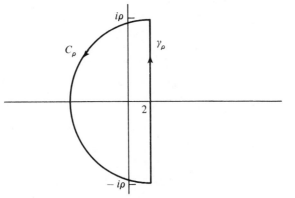

FIGURE 8.8

The integral over C_ρ is bounded by

$$\max_{\pi/2 \le \theta \le 3\pi/2} \frac{\left| e^{(2 + \rho \cos \theta)t} e^{i\rho t \sin \theta} \right|}{(\rho - 2)^4 - 1} \pi\rho = \frac{e^{2t} \pi\rho}{(\rho - 2)^4 - 1} \qquad (\rho > 3),$$

which goes to zero as $\rho \to \infty$.

Now the integrand has four simple poles at ± 1, $\pm i$, all of which eventually lie inside the semicircular contour of Fig. 8.8; in fact,

$$\frac{e^{zt}}{z^4 - 1} = \frac{e^{zt}}{(z - 1)(z + 1)(z - i)(z + i)}.$$

Hence

$$I = \lim_{\rho \to \infty} I_\rho = 2\pi i [\text{Res}(1) + \text{Res}(-1) + \text{Res}(i) + \text{Res}(-i)]$$

$$= 2\pi i \left[\frac{e^t}{2(1 - i)(1 + i)} + \frac{e^{-t}}{(-2)(-1 - i)(-1 + i)} \right.$$

$$\left. + \frac{e^{it}}{(i - 1)(i + 1)(2i)} + \frac{e^{-it}}{(-i - 1)(-i + 1)(-2i)} \right]$$

$$= \pi i \sinh t - \pi i \sin t.$$

So the inverse transform is

$$F(t) = \frac{1}{2\pi i} I = \frac{\sinh t - \sin t}{2}. \qquad \blacksquare$$

402

Chapter 8
Fourier Series and
Integral
Transforms

Exercises 8.3

1. Compute the Laplace transforms of the following functions.

   (a) $F(t) = 3 \cos 2t - 8e^{-2t}$

   (b) $F(t) = 2 - e^{4t} \sin \pi t$

   (c) $F(t) = \begin{cases} 1, & t < 1 \\ 0, & t \geq 1 \end{cases}$

   (d) $F(t) = \begin{cases} 0, & t < 1 \\ 1, & 1 \leq t \leq 2 \\ 0, & t > 2 \end{cases}$

   (e) $F(t) = \sin^2 t$

   (f) $F(t) = \dfrac{1}{\sqrt{t}}$

   [HINT: Let $\chi = \sqrt{st}$ and use the fact that $\int_0^\infty e^{-\chi^2} d\chi = \frac{1}{2}\sqrt{\pi}$.]

2. Find the inverse transform of the following functions.

   (a) $\dfrac{1}{s^2 + 4}$

   (b) $\dfrac{4}{(s-1)^2}$

   (c) $\dfrac{s+1}{s^2 + 4s + 4}$

   (d) $\dfrac{1}{s^3 + 3s^2 + 2s}$

   (e) $\dfrac{s+3}{s^2 + 4s + 7}$

3. Prove Theorem 5.

4. Verify the inversion formula for the following functions.

   (a) $F(t) = e^{-t}$

   (b) $F(t) \equiv 1$

   (c) $F(t) = \dfrac{1}{t^2 + 1}$

5. The effect of a time delay in a physical system is described mathematically by replacing a function $f(t)$ by the *delayed function*

   $$f_\tau(t) = \begin{cases} 0, & 0 \leq t < \tau, \\ f(t - \tau), & \tau \leq t < \infty. \end{cases}$$

   Show that $\mathcal{L}\{f_\tau(t)\} = e^{-\tau s}\mathcal{L}\{f(t)\}$. [Compare Prob. 1(c) and (d).]

6. Use the Laplace transform to solve the following initial-value problems.

    (a) $\dfrac{df}{dt} - f = e^{3t}$, $\quad f(0) = 3$

    (b) $\dfrac{d^2f}{dt^2} - 5\dfrac{df}{dt} + 6f = 0$, $\quad f(0) = 1, f'(0) = -1$

    (c) $\dfrac{d^2f}{dt^2} - \dfrac{df}{dt} - 2f = e^{-t}\sin 2t$, $\quad f(0) = 0, f'(0) = 2$

    (d) $\dfrac{d^2f}{dt^2} - 3\dfrac{df}{dt} + 2f = \begin{cases} 0, & 0 \le t \le 3 \\ 1, & 3 \le t \le 6 \\ 0, & t > 6 \end{cases}$, $\quad f(0) = 0, f'(0) = 0$

    [HINT: See Prob. 5.]

7. In control theory the differential equation

    $$a_n \frac{d^n f}{dt^n} + a_{n-1} \frac{d^{n-1} f}{dt^{n-1}} + \cdots + a_1 \frac{df}{dt} + a_0 f = u$$

    is interpreted as a relation between the (known) input $u(t)$ and the (unknown) output $f(t)$.

    (a) Show that the Laplace transforms $U(s)$ and $F(s)$ of the input and output are related by

    $$F(s) = \frac{U(s)}{a_n s^n + \cdots + a_1 s + a_0} + \frac{P(s)}{a_n s^n + \cdots + a_1 s + a_0},$$

    where $P(s)$ is a polynomial in s whose coefficients depend on the a_i and the initial values of $f(t)$ and its derivatives. The coefficient of $U(s)$ in this expression is called the *transfer function* of the system.

    (b) Show that every solution of the equation with $u(t) \equiv 0$ will go to zero as $t \to +\infty$ if all the poles of the transfer function are simple and lie in the left half-plane. (In other words, the system is *stable*.)

8. Identify the transfer functions and determine the stability of the differential equations in Prob. 6.

9. Consider the mass-spring system shown in Fig. 8.9. Each spring has the same natural (unstretched) length L, but when it is compressed

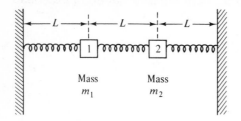

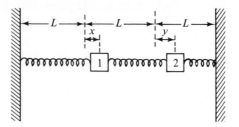

FIGURE 8.9   FIGURE 8.10

or elongated it exerts a force proportional to the amount of compression or elongation (Hooke's law); the constant of proportionality is denoted by K. If we let x and y be the respective displacements of the masses m_1 and m_2 from equilibrium, we have the situation depicted in Fig. 8.10 (for x and y both positive).

(a) By writing Newton's law (force equals mass times acceleration) for each mass, derive the equations of motion

$$m_1 \frac{d^2x}{dt^2} = -Kx - K(x - y),$$

$$m_2 \frac{d^2y}{dt^2} = -Ky + K(x - y).$$

(b) Setting $K = 1$, $m_1 = m_2 = 1$, use Laplace transforms to find the solutions $x(t)$, $y(t)$ of these equations if the masses are released from rest (zero velocity) with each of the following initial displacements:

(i) $x(0) = 1$, $y(0) = -1$.
(ii) $x(0) = 1$, $y(0) = 1$.
(iii) $x(0) = 1$, $y(0) = 0$.

(c) For which initial conditions in part (b) is the *system's* response periodic [note that this requires *both* $x(t + T) = x(t)$ *and* $y(t + T) = y(t)$]? Can you visualize these responses? (They are called *normal modes* for the system.)

SUMMARY

The chief value of sinusoidal analysis in applied mathematics lies in the fact that when a linear system is driven by a function $Ae^{i\omega t}$ its response takes the same form, under the proper circumstances. To take advantage of this we have to be able to express a given function of a real variable as a superposition of sinusoids.

When the function F is periodic, say of period L, and satisfies certain continuity conditions, the appropriate decomposition is given by the Fourier series

$$F(t) = \sum_{n=-\infty}^{\infty} c_n e^{in2\pi t/L},$$

with coefficients defined by

$$c_n = \frac{1}{L} \int_{-L/2}^{L/2} F(t) e^{-in2\pi t/L} \, dt.$$

If F is not periodic but $|F|$ is integrable (and the continuity conditions still hold), then the decomposition takes the form

$$F(t) = \int_{-\infty}^{\infty} G(\omega) e^{i\omega t} \, d\omega,$$

where G is the Fourier transform of F, defined by

$$G(\omega) = \frac{1}{2\pi} \int_{-\infty}^{\infty} F(t) e^{-i\omega t} \, dt.$$

"Direct" differentiation (i.e., termwise or under the integral sign) of both representations can be justified in many circumstances, and since this operation merely amounts to multiplication by $i\omega$, the solution of differential equations can often be greatly simplified by employing Fourier analysis.

To handle initial conditions and transients it is convenient to use the Laplace transform of F,

$$\mathscr{L}\{F\}(s) = \int_0^{\infty} F(t) e^{-st} \, dt.$$

The Laplace transform can be thought of as the Fourier transform of a function which is "turned on" at $t = 0$ [i.e., $F(t) = 0$ for $t < 0$], with $i\omega$ replaced by s. The effect of the initial conditions is displayed in the formulas for the derivatives $F^{(n)}(t)$:

$$\mathscr{L}\{F^{(n)}\}(s) = s^n \mathscr{L}\{F\}(s) - s^{n-1}F(0) - s^{n-2}F'(0) - \cdots - F^{(n-1)}(0).$$

Hence the Laplace transform is the appropriate tool for solving initial-value problems.

Another advantage of the Laplace transform is its ability to handle certain nonintegrable functions. There is an inversion formula for recovering $F(t)$ from $\mathscr{L}\{F\}(s)$, but the use of tables is frequently more convenient.

SUGGESTED READING

The references mentioned at the end of Chapter 3 under the heading "Sinusoidal Analysis" may also be useful for a further study of the material in this chapter. In addition, the following texts deal with the subjects from a more rigorous viewpoint:

Fourier Series

[1]   BACHMAN, G., and NARICI, L. *Functional Analysis.* Academic Press, Inc., New York, 1966.

[2]   EDWARDS, R. E. *Fourier Series*, Vols. I and II. Holt, Rinehart and Winston, Inc., New York, 1967.

[3]   FRIEDMAN, A. *Advanced Calculus.* Holt, Rinehart and Winston, Inc., New York, 1971.

Fourier and Laplace Transforms

[4]   CHURCHILL, R. V. *Operational Mathematics.* McGraw-Hill Book Company, New York, 1958.

[5]   JEFFREYS, H., and JEFFREYS, B. *Methods of Mathematical Physics.* Cambridge University Press, New York, 1956.

I

Equations of Mathematical Physics

In this appendix we shall give a sketch of how solutions of Laplace's equation are used to solve certain problems in physics. The derivations will be quite brief and will deal only with the rudiments of the theory. The interested student should consult the standard physics textbooks listed in the references for a more thorough coverage; our treatment is designed for the reader with no background in these areas, to enable him to bridge the gap between complex-variable methods and complex-variable applications.

I.1 ELECTROSTATICS

As is well known, the electrostatic force between two charged particles P_1 and P_2 obeys an inverse square law; specifically, the magnitude of the force is given by (*Coulomb's law*)

$$(1) \qquad |\mathbf{F}| = \frac{kq_1 q_2}{r^2},$$

where r is the distance between the particles, q_1 and q_2 are the charges on the particles, and k is a dimensional constant. The direction of the force is along the line joining the particles, and the force is attractive if the particles are oppositely charged and repulsive for similarly charged particles.

If the position of particle P_1 in three-dimensional space has coordinates (x_1, y_1, z_1) and particle P_2 is located at (x_2, y_2, z_2), then the force $\mathbf{F}$ on particle P_1 is as depicted in Fig. I.1 for like charges. The component

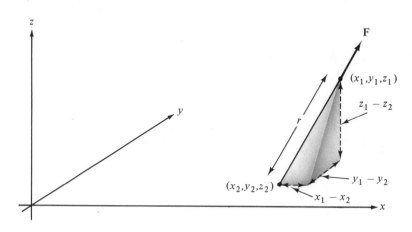

FIGURE I.1

F_x of $\mathbf{F}$ in the x-direction is computed by multiplying Eq. (1) by the cosine of the angle between the x-direction and the line joining the particles; from Fig. I.1, we see that

$$F_x = \frac{kq_1 q_2}{r^2} \frac{x_1 - x_2}{r} = kq_1 q_2 \frac{x_1 - x_2}{r^3}.$$

Similarly, for the components in the y- and z-directions we have

$$F_y = kq_1 q_2 \frac{y_1 - y_2}{r^3},$$

$$F_z = kq_1 q_2 \frac{z_1 - z_2}{r^3}.$$

If other particles $P_3, P_4, \ldots, P_n$ are also present, the forces that they exert on particle P_1 simply add vectorially, or componentwise, so that the total force on P_1 can be expressed as

(2)
$$F_x = \sum_{j=2}^{n} kq_1 q_j \frac{x_1 - x_j}{r_j^3} = q_1 \sum_{j=2}^{n} kq_j \frac{x_1 - x_j}{r_j^3},$$

$$F_y = q_1 \sum_{j=2}^{n} kq_j \frac{y_1 - y_j}{r_j^3},$$

$$F_z = q_1 \sum_{j=2}^{n} kq_j \frac{z_1 - z_j}{r_j^3};$$

here r_j denotes the distance from particle P_1 to particle P_j:

$$r_j = \sqrt{(x_1 - x_j)^2 + (y_1 - y_j)^2 + (z_1 - z_j)^2}.$$

Now let us consider a situation where particles P_2 through P_n are held fixed, while both the location and charge of the first particle are considered as variables. Then from Eqs. (2) we can say (dropping the subscript 1) that the total force on a particle with charge q located at (x, y, z) is given by $q\mathbf{E}(x, y, z)$, where $\mathbf{E}(x, y, z) = (E_x, E_y, E_z)$ is the *electric field* produced by the fixed charges and is given by

$$(3) \qquad E_x(x, y, z) = \sum_{j=2}^{n} \frac{kq_j(x - x_j)}{[(x - x_j)^2 + (y - y_j)^2 + (z - z_j)^2]^{3/2}}$$

with similar formulas for E_y and E_z.

The next step is to consider the function $\phi(x, y, z)$, known as the *electrostatic potential* at (x, y, z) due to the fixed charges:

$$(4) \qquad \phi(x, y, z) \equiv \sum_{j=2}^{n} k \frac{q_j}{r_j}.$$

This function has many interesting physical properties, but the one which concerns us here is the fact that its partial derivatives are the negatives of the components of the electric field; i.e.,

$$\frac{\partial}{\partial x} \phi(x, y, z) = \sum_{j=2}^{n} kq_j \frac{\partial}{\partial x} [(x - x_j)^2 + (y - y_j)^2 + (z - z_j)^2]^{-1/2}$$

$$= \sum_{j=2}^{n} kq_j \left(-\frac{1}{2}\right) [(x - x_j)^2 + (y - y_j)^2 + (z - z_j)^2]^{-3/2} \cdot 2(x - x_j)$$

$$= -E_x,$$

and similarly for $\partial\phi/\partial y$ and $\partial\phi/\partial z$. Obviously, solving for the single function ϕ is easier than solving for the three components of $\mathbf{E}$.

Now in physical configurations where conductors are present, one does not know a priori where the fixed charges are located; the field is produced by the free electrons inside the conductor. Hence expression (4), while still true, is of little value. However, we can say that the electric field $\mathbf{E}$ is zero *inside the conductor*, in the static case, for if it were nonzero, it would produce a motion of the free electrons. In other words, the free electrons inside a conductor arrange themselves so as to cancel the internal field. But then ϕ must be constant inside the conductor, because its derivatives are zero, and we conclude that any conductor is an *equipotential*. In a typical problem one specifies the value of ϕ on each separate conductor and asks for ϕ in the regions external to the conductors.

The last step is to show that ϕ satisfies the three-dimensional Laplace equation

$$\frac{\partial^2 \phi}{\partial x^2} + \frac{\partial^2 \phi}{\partial y^2} + \frac{\partial^2 \phi}{\partial z^2} = 0$$

outside the conductors. This is easy; we have, omitting the summation limits for convenience,

$$\frac{\partial^2 \phi}{\partial x^2} = -\frac{\partial E_x}{\partial x} = -\sum kq_j \frac{\partial}{\partial x}\left(\frac{x - x_j}{r_j^3}\right)$$

$$= -\sum kq_j \left[\frac{1}{r_j^3} - \frac{3(x - x_j)}{r_j^4}\frac{\partial r_j}{\partial x}\right]$$

$$= -\sum kq_j \left[\frac{1}{r_j^3} - \frac{3(x - x_j)}{r_j^4}\frac{1}{2r_j}2(x - x_j)\right]$$

$$= -\sum kq_j \left[\frac{1}{r_j^3} - \frac{3(x - x_j)^2}{r_j^5}\right].$$

Similarly,

$$\frac{\partial^2 \phi}{\partial y^2} = -\sum kq_j \left[\frac{1}{r_j^3} - \frac{3(y - y_j)^2}{r_j^5}\right],$$

$$\frac{\partial^2 \phi}{\partial z^2} = -\sum kq_j \left[\frac{1}{r_j^3} - \frac{3(z - z_j)^2}{r_j^5}\right],$$

so, adding these we get

$$\frac{\partial^2 \phi}{\partial x^2} + \frac{\partial^2 \phi}{\partial y^2} + \frac{\partial^2 \phi}{\partial z^2} = -\sum kq_j \left\{\frac{3}{r_j^3} - \frac{3}{r_j^5}[(x - x_j)^2 + (y - y_j)^2 + (z - z_j)^2]\right\}$$

$$= -\sum kq_j \left(\frac{3}{r_j^3} - \frac{3r_j^2}{r_j^5}\right) = 0.$$

This shows that $\phi(x, y, z)$ is a harmonic function in the regions external to the conductors and hence is determined by its values on the conductors in the manner indicated in Chapter 4.

In two-dimensional problems where there is no z-dependence, ϕ becomes a harmonic function of x and y only. (Notice that a two-dimensional "point charge" can be interpreted as a *line* of charge in three dimensions.)

In closing this section we remark that one usually models (macroscopic) physical situations by assuming a continuous distribution of charge. As a result, the sums in Eqs. (2), (3), and (4) are replaced by integrals, and some aspects of the mathematics become more complicated. In Ref. [1] this is discussed in great detail, and we refer the reader to this source for verification that the potential function ϕ continues to be harmonic in a vacuum.

I.2 HEAT FLOW

The flow of heat in a solid is produced by differences of temperature T in the body. Specifically, if we examine a small ("infinitesimal") rectangle A in, say, the (y, z)-plane, as in Fig. I.2, then it seems reasonable to postu-

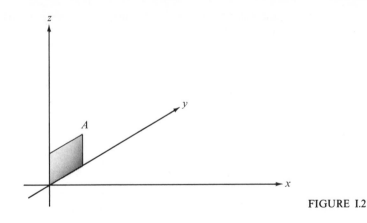

FIGURE I.2

late that the rate at which heat flows through A is proportional to the area of A and to the derivative of T with respect to x. Of course, a positive derivative means that T is increasing, so that heat is driven in the negative x-direction.

Now let's apply this postulate to derive an equation for T. Consider a small cube positioned inside the solid as in Fig. I.3. The rate at which heat flows *into* the cube through the face indicated by the arrow I is given by

$$-k\frac{\partial T(x,\, y,\, z)}{\partial x}h^2.$$

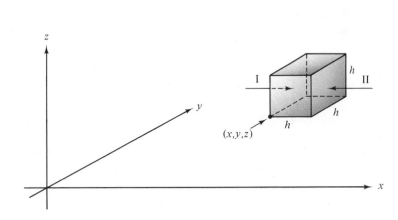

FIGURE I.3

Here k is a dimensional constant measuring the thermal conductivity of the material. Similarly, the rate of heat flow *into* the cube through face II is

$$+k\frac{\partial T(x+h,\, y,\, z)}{\partial x}h^2.$$

Consequently, the total heat flow into the cube in the x-direction is

$$kh^2\left[\frac{\partial T}{\partial x}(x+h, y, z) - \frac{\partial T(x, y, z)}{\partial x}\right],$$

which becomes, using the Mean Value Theorem,

$$kh^3\frac{\partial^2 T}{\partial x^2},$$

where the derivative is evaluated at some intermediate point. We shall eventually let h go to zero, so there is no value in being meticulous here.

Adding the contributions from the other four faces, we see that the rate at which heat flows into the cube is given by

$$(5) \qquad\qquad kh^3\left(\frac{\partial^2 T}{\partial x^2} + \frac{\partial^2 T}{\partial y^2} + \frac{\partial^2 T}{\partial z^2}\right),$$

where the derivatives are evaluated at points inside the cube. Now in a static situation this net influx of heat must be zero; otherwise the temperature would change. Setting (5) equal to zero, dividing by h^3, and taking the limit as $h \to 0$, we conclude that at each point (x, y, z) inside the body

$$\frac{\partial^2 T}{\partial x^2} + \frac{\partial^2 T}{\partial y^2} + \frac{\partial^2 T}{\partial z^2} = 0;$$

i.e., T is a harmonic function. Again, in two-dimensional problems there is no z-dependence, and T is harmonic in x and y only.

I.3 FLUID MECHANICS

The flow of fluids is somewhat similar to heat flow, so the previous derivation will be useful for part of our analysis. Referring to Fig. I.2, we argue that the rate at which a fluid of uniform density flows across the rectangle A should be proportional to the x-component of the fluid velocity $\mathbf{V} = (V_x, V_y, V_z)$ at the rectangle, and to the area of A. Hence, turning now to Fig. I.3, we claim that the flow into the cube through face I is proportional to

$$V_x(x, y, z)h^2$$

and that the inward flow through face II is proportional to

$$-V_x(x+h, y, z)h^2,$$

yielding a net influx in the x-direction proportional to

$$h^2[V_x(x, y, z) - V_x(x+h, y, z)] = -h^3\frac{\partial V_x}{\partial x}$$

(using the Mean Value Theorem again). Adding the contributions of the other faces and setting the total influx equal to zero for the time independent case, we can derive as before (with V_x instead of $\partial T/\partial x$, etc.)

(6)
$$\frac{\partial V_x}{\partial x} + \frac{\partial V_y}{\partial y} + \frac{\partial V_z}{\partial z} = 0.$$

If the problem is two-dimensional and there is no z-dependence, Eq. (6) becomes

(7)
$$\frac{\partial V_x}{\partial x} + \frac{\partial V_y}{\partial y} = 0.$$

Now it turns out to be quite useful in fluid problems to restrict oneself to situations where there are no "whirlpool" tendencies in the flow. Engineers describe this by saying that the *vorticity* is zero. Vorticity is a very involved subject, so here we shall merely be content to say that in this situation one also imposes the condition

(8)
$$\frac{\partial V_x}{\partial y} - \frac{\partial V_y}{\partial x} = 0$$

on the velocity and that one studies *two-dimensional incompressible* (constant density) *irrotational* (zero vorticity) *flows*.

Hopefully the student will remember from vector analysis (Refs. [3] and [4]) that Eq. (7) implies the existence of a function $\phi(x, y)$, called the *stream function* in this context, such that

(9)
$$V_x = \frac{\partial \phi}{\partial y} \quad \text{and} \quad V_y = -\frac{\partial \phi}{\partial x}.$$

(This is reminiscent of the electrostatic situation.) But then Eq. (8) says that ϕ is harmonic:

$$\frac{\partial V_x}{\partial y} - \frac{\partial V_y}{\partial x} = \frac{\partial^2 \phi}{\partial y^2} + \frac{\partial^2 \phi}{\partial x^2} = 0.$$

The curves $\phi(x, y) = \text{constant}$ are everywhere parallel to the velocity field $\mathbf{V}(x, y)$, and they are the *streamlines*. To see this, observe that the vector $(\partial \phi/\partial x, \partial \phi/\partial y)$ is normal to the streamline $\phi(x, y) = \text{constant}$ at the point (x, y). And since Eq. (9) implies that this vector is also perpendicular to $\mathbf{V}$, then $\mathbf{V}$ must be tangent to the streamline.

Consequently we impose the boundary condition that $\phi = \text{constant}$ along the surface of any nonporous obstacle in the flow to indicate that the fluid flows along the surface and not into the body. Hence the problem of finding the two-dimensional incompressible irrotational flow past nonporous obstacles becomes a boundary value problem for a harmonic function and is thus susceptible to complex variable methods.

In closing we remark that Eq. (8) implies the existence of a function $\psi(x, y)$ such that

$$V_x = \frac{\partial \psi}{\partial x} \quad \text{and} \quad V_y = \frac{\partial \psi}{\partial y},$$

and Eq. (7) tells us that ψ, too, is harmonic. In fact, ψ, which is known as the *velocity potential*, is a harmonic conjugate to ϕ. In Ref. [4] this function is dealt with in more detail.

SUGGESTED READING

Electrostatics

[1]   KELLOG, O. D. *Foundations of Potential Theory*. Dover Publications, Inc., New York (reprint).

[2]   PANOFSKY, W. K. H., and PHILLIPS, M. *Classical Electricity and Magnetism*, 2nd ed. Addison-Wesley Publishing Company, Inc., Reading, Mass., 1962.

Heat Flow and Vector Analysis

[3]   DAVIS, H., and SNIDER, A. D. *Introduction to Vector Analysis*, 3rd ed. Allyn and Bacon, Inc., Boston, 1975.

Fluid Mechanics

[4]   OWCZAREK, J. A. *Introduction to Fluid Mechanics*. International Textbook Company, Scranton, Pa., 1968.

II

Table of
Conformal Mappings

The following five pages list some conformal mappings that arise in applications. Notice that corresponding points are labeled by the same letter; thus the point a is mapped to the point A, etc. A more extensive tabulation appears in reference [5] at the end of Chapter 7.

II.1   BILINEAR TRANSFORMATIONS

$z = x + iy$

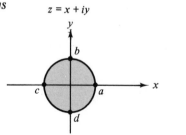

$w = u + iv$

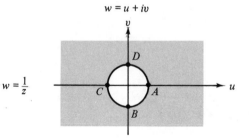

$$w = \frac{1}{z}$$

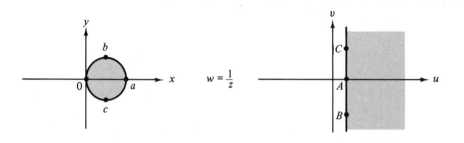

$$w = \frac{1}{z}$$

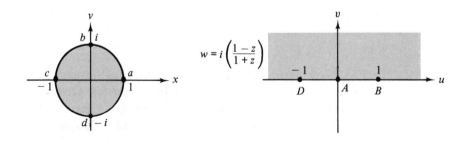

$$w = i\left(\frac{1-z}{1+z}\right)$$

$z = x + iy$

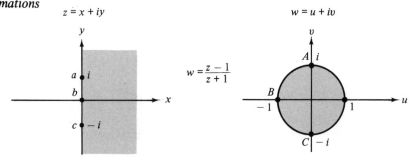

$$w = \frac{z - 1}{z + 1}$$

$w = u + iv$

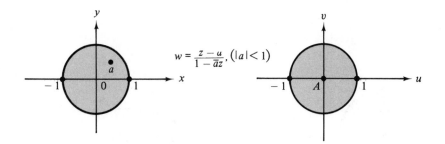

$$w = \frac{z - u}{1 - \bar{a}z}, \ (|a| < 1)$$

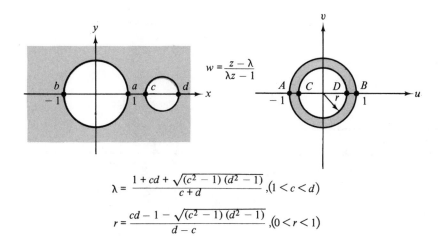

$$w = \frac{z - \lambda}{\lambda z - 1}$$

$$\lambda = \frac{1 + cd + \sqrt{(c^2 - 1)(d^2 - 1)}}{c + d}, (1 < c < d)$$

$$r = \frac{cd - 1 - \sqrt{(c^2 - 1)(d^2 - 1)}}{d - c}, (0 < r < 1)$$

II.2 OTHER TRANSFORMATIONS

$z = x + iy$

$w = u + iv$

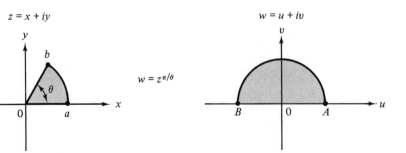

$w = z^{\pi/\theta}$

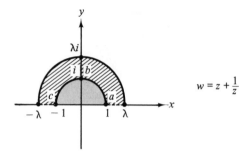

$w = z + \dfrac{1}{z}$

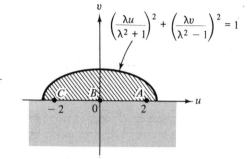

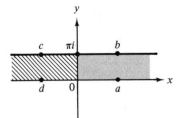

$w = e^z$

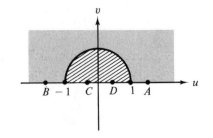

$z = x + iy$ $w = u + iv$

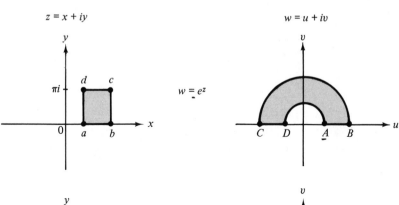

$w = e^z$

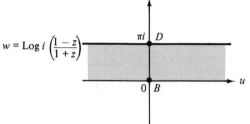

$w = \text{Log } i \left(\dfrac{1 - z}{1 + z} \right)$

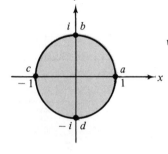

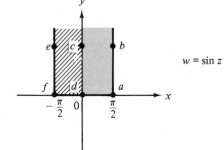

$w = \sin z$

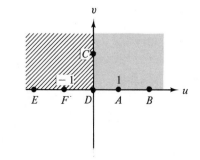

$z = x + iy$

$w = u + iv$

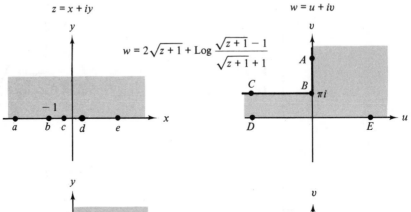

$$w = 2\sqrt{z+1} + \text{Log}\,\frac{\sqrt{z+1}-1}{\sqrt{z+1}+1}$$

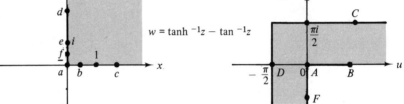

$$w = \tanh^{-1}z - \tan^{-1}z$$

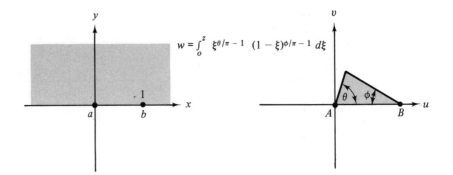

$$w = \int_o^z \xi^{\theta/\pi-1}\,(1-\xi)^{\phi/\pi-1}\,d\xi$$

Answers to Selected Problems

CHAPTER 1

Exercises 1.1, Page 5

5. (b) $3 + 0i = 3$; (d) $6 - 3i$; (f) $1 + i$.

7. $\dfrac{61}{185} - \dfrac{107}{185}i.$

9. $2 + 0i = 2.$

12. $i^{4k} = 1$; $i^{4k+1} = i$; $i^{4k+2} = -1$; $i^{4k+3} = -i.$

15. $a^3 - 3ab^2 + 5a^2 - 5b^2 = a,\; 3a^2b - b^3 + 10ab = b + 3,$ where $z = a + bi.$

2. $-\dfrac{25}{11} - \dfrac{7}{11}i.$

10. (b) circle, center $= 1 - i$, rad $= 3$;

(d) perpendicular bisector of segment joining $z = 1$ and $z = -i$;

(f) ellipse with foci at ± 1;

(g) circle, center $= \frac{9}{8}$, rad $= \frac{3}{8}$;

(j) all points outside the circle $|z| = 6$.

Exercises 1.3, Page 18

6. (a) 1; (b) $5\sqrt{26}$; (d) 1.

10. (b) arg $= 3\pi/4 + 2k\pi$,

polar form: $3\sqrt{2}(\cos 3\pi/4 + i \sin 3\pi/4)$;

(d) arg $= 7\pi/6 + 2k\pi$,

polar form: $4(\cos 7\pi/6 + i \sin 7\pi/6)$;

(f) arg $= 5\pi/3 + 2k\pi$, polar form: $4(\cos 5\pi/3 + i \sin 5\pi/3)$;

(h) arg $= 13\pi/12 + 2k\pi$,

polar form: $\dfrac{\sqrt{14}}{2}(\cos 13\pi/12 + i \sin 13\pi/12)$.

15. (a) $-3\pi/4$; (c) $\pi/2$.

16. (a) F; (b) T; (c) F; (d) T.

Exercises 1.4, Page 24

5. (a) $2(\cos \pi/4 + i \sin \pi/4)$, $2(\cos 3\pi/4 + i \sin 3\pi/4)$,

$2(\cos 5\pi/4 + i \sin 5\pi/4)$, $2(\cos 7\pi/4 + i \sin 7\pi/4)$;

(c) $\cos \pi/8 + i \sin \pi/8$, $\cos 5\pi/8 + i \sin 5\pi/8$,

$\cos 9\pi/8 + i \sin 9\pi/8$, $\cos 13\pi/8 + i \sin 13\pi/8$;

(e) $\sqrt[4]{2}(\cos 3\pi/8 + i \sin 3\pi/8)$, $\sqrt[4]{2}(\cos 11\pi/8 + i \sin 11\pi/8)$.

8. (b) $2 - i, 1 - i$.

9. $1, 1 + i\sqrt{3}, 1 - i\sqrt{3}$.

10. $z = 1/(w - 1)$, $w = \cos 2k\pi/5 + i \sin 2k\pi/5$, $k = 1, 2, 3, 4$.

16. $\sqrt[4]{8}(\cos 5\pi/8 + i \sin 5\pi/8)$; $\sqrt[4]{8}(\cos 13\pi/8 + i \sin 13\pi/8)$.

Exercises 1.5, Page 27

3. b, c, f.

4. b, c.

5. a, c.

8. a, e.

11. all points of S and 0.

18. No.

CHAPTER 2

Exercises 2.2, Page 36

1. (b) $x/(x^2 + y^2) - iy/(x^2 + y^2)$;
    (d) $(2x^2 - 2y^2 + 3)/\sqrt{(x - 1)^2 + y^2} + i4xy/\sqrt{(x - 1)^2 + y^2}$.

3. (a) half-plane Re $w > 5$;   (b) upper half-plane Im $w \geq 0$;
    (c) $|w| \geq 1$;
    (d) circular sector $|w| < 2$, $-\pi < \text{Arg } w < \pi/2$.

4. (b) $\sqrt{2}/3 + 2i/3$; $-\sqrt{2}/3 + 2i/3$.

8. (b) does not converge;   (d) converges to 0.

12. (a) $-8i$;   (c) $6i$;   (e) $2z_0$.

20. $-\frac{1}{2} - i$.

Exercises 2.3, Page 42

6. (b) $-12z(z^2 - 3i)^{-7}$;
    (d) $(z + 2)^2[-5z^2 + (-16 - i)z + 3 - 8i]/(z^2 + iz + 1)^5$.

9. (a) $2 - 3i$;   (c) $-\frac{1}{2} \pm i\sqrt{15}/2$.

10. (b) nowhere analytic;   (d) everywhere analytic;   (f) analytic
    except at $z = 0$;   (g) nowhere analytic.

11. (a) T;   (c) F;   (e) F;   (g) T.

14. $\frac{3}{5}$.

Exercises 2.4, Page 48

3. $g(z) = 3z^2 + 2z - 1$.

5. $f'(z) = 2e^{x^2 - y^2}(x + iy)[\cos(2xy) + i\sin(2xy)]$.

Exercises 2.5, Page 53

2. $ax^2 + bxy - ay^2$.

4. (one example) $\phi(x, y) = r^4 \sin 4\theta = 4x^3y - 4xy^3$.

6. (a) $\phi(x, y) = \text{Re}(z^2 + 5z + 1)$;

   (b) $\phi(x, y) = 2\,\text{Re}\left(\dfrac{z^2}{z + 2i}\right)$.

7. (a) $v = -x + a$; (c) $v = y^2/2 - y - x^2/2 - x + a$;
   (e) $v = \text{Tan}^{-1}(y/x) + a$.

CHAPTER 3

Exercises 3.1, Page 67

6. (b) $i\exp(2)$; (d) $\cos(1)\cosh(1) + i\sin(1)\sinh(1)$; (f) 0.

10. (b) $-2\sin(2z) - (i/z^2)\cos(1/z)$; (d) $3\tan^2(z)\sec^2(z)$;
    (f) $1 - \tanh^2(z) = \text{sech}^2(z)$.

17. (a) $z = ik\pi/2,\ k = 0, \pm 1, \pm 2, \ldots$;
    (b) $z = -i\cdot\ln 3 + 2k\pi,\ k = 0, \pm 1, \pm 2, \ldots$;
    (c) No solution.

Exercises 3.2, Page 74

1. (b) $(\frac{1}{2})\text{Log}(2) + i(7\pi/4 + 2k\pi),\ k = 0, \pm 1, \pm 2, \ldots$;
   (d) $\text{Log}(2) + i(\pi/6)$.

6. (a) $\text{Log}(2) + i(\pi/2 + 2k\pi),\ k = 0, \pm 1, \pm 2, \ldots$;
   (b) $\pm\sqrt[4]{2}\exp(i\pi/8)$;
   (c) $i(2\pi/3 + 2k\pi),\ i(4\pi/3 + 2k\pi),\ k = 0, \pm 1, \pm 2, \ldots$.

13. (a) $\text{Log}(2z - 1)$; (c) $\mathscr{L}_{\pi/2}(2z - 1)$, where $\mathscr{L}_{\pi/2}(re^{i\theta}) = \text{Log } r + i\theta$,
    $\pi/2 < \theta < 5\pi/2$.

14. $w = (1/\pi)\text{Log } z$.

Exercises 3.3, Page 78

1. (a) $\exp(-\pi/2 - 2k\pi)$, $k = 0$, ± 1, ± 2, ...;

    (b) $+1$, $-\frac{1}{2} + i\frac{\sqrt{3}}{2}$, $-\frac{1}{2} - i\frac{\sqrt{3}}{2}$;

    (c) $\exp[-2k\pi^2 + \pi i \text{ Log } 2]$, $k = 0$, ± 1, ± 2, ...;

    (d) $(1 + i)\exp[2k\pi + \pi/4 - (i/2)\text{Log } 2]$, $k = 0$, ± 1, ± 2, ...;

    (e) $-2 + 2i$.

2. (a) 2; (c) $(1 + i)\exp[(i/2)\text{Log } 2 - \pi/4]$.

4. No.

7. $(1 + i)\exp(-\pi/2)$.

12. $z = \pi/4 + k\pi$, $k = 0$, ± 1, ± 2, ...

Exercises 3.4, Page 85

4. (a) $I_s = \dfrac{A}{R_0} \cos(\omega t - \phi_0)$,

    where $R_0 = \left| R + \dfrac{1}{i\omega C} \right| = \sqrt{R^2 + \dfrac{1}{\omega^2 C^2}}$,

    and $\phi_0 = \text{Arg}\left(R + \dfrac{1}{i\omega C} \right) = -\tan^{-1}\left(\dfrac{1}{\omega RC} \right)$.

    (b) $I_s = \dfrac{A\sqrt{\omega^2 L^2 + (\omega^2 RCL - R)^2}}{\omega RL} \cos[\omega(t + \beta) + \phi_0]$

    where $\phi_0 = \tan^{-1}\left(\dfrac{\omega^2 RCL - R}{\omega L} \right)$.

    (c) $I_s = \dfrac{A\omega_1}{\omega_1^4 + 4} \sqrt{16 + 4\omega_1^2 + 4\omega_1^4 + \omega_1^6} \, \cos(\omega_1 t + \phi_1)$

    $\qquad + \dfrac{B\omega_2}{\omega_2^4 + 4} \sqrt{16 + 4\omega_2^2 + 4\omega_2^4 + \omega_2^6} \, \sin(\omega_2 t + \phi_2)$,

    where $\phi_1 = \tan^{-1}\left(\dfrac{4}{\omega_1^3 + 2\omega_1} \right)$, $\phi_2 = \tan^{-1}\left(\dfrac{4}{\omega_2^3 + 2\omega_2} \right)$.

CHAPTER 4

Exercises 4.1, Page 100

3. (a) $z(t) = (1 + i) + t(-3 - 4i)$, $0 \le t \le 1$;

    (b) $z(t) = 2i + 4e^{-it}$, $0 \le t \le 2\pi$;

(c) $z(t) = Re^{it}$, $\pi/2 \leq t \leq \pi$;

(d) $z(t) = t + it^2$, $1 \leq t \leq 3$.

5. Yes.

7. 8.

8. $\Gamma : z(t) = \begin{cases} (-2 + 2i) + t(1 - 2i), 0 \leq t \leq 1, \\ e^{-it\pi}, 1 \leq t \leq 2 \end{cases}$

$-\Gamma : z(t) = \begin{cases} e^{it\pi}, -2 \leq t \leq -1, \\ (-2 + 2i) - t(1 - 2i), -1 \leq t \leq 0. \end{cases}$

12. 15π.

Exercises 4.2, Page 110

4. (b) $(1 + i)\sinh 2$; (d) $1/(2i) - 1/(8 + 2i)$

5. $4\pi i$.

6. (a) $8\pi i$; (b) $-8\pi i$; (c) $-24\pi i$.

8. $13/10 + i/6$.

11. (a) $3 + i$; (b) $3 + i$; (c) $3 + i$.

12. $-2i$.

Exercises 4.3, Page 117

2. (a) $-3 + 2i$; (c) $i\pi$; (e) $\dfrac{-i \sinh^3 1}{3}$;

(g) $-\dfrac{\sqrt{2}}{3} - \dfrac{2}{3}(\sqrt{\pi})^3 + \dfrac{i\sqrt{2}}{3}$;

(i) $\pi/4 - (\frac{1}{2})\arctan 2 + (i/4)\text{Log } 5$.

Exercises 4.4, Page 139

2. $z(s, t) = (2 - s)\cos 2\pi t + i(3 - 2s)\sin 2\pi t$, $0 \leq s, t \leq 1$.

4. (a) is; (b) is; (c) is not; (d) is; (e) is.

8. (a) is; (b) is not; (c) is; (d) is;
   (e) is not; (f) is.

13. (a) π; (b) 0; (c) $-\pi$.

14. $-4\pi i$.

15. 0.

Exercises 4.5, Page 151

1. 0.

3. (a) $-2\pi i$; (b) $3\pi i e^{3/2}/2$; (c) $2\pi i/9$; (d) $10\pi i$;
(e) $-2e\pi i$; (f) $-i\pi/2$.

4. (a) $\pi(1 + 2i)/2$; (b) $-\pi(1 + 2i)/2$; (c) 0.

5. $G(1) = 4\pi i$, $G'(i) = -2\pi(2 + i)$, $G''(-i) = 4\pi i$.

6. π/e.

7. $-2\pi i/9$.

Exercises 4.6, Page 158

7. $f(z) \equiv i$ is the only solution.

15. $9\sqrt{2}/8$.

Exercises 4.7, Page 165

11. 3.

13. $\dfrac{1}{\pi}\left[\tan^{-1}\left(\dfrac{1-x}{y}\right) - \tan^{-1}\left(\dfrac{-1-x}{y}\right)\right]$ for $y > 0$.

Exercises 5.1, Page 175

1. (a) $9/10 + i3/10$; (b) $3(1 - i)$; (c) $3/5$;
(d) $(-2 + i)/(5 \cdot 2^{13})$; (e) $9/8$; (f) -1.

2. 15 terms are sufficient.

8. (a) diverges; (b) converges; (c) diverges;
(d) diverges; (e) converges.

12. (a) $|z| < 1$; (b) $|z - i| < 2$; (c) all z;
(d) $|z + 5i| < 1$.

Exercises 5.2, Page 185

2. (a) entire plane; (b) entire plane; (c) entire plane;
   (d) $|z - i| < \sqrt{2}$; (e) $|z| < 1$; (f) entire plane.

5. (a) $\sum_{j=0}^{\infty} (-1)^j z^j$, $|z| < 1$;

   (c) $\sum_{j=0}^{\infty} \frac{(-1)^j 3^{2j+1} z^{2j+4}}{(2j+1)!}$, entire plane;

   (e) $i + \sum_{j=1}^{\infty} \frac{2}{(1-i)^{j+1}} (z - i)^j$, $|z - i| < \sqrt{2}$;

   (g) $\sum_{j=1}^{\infty} j z^j$, $|z| < 1$.

7. $\text{Log}\left(\frac{1+z}{1-z}\right) = 2 \sum_{j=0}^{\infty} \frac{z^{2j+1}}{2j+1}$, $|z| < 1$.

11. (a) $1 + z - \frac{z^3}{3} + \cdots$; (b) $-1 - 2z - \frac{5z^2}{2} + \cdots$;

    (c) $1 + \frac{z^2}{2} + \frac{5z^4}{24} + \cdots$; (d) $z - \frac{z^3}{3} + \frac{2z^5}{15} + \cdots$.

12. $f(z) = \frac{z + z^2}{(1 - z)^3}$.

19. 9 terms.

Exercises 5.3, Page 193

3. (b) $|z - 1| = \frac{1}{2}$; (d) $|z - i| = 3$; (f) $|z| = 2$.

4. No, since $|2 + 3i| > |3 - i|$.

5. (a) $6! 6^3/3^6$; (b) $2\pi i$; (c) 0; (d) 0.

8. $z + z^3/3 + z^5/10 + \cdots$.

14. (a) $f(z) = \sum_{j=0}^{\infty} \frac{z^{2j}}{2^j(j!)}$; (c) $f(z) = \sum_{j=0}^{\infty} (j + 1) z^{2j}$.

Exercises 5.4, Page 201

1. (a) $\lim \sup = 2$; (b) $\lim \sup = \infty$;
   (c) $\lim \sup = 0$; (d) $\lim \sup = \infty$.

4. (a) 2; (c) $\frac{1}{3}$; (e) 1; (f) 1.

6. (a) R; (b) R^4; (c) $\sqrt{R}$; (d) R; (e) ∞ if $R > 0$.

10. $f(z) = a_0 + z + z^2 + z^4 + z^8 + \cdots$, a_0 arbitrary.

Exercises 5.5, Page 209

1. No.

2. (a) $\sum\limits_{n=-1}^{\infty} (-1)^{n+1} z^n$; (b) $\sum\limits_{n=2}^{\infty} (-1)^n z^{-n}$; (c) $-\sum\limits_{n=-1}^{\infty} (z+1)^n$;

(d) $\sum\limits_{n=2}^{\infty} (z+1)^{-n}$.

3. (a) $\sum\limits_{j=1}^{\infty} \frac{1}{3}\left[(-1)^j - \frac{1}{2^j}\right] z^j$; (b) $\frac{1}{3} \sum\limits_{j=0}^{\infty} \frac{(-1)^j}{z^{j+1}} - \frac{1}{3} \sum\limits_{j=0}^{\infty} \left(\frac{z}{2}\right)^j$;

(c) $\sum\limits_{j=1}^{\infty} \frac{1}{3}[(-1)^{j-1} + 2^j] z^{-j}$.

5. $\dfrac{5/4}{(z-4)^3} - \dfrac{1/16}{(z-4)^2} + \dfrac{1/64}{z-4} + \sum\limits_{n=0}^{\infty} \dfrac{(-1)^{n+1}}{4^{n+4}} (z-4)^n$.

6. $z^2 - \dfrac{1}{18} + \sum\limits_{j=2}^{\infty} \left(-\dfrac{1}{9}\right)^j \dfrac{z^{2-2j}}{(2j)!}$.

7. (a) $\dfrac{1}{z^2} + \dfrac{1}{z^3} + \dfrac{3}{2z^4} + \cdots$; (c) $\dfrac{1}{z} + \dfrac{z}{6} + \dfrac{7z^3}{360} + \cdots$.

8. $\frac{1}{2} < |z| < 2$.

Exercises 5.6, Page 217

1. (a) pole of order 2 at 0,
   removable singularity at -1;
   (c) simple poles at $z = \pm i$;

   (e) simple poles at $z = \pm(2n+1)\dfrac{\pi}{2}$, $n = 0, 1, 2, \ldots$;

   (g) removable singularity at 0.

2. 8.

4. (a) F; (b) T; (c) T; (d) F; (e) T.

7. essential.

9. Yes, $e^{1/z}$ along negative real axis.

12. $-m$.

Exercises 5.7, Page 223

1. (a) essential singularity; (c) analytic;
(e) pole of order 2;
(g) essential singularity; (i) analytic.

4. (a) $1 + 2 \sum\limits_{n=1}^{\infty} (-1)^n/z^n$; (c) $\sum\limits_{n=0}^{\infty} i^n/z^{3n+3}$

Exercises 5.8, Page 232

3. $f(z) = z^2$.

4. Yes, $\sin\left(\dfrac{1+z}{1-z}\right) = 0$ for $z = (n\pi - 1)/(n\pi + 1)$, $n = 1, 2, 3, \ldots$.

5. all α except $0 \le \alpha < 1$.

8. (a) No; (b) Yes; (c) Yes;
(d) No; (e) Yes; (f) Yes.

Exercises 6.1, Page 243

1. (a) $\text{Res}(2) = e^6$; (b) $\text{Res}(1) = -2$, $\text{Res}(2) = 3$;
(c) $\text{Res}(0) = 0$; (d) $\text{Res}(-1) = -6$;
(e) $\text{Res}(0) = 1$, $\text{Res}(-1) = -5/2e$; (f) $\text{Res}(0) = \frac{1}{3}$;
(g) $\text{Res}[\pm(2n+1)\pi/2] = -1$, $n = 0, 1, 2, \ldots$;
(h) $\text{Res}(n\pi) = (-1)^n(n\pi - 1)$, $n = 0, \pm 1, \pm 2, \ldots$.

2. $\text{Res}(1) = -2$.

3. (a) $\pi i \sin 2$; (b) $\pi i(e^2 - 1)/4$; (c) $-8\pi i$;

(d) $\pi i \left[\dfrac{(2 - 5i)e^{2i}}{58} - \dfrac{12 - 5i}{50} \right]$; (e) $\pi i/3$;

(f) 0; (g) 0.

4. $2\pi i$.

Exercises 6.2, Page 247

1. $2\pi/\sqrt{3}$.　　2. $8\pi/\sqrt{21}$.

3. $\pi\sqrt{2}$.　　4. $3\pi\sqrt{5}/25$.

Exercises 6.3, Page 256

4. $\pi/6$.　　5. $-\pi/27$.

6. $\pi/6$.　　7. $3\pi\sqrt{2}/16$.　　11. $2\pi\sqrt{3}/9$.

Exercises 6.4, Page 264

4. $(\pi \sin 3)/2e^3$.

6. $\dfrac{\pi}{3e}\left(1 - \dfrac{1}{2e}\right)$.

8. $i\pi/e^6$.

Exercises 6.5, Page 272

1. (a) $i\pi/2$;　(c) 0.

5. $-\pi/2$.

6. $\dfrac{\pi}{5}(\cos(1) - e^{-2})$.

7. $\pi[\sin(1) - 2\sin(2)]$.

11. $3\pi/4$.

12. $-\pi\cot(a\pi)$.

Exercises 6.7, Page 290

1. a, c, e, and f.

2. 1.

8. all roots.

Exercises 7.1, Page 299

3. (a) $\phi(z) = \text{Arg}(z - 2) - \text{Arg}(z + 1)$;

   (b) $\phi(z) = a_4 + \sum_{k=1}^{3} \frac{1}{\pi}(a_k - a_{k+1})\text{Arg}(z - x_k)$.

5. $\phi(x, y) = \dfrac{1 + y}{x^2 + (1 + y)^2}$.

Exercises 7.2, Page 307

1. (a) 1; (b) 1; (c) 2.

6. the pure imaginary constant functions.

11. (a) the upper half-plane: Im $w > 0$;

    (b) the whole plane minus the logarithmic spiral $\rho = e^\phi$, $-\infty \leq \phi < \infty$;

    (c) $\{w: |w| < 1, \text{Im } w > 0\}$;

    (d) $\{w: |w| > 1, \text{Im } w > 0\}$;

    (e) the upper half-annulus $\{w: e < |w| < e^2, \text{Im } w > 0\}$;

    (f) $\{w: |w| > 1\}$ and $\{w: |w| < 1\}$.

12. (a) the upper half-plane: Im $w > 0$; (b) fourth quadrant;

    (c) the whole plane minus the real intervals $(-\infty, -1], [1, \infty)$;

    (d) the interior of the ellipse $(u^2/\cosh^2 1) + (v^2/\sinh^2 1) = 1$ excluding the real segments $[-\cosh 1, -1], [1, \cosh 1]$.

Exercises 7.3, Page 319

1. $w = 3iz + 5$.

2. (a) $\{w: |w - 2 + 2i| \leq 1\}$; (b) $\{w: |w - 6i| \leq 3\}$;

   (c) $\{w: \text{Re}(w) \leq 1/2\}$; (d) $\{w: \text{Re}(w) \geq 3/2\}$;

   (e) $\{w: |w - 2/3| \leq 1/3\}$.

3. (a) $w = iz$; (b) $w = (2z)/(z + 1)$; (c) $w = (z + i)/(z - 1)$;

   (d) $w = (z + 1)/(z - 1)$.

4. $w = e^{3\pi i/4}\left(\dfrac{z + i}{z - 1}\right)$.

7. the region exterior to both of the circles C_1: $|w - (1 - i)/2| = 1/\sqrt{2}$ and C_2: $|w - (1 + i)/2| = 1/\sqrt{2}$.

9. lower half-disk $\{w: |w| < 1, \text{Im } w < 0\}$.

10. $w = -[(1 - z)/(1 + z)]^2$.

11. $w = \exp\left(4\pi\left[-i\left(\dfrac{1}{z-2} + \dfrac{1}{4}\right)\right]\right)$.

Exercises 7.4, Page 328

1. $z - 2$.   2. 0.

3. (a) $\dfrac{4 - 3i}{25}$;   (b) $\dfrac{7 - i}{6}$;   (c) $\dfrac{5 - 2i}{3}$.

5. No.   6. $\lambda > 0$.

13. $w = \dfrac{i - 2z}{2 + iz}$.

16. $w = \dfrac{az + b}{cz + d}$, where a, b, c, d are real and $ad - bc > 0$.

Exercises 7.5, Page 340

2. $w = A(z - x_1)^2 - 1$, where $A < 0$.

3. $w = -\dfrac{i}{\pi}\sin^{-1} z + i/2$.

4. $w = \dfrac{1}{\pi}\text{Log}\left(\dfrac{z - x_2}{x_2 - x_1}\right)$, where $x_1 < x_2$.

5. $w = -\dfrac{2}{\pi}(\sin^{-1} z + z\sqrt{1 - z^2})$.

6. $w = 1 + \dfrac{i}{\pi}[\sqrt{z^2 - 1} + \log(z + \sqrt{z^2 - 1})]$.

Exercises 7.6, Page 353

1. $\phi(x, y) = \dfrac{2}{\pi}\text{Arg}\left(\dfrac{1 + z}{1 - z}\right)$.

2. $T(x, y) = \dfrac{2}{\pi} \text{Arg}\left(\dfrac{1 + z^2}{1 - z^2}\right) - 1.$

4. $\phi(x, y) = \dfrac{1}{\pi} \text{Arg}\left(\dfrac{\sqrt{z^2 + 1} - 1}{\sqrt{z^2 + 1} + 1}\right).$

6. $\phi(x, y) = 2 \text{Re}\left(\dfrac{z}{z - 2}\right).$

7. (a) $T(x, y) = -\dfrac{1}{\pi} \text{Arg}[\sin^2(\pi i z)];$

   (b) $T(x, y) = \dfrac{2}{\pi} \text{Arg}(e^{\pi z} + 1) - \dfrac{1}{\pi} \text{Arg}(e^{\pi z}).$

8. $\phi(x, y) = -3 + \dfrac{6}{\pi} \text{Arg}\left(\dfrac{e^{-i\pi/6} + z}{e^{i\pi/6} - z}\right).$

9. $\phi(x, y) = (\text{Log } r)^{-1} \text{Log}|(4 - \lambda z)/(4\lambda - z)|,$ where
   $\lambda = (19 + \sqrt{105})/16$ and $r = (13 - \sqrt{105})/8.$

Exercises 7.7, Page 362

2. isotherms: $z(t) = g(a + it),\ t \geq 0,\ -\dfrac{\pi}{2} < a < \dfrac{\pi}{2},$ where

   $g(w) = \dfrac{1}{2} + \dfrac{w - \cos w}{\pi}$ maps the half-strip $-\dfrac{\pi}{2} < u < \dfrac{\pi}{2},\ v > 0,$

   onto the given region.

3. $T(z) = \dfrac{2}{\pi} \text{Re}[\sin^{-1}(z^2)].$

4. $T(z) = -\dfrac{20}{\pi} \text{Re}[\sin^{-1}(-e^{-\pi z})].$

7. streamlines: $\text{Im}(z^{\pi/\alpha} + z^{-\pi/\alpha}) = \text{constant}.$

Exercises 8.1, Page 381

4. (a) $F(t) = \dfrac{i}{8}[-3e^{it} + 3e^{-it} + e^{3it} - e^{-3it}];$

   (b) $c_n = \dfrac{12(-1)^n}{\pi(9 - 4n^2)(1 - 4n^2)};$

   (c) $c_n = \dfrac{2(-1)^n}{n^2}$ for $n \neq 0,\ c_0 = \dfrac{\pi^2}{3};$

(d) $c_0 = 0$, $c_n = (-1)^n \left[\dfrac{i\pi}{n} - \dfrac{2i}{\pi n^3} \right] + \dfrac{2i}{\pi n^3}$ $(n \neq 0)$.

5. a, b.

8. (a) $f(t) = \dfrac{1}{8} + \dfrac{\cos 2t}{2} - \dfrac{\cos 4t}{104}$;

(b) $f(t) = \dfrac{\pi^2}{3} + \displaystyle\sum_{\substack{n=-\infty \\ n \neq 0}}^{\infty} \dfrac{c_n e^{int}}{1 + in - n^2}$,

where c_n is defined as in 4(c) above.

(c) $f(t) = \displaystyle\sum_{\substack{n=-\infty \\ n \neq 0}}^{\infty} \dfrac{c_n e^{int}}{2 + 4in - n^2}$, where c_n is as in Example 8.4.

Exercises 8.2, Page 392

1. (a) $G(\omega) = \dfrac{1}{\pi} \left(\dfrac{1}{1 + \omega^2} \right)$;

(b) $G(\omega) = \dfrac{1}{2\sqrt{\pi}} e^{-\omega^2/4}$;

(c) $G(\omega) = \dfrac{-i\omega e^{-\omega^2/4}}{4\sqrt{\pi}}$;

(d) $G(\omega) = \begin{cases} \frac{1}{2} & \text{if} \quad |\omega| < 1 \\ 0 & \text{if} \quad |\omega| > 1 \\ \frac{1}{4} & \text{if} \quad \omega = \pm 1 \end{cases}$;

(e) $G(\omega) = \begin{cases} \dfrac{i}{2} \sin \omega & \text{if} \quad |\omega| \leq \pi \\ 0 & \text{if} \quad |\omega| \geq \pi. \end{cases}$

2. (a) $G(\omega) = \dfrac{e^{-|\omega|/\sqrt{2}}}{2\sqrt{2}} \left(\cos \dfrac{|\omega|}{\sqrt{2}} + \sin \dfrac{|\omega|}{\sqrt{2}} \right)$;

(b) $G(\omega) = -\dfrac{ie^{-|\omega|/\sqrt{2}}}{2} \sin \dfrac{\omega}{\sqrt{2}}$;

(c) $G(\omega) = \dfrac{1}{2\sqrt{\pi}} e^{-\omega^2/4}$.

4. (a) $f(t) = \int_{-\infty}^{\infty} \dfrac{e^{-\omega^2/4}}{2\sqrt{\pi}} \left(\dfrac{1}{-\omega^2 + i\omega + 1} \right) e^{i\omega t} \, d\omega$;

(b) $f(t) = \int_{-\infty}^{\infty} \dfrac{1}{2\pi(\omega + i)} \left(\dfrac{e^{i\omega t}}{-\omega^2 + 4i\omega + 1} \right) d\omega$;

(c) $f(t) = \int_{-\infty}^{\infty} \dfrac{\sin \omega}{\omega \pi} \left(\dfrac{e^{i\omega t}}{-\omega^2 + 2i\omega + 3} \right) d\omega$.

Exercises 8.3, Page 402

1. (a) $\dfrac{3s}{s^2 + 4} - \dfrac{8}{s + 2}$;

(b) $\dfrac{2}{s} - \dfrac{\pi}{(4 - s)^2 + \pi^2}$;

(c) $\dfrac{1}{s}(1 - e^{-s})$;

(d) $\dfrac{1}{s}(e^{-s} - e^{-2s})$;

(e) $\dfrac{2}{s(s^2 + 4)}$;

(f) $\sqrt{\dfrac{\pi}{s}}$.

2. (a) $\frac{1}{2} \sin(2t)$;
    (b) $4te^t$;
    (c) $e^{-2t} - te^{-2t}$;
    (d) $\frac{1}{2}[1 - 2e^{-t} + e^{-2t}]$;

(e) $e^{-2t}\left(\cos\sqrt{3}\,t + \dfrac{1}{\sqrt{3}} \sin \sqrt{3}\,t \right)$.

6. (a) $f(t) = \frac{1}{2}e^{3t} + \frac{5}{2}e^t$;
    (b) $f(t) = 4e^{2t} - 3e^{3t}$;
    (c) $f(t) = \frac{28}{39}e^{2t} - \frac{5}{6}e^{-t} - \frac{1}{13}e^{-t} \sin 2t + \frac{3}{26}e^{-t} \cos 2t$;

(d) $f(t) = \begin{cases} 0 & \text{if } 0 \le t \le 3 \\ \frac{1}{2} + \frac{1}{2}e^{2t-6} - e^{t-3} & \text{if } 3 \le t \le 6 \\ \frac{1}{2}e^{2t-6} - \frac{1}{2}e^{2t-12} + e^{t-6} - e^{t-3} & \text{if } t \ge 6. \end{cases}$

9. (b) (i) $x(t) = \cos \sqrt{3}\,t$, $y(t) = -\cos \sqrt{3}\,t$;
    (ii) $x(t) = y(t) = \cos t$;
    (iii) $x(t) = \frac{1}{2}(\cos t + \cos \sqrt{3}\,t)$, $y(t) = \frac{1}{2}(\cos t - \cos \sqrt{3}\,t)$.
    (c) (i), (ii).

Index